PREDICTING BREEDING VALUES
with Applications in Forest Tree Improvement

FORESTRY SCIENCES

Volume 33

This is Journal Series No. R00051 of the Institute of Food and Agricultural Sciences of the University of Florida, Gainesville, FL, 32611.

For a list of titles in this series see final page of this volume.

Predicting Breeding Values

with Applications in Forest Tree Improvement

by

TIMOTHY L. WHITE

and

GARY R. HODGE

Department of Forestry, University of Florida, Gainesville, U.S.A.

KLUWER ACADEMIC PUBLISHERS

DORDRECHT / BOSTON / LONDON

Library of Congress Cataloging in Publication Data

White,.Timothy L., 1951-
Predicting breeding values with applications in forest tree improvement / Timothy L. White, Gary R. Hodge.
p. cm. -- (Forestry sciences ; 33)
Includes bibliographical references.
ISBN 0-7923-0460-8 (U.S.)
1. Trees--Breeding. 2. Trees--Selection. 3. Forest genetics. 4. Breeding. 5. Quantitative genetics. I. Hodge, Gary R., 1958- . II. Title. III. Series.
SD399.5.W47 1989
631.5'2--dc20 89-19903

ISBN 0-7923-0460-8

Published by Kluwer Academic Publishers,
P.O. Box 17, 3300 AA Dordrecht, The Netherlands.

Kluwer Academic Publishers incorporates
the publishing programmes of
D. Reidel, Martinus Nijhoff, Dr W. Junk and MTP Press.

Sold and distributed in the U.S.A. and Canada
by Kluwer Academic Publishers,
101 Philip Drive, Norwell, MA 02061, U.S.A.

In all other countries, sold and distributed
by Kluwer Academic Publishers Group,
P.O. Box 322, 3300 AH Dordrecht, The Netherlands.

printed on acid free paper

Printed in The Netherlands

To Mary, Dorothy and Suzanne

and Terry, Daniel and Megan

Contents

Preface

In most breeding programs of plant and animal species, genetic data (such as data from field progeny tests) are used to rank parents and help choose candidates for selection. In general, all selection processes first rank the candidates using some function of the observed data and then choose as the selected portion those candidates with the largest (or smallest) values of that function. To make maximum progress from selection, it is necessary to use a function of the data that results in the candidates being ranked as closely as possible to the true (but always unknown) ranking.

Very often the observed data on various candidates are messy and unbalanced and this complicates the process of developing precise and accurate rankings. For example, for any given candidate, there may be data on that candidate and its siblings growing in several field tests of different ages. Also, there may be performance data on siblings, ancestors or other relatives from greenhouse, laboratory or other field tests. In addition, data on different candidates may differ drastically in terms of quality and quantity available and may come from varied relatives. Genetic improvement programs which make most effective use of these varied, messy, unbalanced and ancestral data will maximize progress from all stages of selection.

In this regard, there are two analytical techniques, best linear prediction (BLP) and best linear unbiased prediction (BLUP), which are quite well-suited to predicting genetic values from a wide variety of sources, ages, qualities and quantities of data. These techniques were largely developed by the late C.R. Henderson and his colleagues for predicting breeding values of dairy cows from massive amounts of extremely messy data. BLUP is now widely used in the United States and other countries in the dairy industry. These analytical techniques (BLP and BLUP) are equally well-suited to breeding programs of other plant and animal species that generate messy genetic data, but have <u>not</u> been widely used. While these techniques are well-documented in the literature, many of the papers require a fair amount of mathematical sophistication and often deal with the extraordinary computational problems encountered in dairy cow breeding programs.

The purpose of this book is to describe these techniques of predicting genetic values in a manner requiring a minimum of prerequisite background (one year of statistical methods and a course in quantitative genetics). While we have tried to be somewhat rigorous in the presentation, we try to stress concepts and the intuitively pleasing manner with which these techniques handle different types of data.

There are three analytical techniques discussed in this book: BLP, selection index and BLUP. The common thread of these techniques is that in the derivation of their analytical formulas, genetic values are considered random variables to be predicted rather than fixed effects (i.e. constants) to be estimated. While this seems subtle at first, it results in quite different analytical methods and formulas than often employed in breeding programs.

While these three methods are similar in some regards (in terms of treating genetic values as random variables and in terms of some characteristics in the way they handle data), BLP, selection index and BLUP differ from one another in some other critical underlying assumptions. Thus, the appropriate technique to use for any given problem depends on which set of assumptions of the three methods is most suitable in that situation. For this reason, we try to be quite explicit in describing the underlying assumptions, problems and issues involved in the application of all three techniques.

The book is intended for students and scientists in quantitative genetics and for breeders in genetic improvement programs. While the examples and applications given are necessarily limited to our experiences in forest genetics, we hope the book finds appeal in other disciplines. The concepts are wholly general.

The first three chapters develop some background material in matrix algebra (Chapter 1), statistics (Chapter 2) and progeny testing (Chapter 3) that is useful in the ensuing discussions of BLP (Chapters 4 - 8), selection index (Chapters 9 - 10), and BLUP (Chapters 11 - 12). Chapters 4, 9 and 11 develop the theory and properties of BLP, selection index and BLUP, respectively. These chapters are the most difficult mathematically; however, they can be skimmed lightly the first reading without disrupting continuity. These theory chapters are included for completeness and for later reference. The other chapters dealing with each technique rely heavily on numerical examples to 1) demonstrate how to apply the technique to real data sets and 2) develop intuitive concepts about how the technique handles data (e.g. how different qualities of data get different weights in predicting genetic values). In particular, Chapters 5 and 6 develop many concepts about BLP that also apply to the other two techniques. We view these as key chapters to read even if primary interest is on selection index or BLUP.

The treatment of these three techniques in this book (particularly BLUP) is certainly not exhaustive. Rather, we try to develop concepts and demonstrate applications to motivate and allow use of these techniques by other geneticists. Errors in the presentation inevitably exist and we would appreciate them being brought to our attention. Also,we are still learning about new applications and ramifications of these techniques and would appreciate discussing them with other scientists.

Many people have helped us in the preparation of this book and we are deeply grateful. Drs. Mike DeLorenzo and Charles Wilcox stimulated our original interest and were extremely helpful in our early applications of BLP to forestry problems. The 40 students in two shortcourses (the USDA Forest Service course in Corvallis, Oregon, and the Tree Breeding course in Canberra, Australia) made tremendous contributions to earlier drafts of the manuscript and helped in many other ways. Two reviewers, Dr. Dale van Vleck and Dr. Sharon Friedman, read the entire final draft. They provided simply excellent review comments for which we are grateful. We also thank other reviewers that read one to several chapters: Dr. Floyd Bridgwater, Mr. Dudley Huber, Dr. Steve McKeand, Ms. Sonia De Souza and Dr. Claire Williams. We also want to thank the 15 industrial and state agencies that are the members of the Cooperative Forest Genetics Research Program. Their support and patience made this possible. Finally, we want to thank Mr. Greg Powell for everything he does: managing and analyzing the data, keeping the computers running, picking up the slack and steering a steady course even in high winds.

Chapter 1
Matrix Algebra

INTRODUCTION

A matrix is a two dimensional rectangular array of numbers with rows and columns (much like a spreadsheet or table of numbers). Matrix algebra is a shorthand method of writing complicated algebraic formulas. The use of matrix algebra greatly simplifies the explanation and calculation of many statistical methods such as regression, and is a necessity for explaining the prediction methods discussed in Chapters 4-12. In this chapter, we present a very brief and informal introduction to matrix algebra. We discuss only those topics needed for use in later chapters and therefore, this is by no means a complete introduction.

While matrix algebra allows complicated equations to be written and manipulated very compactly, each symbol used to represent a matrix actually represents many numbers contained in that matrix. Therefore, while the written representation of a matrix equation is very simple, the physical meaning of that equation can be quite complex (and sometimes hard to understand). Also, because matrices are arrays of numbers and not just single numbers, they have a different set of algebraic properties. It is our experience that the only way to become comfortable with matrix representations of equations is to work through them again and again. Therefore, in this chapter, we make heavy use of numerical examples and demonstrations of the concepts and "rules" of matrix algebra. Proofs are given only when they seem instructive to illustrate concepts.

Readers familiar with matrix algebra can skip over this brief introduction. For readers wishing more detailed discussions of matrix and linear algebra, the books below are listed in increasing order of difficulty (our presentation relies most heavily on numbers 1-3 below).

1. Neter and Wasserman 1974. Chapter 6 provides a very introductory account of definitions and matrix arithmetic.
2. Searle 1966. The entire book is devoted to matrix algebra as commonly applied in biological sciences and statistics.
3. Noble and Daniel 1977. This is a standard and readable textbook on linear algebra. Their geometric interpretations are especially instructive.
4. Eves 1966. This paperback book provides extra depth on some topics.
5. Halmos 1958. This is a more advanced text on the geometry of vector spaces.

BASIC DEFINITIONS

Matrix

A matrix is a rectangular array of numbers arranged in rows and columns. Throughout this book, matrices will be represented by boldface capital letters such as **A** or **B**. The numbers in a matrix are called the **elements** of the matrix and are indexed as to their exact position in the matrix by two subscripts. For example, a_{ij} would indicate the element in matrix **A** that occupies the position in the ith row and j^{th} column. The row subscript is always designated first. The matrix **A** can be represented in complete generality by **A** = $\{a_{ij}\}$ where $\{a_{ij}\}$ represents the collection of all elements. If **A** has r rows and c columns, then
A is written as:

$$\mathbf{A} = \begin{bmatrix} a_{11} & a_{12} & \cdots & a_{1j} & \cdots & a_{1c} \\ a_{21} & a_{22} & \cdots & a_{2j} & \cdots & a_{2c} \\ \vdots & \vdots & & \vdots & & \vdots \\ a_{i1} & a_{i2} & \cdots & a_{ij} & \cdots & a_{ic} \\ \vdots & \vdots & & \vdots & & \vdots \\ a_{r1} & a_{r2} & \cdots & a_{rj} & \cdots & a_{rc} \end{bmatrix}$$

Note that there are r rows and c columns. Every element in the first row has a 1 as the first subscript and a_{1c} represents the number (i.e. element) in the first row and last column (since there are c columns). If r = c, then the matrix is said to be **square**. The **dimension** of a matrix is expressed as r x c, i.e. the number of rows by the number of columns. Two matrices are said to be **equal** if and only if the corresponding elements in both matrices are the same. That is, to be equal, the two matrices **A** and **B** must have the same dimensions and have $a_{ij} = b_{ij}$ for all i and j.

> **Example 1.1**: Suppose that open-pollinated seed from four parents were planted in three different progeny tests (i.e. at three test sites). After measuring each seedling and calculating the family mean for each parent at each test site, the family means for say 10-year height (in feet) could be represented in a data matrix, **Y** , as follows:

	Columns = Test Sites		
Rows = Parents	j=1	j=2	j=3
i=1	25	36	22
i=2	22	33	18
i=3	33	44	20
i=4	15	22	18

The first element, y_{11}, is the average 10-year height of family 1 at test location 1. It appears that site 2 is a better site (column 2 has higher average heights for all families) and that parents (rows) 3 and 4 are the fastest and slowest growing families across all three test sites. In this case, r=4 and c=3 and this matrix has dimension 4 x 3.

Vector

A vector is a special kind of matrix which has either only a single row (r=1) or a single column (c=1). There are two kinds of vectors: a **row vector** has a single row and dimension 1 x c; a **column vector** has a single column and dimension r x 1. In this book, vector names will be designated with boldface small letters such as **y**, etc. When not explicitly stated, a vector is understood to be a column vector. All the rules of matrix algebra apply to vectors since they are just special types of matrices.

Example 1.2: From the data matrix, **Y** in Example 1.1, each of the four rows can be considered a 1 x 3 row vector. So, $\mathbf{y}_1 = [\ 25 \ \ 36 \ \ 22\ ]$ is a row vector containing the family means for parent 1 from all three test sites. As is done below, it is sometimes preferable to place the data from all parents into a single vector. Note that since the type of vector is not stated, it is a column vector. As an example, we order the family means from Example 1.1 so that all of the data from parent 1 comes first, followed by that from parent 2, etc. Then the 12 x 1 data vector **y** is:

$$\mathbf{y} = \left[\begin{array}{c} 25 \\ 36 \\ 22 \\ 22 \\ 33 \\ 18 \\ 33 \\ 44 \\ 20 \\ 15 \\ 22 \\ 18 \end{array}\right] \begin{array}{l} \\ \} \text{ The first three elements are for parent 1.} \\ \\ \\ \} \text{ The second three elements are for parent 2.} \\ \\ \\ \} \text{ The next three elements are for parent 3.} \\ \\ \\ \} \text{ The final three elements are for parent 4.} \\ \\ \end{array}$$

Scalar

The smallest matrix is just a single number and is called a **scalar**. This special type of matrix has both a single row (r=1) and a single column (c=1) and therefore has dimension 1 x 1. Since scalars are both numbers and the smallest matrices, they follow all the rules of both real-number algebra and matrix algebra. Scalars will be designated with non-boldface type in this text.

Identity Matrix and Null Matrix

The identity matrix is a special type of square matrix which is denoted as **I**. It can be of any size (as long as it is square) and has ones for the diagonal elements and zeros for all the off-diagonal elements. Identity matrices serve a special role in matrix algebra. The size of an identity matrix is often designated as a subscript; for example, $\mathbf{I}_n$ is an n x n identity matrix.

A null matrix is a matrix of any dimension (not necessarily square) whose elements are all zeros. Null matrices are denoted by boldface zeros, i.e. **0**. The 2 x 2 identity matrix and null matrix are:

$$\mathbf{I}_2 = \begin{bmatrix} 1 & 0 \\ 0 & 1 \end{bmatrix} \qquad \mathbf{0} = \begin{bmatrix} 0 & 0 \\ 0 & 0 \end{bmatrix}$$

MATRIX OPERATIONS

Transpose

The transpose operation interchanges the rows and columns of a matrix. The transpose of matrix **A** is denoted by **A'**. The first column of **A** becomes the first row of **A'**, the second column of **A** becomes the second row of **A'**, etc. Formally, for a matrix $\mathbf{A} = \{a_{ij}\}$, $\mathbf{A}' = \{a_{ji}\}$. So the element in the ith row and j^{th} column of **A** is in the ith column and j^{th} row of **A'**. If the dimension of **A** is r x c, then the dimension of **A'** is necessarily c x r.

The transpose operation can also be performed on vectors and scalars. The transpose of a column vector with dimension r x 1 is a row vector with dimension 1 x r. Correspondingly, the transpose of a row vector with dimension 1 x c is a column vector with dimension c x 1. The transpose of a scalar is just the scalar itself.

Example 1.3: The transpose of the 4 x 3 data matrix, **Y**, in Example 1.1 is obtained by writing the columns of **Y** as the rows of **Y'**. Thus, the first column of **Y** becomes the first row of **Y'**, etc. The transpose has dimension 3 x 4 and is shown below:

$$\mathbf{Y}' = \begin{bmatrix} 25 & 22 & 33 & 15 \\ 36 & 33 & 44 & 22 \\ 22 & 18 & 20 & 18 \end{bmatrix}$$

For a square matrix, when $\mathbf{A} = \mathbf{A}'$, the matrix is called **symmetric**. Note that this means that the $a_{ij} = a_{ji}$ for all i and j. For example, $a_{12} = a_{21}$. Thus, the off-diagonal elements of a symmetric matrix have a special relationship in that the below diagonal elements can be obtained be "folding" the above diagonal elements down across the diagonal.

Addition and Subtraction

To conform for addition or subtraction, two matrices must have the same dimensions. That is, they must have both the same number of rows and the same number of columns. If two matrices do not have the same dimensions, then they do not conform for addition or subtraction and can neither be added nor subtracted. The sum or difference of two conforming matrices is another matrix whose elements are the sums or differences of the corresponding elements of the two matrices being added or subtracted. In general, if **A** and **B** have dimensions r x c, then **A** + **B** = **C** and **C** has dimension r x c. The elements of **C** are $a_{ij} + b_{ij}$ for all i and j. Subtraction works in the same fashion.

Example 1.4: Suppose that we have both the 10-year heights and the 5-year heights (in feet) for the 4 open-pollinated families described in Example 1.1. If the data matrices containing the 5- and 10-year family means from each test site are specified as **X** and **Y**, then the difference matrix, **Z**, represents the average growth (on a family mean basis) that occurred on each site between the fifth and tenth years.

If

$$\mathbf{X} = \begin{bmatrix} 15 & 25 & 10 \\ 10 & 15 & 5 \\ 30 & 29 & 14 \\ 8 & 12 & 9 \end{bmatrix} \quad \text{and} \quad \mathbf{Y} = \begin{bmatrix} 25 & 36 & 22 \\ 22 & 33 & 18 \\ 33 & 44 & 20 \\ 15 & 22 & 18 \end{bmatrix}$$

then **Y** - **X** = **Z** and

$$\mathbf{Z} = \begin{bmatrix} 10 & 11 & 12 \\ 12 & 18 & 13 \\ 3 & 15 & 6 \\ 7 & 10 & 9 \end{bmatrix}$$

Matrix Multiplication

Multiplication of a matrix by a scalar: The product of a scalar multiplied by a matrix, **A**, is defined as a new matrix, **B**, that has the same dimension as **A** and whose elements are obtained by multiplying each element of **A** by the scalar. Thus, for k**A** = **B**, **B** = $\{b_{ij}\} = \{k \cdot a_{ij}\}$ for all i and j.

Multiplication of a matrix by a matrix: The multiplication of two matrices is more complicated than any of the operations we have discussed so far. Matrix multiplication has a very special definition and it takes practice to make it a routine operation. In general, the product of two matrices is written **AB** = **C** and the ijth element of **C** (i.e. the element in the ith row and jth column) is obtained as the sum of products of the ith row of **A** and the jth column of **B.**

So, the element in the second row, third column of **C** is obtained as the sum of products between elements of the second row of **A** and third row of **B**:

$$c_{23} = a_{21}b_{13} + a_{22}b_{23} + a_{23}b_{33} + \ldots + a_{2m}b_{m3}$$

where the dimensions of **A** and **B** are r x m and m x c, respectively. Think of moving along the second row of **A** and simultaneously down the third column of **B** summing the products of the corresponding elements. Note that the first subscript on "a" is always 2 and the second subscript on "b" is always 3 since we are obtaining the element for the second row, third column of the product matrix, **C**. Also, the second subscript on "a" equals the first subscript on "b" for each elemental product because we are moving simultaneously along the row and down the column.

This latter consideration means that to conform for multiplication **A** must have the same number of columns as **B** has rows. If this condition is not met, then the two matrices can **not** be multiplied. Also, dimensional analysis can be used to ascertain the dimension of the product matrix. If **A** is r x m and **B** is m x c, then **C** = **AB** is r x c. Note that the product **BA** is not defined. In the first case, **B** is **premultiplied** by A, while in the second case **B** is **postmultiplied** by **A**. The order of matrix multiplication will usually give different results and multiplication may only be defined in one order. We discuss this more in the section "Algebraic Properties" .

Example 1.5: Continuing with the **X** and **Y** matrices (defined in Example 1.4) of family averages of 5- and 10-year heights, note that **X** and **Y** can not be multiplied in either order. They do not conform for multiplication because in neither order does the number of columns of the first matrix equal the number of rows in the second matrix. There are two products that may have biological meaning that are defined: **XY'** and **X'Y**. The resultant matrix of **X'Y** has dimension (3 x 4) X (4 x 3) => 3 x 3 and is shown below:

$$\underset{\mathbf{X'}}{\begin{bmatrix} 15 & 10 & 30 & 8 \\ 25 & 15 & 29 & 12 \\ 10 & 5 & 14 & 9 \end{bmatrix}} \underset{\mathbf{Y}}{\begin{bmatrix} 25 & 36 & 22 \\ 22 & 33 & 18 \\ 33 & 44 & 20 \\ 15 & 22 & 18 \end{bmatrix}} = \underset{\mathbf{Z}}{\begin{bmatrix} 1705 & 2366 & 1254 \\ 2092 & 2935 & 1616 \\ 957 & 1339 & 752 \end{bmatrix}}$$

The element in the second row, third column of **Z** is calculated as:

$$1616 = (25 \cdot 22) + (15 \cdot 18) + (29 \cdot 20) + (12 \cdot 18).$$

Special cases of matrix multiplication: Two special cases of matrix multiplication are stated here and then left as problems for the reader to verify. The multiplication of any matrix, **A**, times the identity matrix of the appropriate dimension is just **A**. That is, $\mathbf{AI}_c = \mathbf{I}_r\mathbf{A} = \mathbf{A}$ where **A** is r x c and $\mathbf{I}_c$ and $\mathbf{I}_r$ are the c x c and r x r identity matrices, respectively.

Another often used special result is that **A'A** and **AA'** are symmetric (and therefore square) matrices regardless of the dimension of **A**. Note that the dimension of the resultant product matrix depends on the order of the multiplication.

Inverse of a Matrix

The inverse of a matrix, **A**, is another matrix, denoted $\mathbf{A}^{-1}$, that when multiplied by the original matrix results in the identity matrix. Thus, $\mathbf{AA}^{-1} = \mathbf{I}$ if and only if $\mathbf{A}^{-1}$ is the inverse of **A**. Note that by the same definition **A** is the inverse of $\mathbf{A}^{-1}$. The inverse of a matrix is analogous to the reciprocal of a scalar which when multiplied by the original scalar equals 1. The inverse matrix is used in matrix algebra in manner similar to the way reciprocals are used to solve equations. For example using scalars, if ab = c, we can solve for b by multiplying both sides of the equation by 1/a (or equivalently by a^{-1}). Then, $b = ca^{-1} = c/a$. Analogously in matrix algebra, if $\mathbf{AB} = \mathbf{C}$ and if an inverse of **A** exists, then to solve for **B** premultiply both sides of the equation by $\mathbf{A}^{-1}$. So, $\mathbf{A}^{-1}\mathbf{AB} = \mathbf{IB} = \mathbf{B} = \mathbf{A}^{-1}\mathbf{C}$. So, the solution for **B** is $\mathbf{B} = \mathbf{A}^{-1}\mathbf{C}$. The order of multiplication is important in matrix algebra and in this case we consistently used premultiplication.

To have an inverse, a matrix must be square, but not even all square matrices have inverses. For an inverse to exist, the matrix must be of full rank (also called non-singular). When a matrix is of full rank (i.e. non-singular), an inverse always exists and that inverse is unique. We do not discuss this concept of singularity any further since most matrices we encounter have inverses. When an inverse does exist, multiplication of the matrix and the inverse is commutative. That is, $\mathbf{AA}^{-1} = \mathbf{A}^{-1}\mathbf{A} = \mathbf{I}$. For more detail on inverses, see Neter and Wasserman (1974) or Noble and Daniel (1977).

In Chapters 11 and 12, singular matrices are encountered in the discussion of Best Linear Unbiased Prediction; these do not have true inverses as we have just discussed. That is, for a singular matrix, **S**, no matrix can be found such that $\mathbf{SS}^{-1} = \mathbf{I}$; $\mathbf{S}^{-1}$ does not exist. In this case it is possible to find something called a generalized inverse. This is often shortened to G-inverse and for a matrix **S** is designated $\mathbf{S}^{-}$. Generalized inverses do not have the most desirable property of true inverses (i.e. $\mathbf{SS}^{-}$ does not equal **I**). However, a G-inverse may have some of the other desirable properties of true inverses. There are several different types and definitions of G-inverses corresponding to which properties they possess (see Searle 1966; Graybill 1976, p. 24). And for any singular matrix, many different G-inverses can be found that possess different sets of desirable properties. For this book, a matrix will be termed a generalized inverse of **S** if $\mathbf{SS^-S} = \mathbf{S}$. For any matrix, **S**, it is always possible to find a G-inverse that satisfies this property.

The computation of an inverse of a matrix can become quite laborious. Even for relatively small matrices (such as 5 x 5), the computations can be overwhelming if done by hand. We do not discuss how to compute the inverse of a matrix, but rather rely on the fact that all matrix algebra computer programs have an inverse function. Computer solutions of inverses can sometimes suffer from rounding off problems and it is always good to check the solution by multiplying the calculated inverse by the original matrix to see if the product is the identity matrix. Most matrix algebra computer programs also have a G-inverse function and any calculated G-inverse should be checked for rounding errors by verifying that $\mathbf{SS^-S} = \mathbf{S}$.

ALGEBRAIC PROPERTIES

Commutative Property

The commutative property holds for addition: As long as the two matrices conform for addition (i.e. have the same dimension), then $\mathbf{A} + \mathbf{B} = \mathbf{B} + \mathbf{A}$. This is because addition deals with each pair of corresponding elements and $a_{ij} + b_{ij} = b_{ij} + a_{ij}$ for each and every i and j.

The commutative property does not hold for multiplication: As discussed previously, the order in which two matrices are multiplied affects the product matrix. In fact, unless the two matrices are square and of the same dimension, multiplication is not even defined in both orders. So, in general, **AB** does not equal **BA**.

There are four special cases where multiplication is commutative. Three of the special cases involve square matrices only. For an r x r square matrix **A**, multiplication by the r x r identity matrix, the r x r null matrix and the r x r inverse (if a unique inverse exists) are commutative operations. That is:

$$\mathbf{I_rA} = \mathbf{AI_r} = \mathbf{A} \qquad \textbf{Eq. 1.1}$$

and

$$\mathbf{0_r A} = \mathbf{A 0_r} = \mathbf{0_r} \qquad \text{Eq. 1.2}$$

and

$$\mathbf{AA^{-1}} = \mathbf{A^{-1}A} = \mathbf{I_r}.$$

The fourth special case involves multiplication of a matrix by a scalar and by the definition of that operation given above, it can be seen that $\mathbf{kA} = \mathbf{Ak}$.

Associative Law

The associative law holds for addition: Again because addition works with simple sums of corresponding pairs of elements, the associative law holds and thus:

$$\mathbf{A} + \mathbf{B} + \mathbf{C} = (\mathbf{A} + \mathbf{B}) + \mathbf{C} = \mathbf{A} + (\mathbf{B} + \mathbf{C}). \qquad \text{Eq. 1.3}$$

It is understood that all matrices must conform for addition.

For subtraction, an analogous expression to that for subtracting real numbers holds for subtracting matrices: $(\mathbf{A} - \mathbf{B}) - \mathbf{C} = \mathbf{A} - (\mathbf{B} + \mathbf{C})$.

The associative law holds for multiplication: Providing the matrices conform for multiplication, the associative law also holds for multiplication. Thus:

$$\mathbf{ABC} = (\mathbf{AB})\mathbf{C} = \mathbf{A}(\mathbf{BC}). \qquad \text{Eq. 1.4}$$

This is proven by showing that each element in the final product matrix is the same regardless of the grouping for multiplication. Thus, **A** can first be postmultiplied by **B** and then the intermediate result postmultiplied by **C**. Or, **B** can first be postmultiplied by **C** and the intermediate result premultiplied by **A**.

The distributive law holds: If **B** and **C** conform for addition and **A** conforms with them for multiplication, then

$$\mathbf{A}(\mathbf{B} + \mathbf{C}) = \mathbf{AB} + \mathbf{AC}\,. \qquad \text{Eq. 1.5}$$

This law is also proven by showing that every element in the resultant matrix is the same for the expressions on both sides of Equation 1.5.

Reflexive Law for Transposition

Taking the transpose of a transpose returns the original matrix. This is seen by tracing the elements through the operations and leads to several properties (theorems) stated below without proof (the matrices must conform for each designated operation).

1. Transposing a matrix that is itself already the transpose of an original matrix results in the original matrix.

$$(\mathbf{A}')' = \mathbf{A} . \qquad \textbf{Eq. 1.6}$$

2. Transposing a matrix that is the sum of two matrices is equivalent to first transposing each matrix and then summing them.

$$(\mathbf{A} + \mathbf{B})' = \mathbf{A}' + \mathbf{B}' . \qquad \textbf{Eq. 1.7}$$

3. Transposing a matrix that is the product of two matrices is equivalent to transposing each matrix and then multiplying them in reverse order.

$$(\mathbf{AB})' = \mathbf{B}'\mathbf{A}' . \qquad \textbf{Eq. 1.8}$$

Properties of Inverses

Three properties of the inverse operation are shown below. The proofs are given because they are instructive of the manner in which many proofs in matrix algebra are developed. Since we have defined the inverse operation only for square matrices, **A**, **B**, and **C** (below) are all square and of the same dimension.

1. Transposing the inverse of a matrix is equivalent to taking the inverse of the transpose.

   Proof:

$$(\mathbf{A}')^{-1}\mathbf{A}' = \mathbf{I} \qquad \text{(by definition of inverse of } \mathbf{A}')$$

and

$$[(\mathbf{A}')^{-1}\mathbf{A}']' = \mathbf{I}' = \mathbf{I} \qquad \text{(since } \mathbf{I} \text{ is symmetric)}$$

so

$$\mathbf{A}[(\mathbf{A}')^{-1}]' = \mathbf{I} \qquad \text{(by Eq. 1.8).}$$

Thus, we have two equivalent expressions for the inverse of **A** such that

$$\mathbf{A}^{-1} = [(\mathbf{A}')^{-1}]'$$

and taking the transpose of both sides of this expression completes the proof:

$$(\mathbf{A}^{-1})' = (\mathbf{A}')^{-1} . \qquad \textbf{Eq. 1.9}$$

2. Taking the inverse of a matrix which results from multiplying several matrices is equivalent to first inverting each matrix and then multiplying the inverses in reverse order.

Proof:

$$\mathbf{ABC}(\mathbf{ABC})^{-1} = \mathbf{I} \qquad \text{(by definition of inverse of } \mathbf{ABC})$$

and it is also true that

$$\mathbf{ABC}(\mathbf{C}^{-1}\mathbf{B}^{-1}\mathbf{A}^{-1}) = \mathbf{I}$$

because

$$\mathbf{ABCC}^{-1}\mathbf{B}^{-1}\mathbf{A}^{-1} = \mathbf{ABIB}^{-1}\mathbf{A}^{-1} = \mathbf{AIA}^{-1} = \mathbf{I}.$$

Thus, we have two expressions that both serve as the inverse to **ABC** and since the inverse of a full rank matrix is unique, these expressions must be equivalent. Therefore,

$$(\mathbf{ABC})^{-1} = \mathbf{C}^{-1}\mathbf{B}^{-1}\mathbf{A}^{-1} . \qquad \textbf{Eq. 1.10}$$

3. The inverse of an inverse is the original matrix.

Proof:

$$\mathbf{A}^{-1}(\mathbf{A}^{-1})^{-1} = \mathbf{I}. \qquad \text{(By definition of inverse of } \mathbf{A}^{-1}\text{)}$$

Premultiply both sides by **A**, then

$$\mathbf{A}\mathbf{A}^{-1}(\mathbf{A}^{-1})^{-1} = \mathbf{AI} = \mathbf{A}$$

and so

$$\mathbf{I}\,(\mathbf{A}^{-1})^{-1} = \mathbf{A} = (\mathbf{A}^{-1})^{-1} .$$

Thus,

$$(\mathbf{A}^{-1})^{-1} = \mathbf{A} . \qquad \textbf{Eq. 1.11}$$

SUMMARY AND CONCLUSIONS

The discussion of matrix algebra presented in this chapter is both brief and informal. In our experience, the best way to become comfortable with the operations and algebraic rules associated with matrix manipulations is to take very simple matrices and verify each property or rule by hand calculations.

With the larger matrices in later chapters, it is not feasible to carry through all calculations by hand. In one sense, the beauty of matrix algebra is that these large algebraic expressions can be compactly represented. On the other hand, it is more difficult to understand the operations intuitively. To facilitate understanding, it is important for the reader go through several steps for each matrix representation in later chapters:

1) Try to completely specify all elements in one or two rows of each matrix in an expression to make sure you understand where the elements of each matrix really come from,
2) Understand the units (e.g. percent, feet, meters, degrees, etc.) for each element you specify,
3) Determine the dimension of each matrix (e.g. the matrix in Example 1.1 is 4 x 3 because there are data for four parents and three test sites),
4) Make sure all matrices conform for the operations specified in the expression (e.g. if the expression is **AB** = **C** then the number of columns of **A** must equal the number of rows of **B**),
5) Conduct a dimensional analysis (see Problem 1.1) to determine the dimension of the final product, and
6) When possible, try to actually conduct the operations specified in the expression for a limited number of rows and columns (this will not always be possible, e.g. for the inverse operation).

When expressions become complicated, it can still be difficult to understand, intuitively, the way the operations are handling the data. We often create small example matrices and calculate the expression several times each time varying some numbers in the matrices to see the effects on the final result.

After some practice, the operations, properties and rules of matrix algebra become familiar. The compactness with which algebraic expressions are represented greatly facilitates the description of the prediction methods in this book.

PROBLEMS

Problem 1.1: Demonstrate by dimensional analysis (i.e. by assigning general dimensions and finding the dimensions of the product), the following facts:

a. A row vector postmultiplied by a column vector is a scalar.

b. A column vector postmultiplied by a row vector is a matrix.

c. A matrix postmultiplied by a column vector is a column vector.

d. A matrix premultiplied by a row vector is a row vector.

Problem 1.2: Using **X** and **Y** from Example 1.5, calculate by hand, **Z** = **XY'**.

Problem 1.3: Using **Y** from Example 1.1, verify (i.e. do the calculations by hand):

$\mathbf{I}_4\mathbf{Y} = \mathbf{Y}\mathbf{I}_3 = \mathbf{Y}$, where $\mathbf{I}_4$ and $\mathbf{I}_3$ are 4 x 4 and 3 x 3 identity matrices.

Problem 1.4: Use a computer program when needed and use the **X** matrix from Example 1.4.

a. What are the dimensions of **X'X** and **XX'** ?

b. Calculate **Z** = **X'X** and **W** = **XX'**.

c. Describe in words the 1) meaning of the diagonal elements of **Z** and **W** and 2) the relationships of the off-diagonals in each matrix.

Problem 1.5: Use a computer program and the matrices given below for this problem.

$$\mathbf{A} = \begin{bmatrix} 24 & 18 & 2000 \\ 29 & 58 & 4 \\ 99 & 39 & 945 \end{bmatrix} \qquad \mathbf{B} = \begin{bmatrix} 88 & 98 & 33 \\ 5 & 0 & 1 \\ 32 & 14 & 5 \end{bmatrix}$$

$$\mathbf{C} = \begin{bmatrix} 15 & 18 & 23 & 66 \\ 49 & 23 & 0 & .3 \\ 14 & 12 & 0 & 18 \end{bmatrix}$$

a. Calculate **(ABC)'**.

b. Calculate **C'B'A'**.

c. Calculate **(BC)'A'**.

d. Calculate **C'E** where **E** = **(AB)'**.

e. Compare the results from a,b,c and d.

Problem 1.6: Use a computer program and the matrices from Problem 1.5 for this problem.

a. Calculate $\mathbf{A}^{-1}$.

b. Calculate $\mathbf{AA}^{-1}$.

c. Calculate $\mathbf{AA}^{-1}\mathbf{A}$.

d. Calculate $(\mathbf{A}^{-1})^{-1}$.

e. Calculate $(\mathbf{A}^{-1})^{-1}\mathbf{A}^{-1}$.

f. Calculate $(\mathbf{A}^{-1})'$ and $(\mathbf{A}')^{-1}$.

g. Describe each of the above calculations or properties in words.

Problem 1.7: Use a computer program and **A** and **B** from Problem 1.5 for this problem.

a. Calculate $(\mathbf{AB})^{-1}$.

b. Calculate $\mathbf{AB}(\mathbf{AB})^{-1}$.

c. Calculate $\mathbf{B}^{-1}\mathbf{A}^{-1}$.

d. Calculate $\mathbf{ABB}^{-1}\mathbf{A}^{-1}$.

Problem 1.8: Use a computer program and **L** and **M** below for this problem.

$$\mathbf{L} = \begin{bmatrix} 14 & 22 & 18 \\ 41 & 33 & 19 \\ 69 & 77 & 55 \end{bmatrix} \qquad \mathbf{M} = \begin{bmatrix} 14 & 22 & 81 \\ 41 & 33 & 19 \\ 69 & 77 & 55 \end{bmatrix}$$

a. Calculate the inverses of **L** and **M**.

b. Verify that these are inverses by multiplying to get **I**.

c. Calculate the G-inverses of **L** and **M**.

d. Multiply **L** and **M** by their G-inverses. Are the products equal to the identity matrix?

e. Calculate $\mathbf{LL}^{-}\mathbf{L}$ and $\mathbf{MM}^{-}\mathbf{M}$. What are the products?

Chapter 2
Statistics

INTRODUCTION

Consider a 40-acre stand of Douglas-fir trees that contains 20,000 trees each of which is 15 years old. Suppose we wish to know the average height of the stand to both assess the site quality and determine when the stand will be ready for harvest. Instead of measuring the whole stand, we opt to save time and money by measuring a random sample of trees from the stand. Based on our **observed sample** of trees, we then make inferences about the entire **population** (i.e. the entire 20,000 trees).

There are a set of **parameters** that we wish to estimate which are associated with the population. These parameters characterize the population of trees of interest. For example, the population of trees has an average height and a variance associated with the distribution of heights. These parameters are unknown to us, but are fixed constants. They are <u>not</u> random variables. For example, the true average height may be 30 feet. This is an unknown constant that we wish to estimate from observations on a sample of trees from the stand.

Now suppose that a random sample of ten trees is drawn from the population of 20,000. Associated with this sample is a set of **statistics** that will be used to estimate the population parameters. For example, if the sample mean is 33.8 feet for the ten trees, then an estimate of the overall population mean is 33.8. In this case, the sample estimate is 3.8 feet above the true population mean, but in general the true population mean will never be known and we will never know the exact error associated with our estimates. A key concept is that statistics are **random variables** that can take on many different values. If a second sample of ten trees is drawn from the population, there is very little likelihood that the sample mean will again be 33.8. In fact, if a large number of 10-tree samples are drawn and their sample means calculated, then we could plot a frequency distribution of these sample means to determine the probability (i.e. frequency) of obtaining a sample mean of 33.8 when the true mean is in fact 30.

The formal definition of a **random variable** is a variable that assumes each of its possible values with some definite (but not necessarily known) probability. A **constant** has only one possible value (which may not be known) and therefore assumes this value with probability equal to one.

In forest genetics and tree improvement there are many, many parameters that geneticists are interested in estimating: means, variance and covariance components, heritabilities, etc. These parameters are unknown constants associated with some inference population (such as the population of first generation selections made in a given tree improvement program). Geneticists use experiments (such as progeny tests) to estimate these parameters, but these parameters are never known exactly. The estimates developed for any particular parameter will have more or less error associated with it depending on the quality of the experiment used to develop the estimate. In this text, we denote the estimates of a parameter by placing a ^ over the top of the parameter symbol.

The purpose of this chapter is to develop some statistical concepts and notation that will be used in later chapters. In particular, the algebra associated with certain operations (i.e. expected value, variance and covariance) on random variables is discussed both for random scalars and random vectors. We also discuss some concepts of linear statistical models often used in forest genetics experiments.

EXPECTED VALUE, VARIANCE AND COVARIANCE

Throughout the definitions of expected value, variance and covariance, the random variable, y, is assumed to be a discrete random variable which assumes a finite number of outcomes. For continuous random variables that can assume an infinite number of outcomes, the summation sign in the definition is replaced by an integral. All of the concepts of algebra for expected values, variances and covariances hold for both discrete and continuous random variables. Other discussions of these concepts are: Neter and Wasserman (p. 1-6, 1974), Mendenhall and Schaeffer (p. 178-190, 1973) and Brunk (Chapter 5, 1975).

The definitions and computational formulas presented here are for the true parameters (not estimates of the parameters) and assume that the entire population can be enumerated. While this is almost never the case, the purpose here is to develop the algebraic rules of working with linear transformations of random variables. Computational formulas for sample estimates of means, variances and covariances can be found in any elementary statistics book.

Expected Value

The expected value of a random variable, y, is denoted either E(y) or μ_y and is simply the true weighted average of that random variable. That is,

$$E(y) = \mu_y = \Sigma\, y_i f(y_i) \qquad \textbf{Eq. 2.1}$$

where Σ indicates the summation over all possible values of y and $f(y_i)$ is the probability of each occurrence of y.

This is a weighted average of y because before being summed, each value of y is weighted by its frequency (i.e. probability) of occurrence. The summation is over the whole population of y, not just a sample. Therefore, the expected value of y is a constant, not a random variable. So, while y is a random variable that can take on many values (y_i takes on each of its i values with probability $f(y_i)$), the expected value of y is a constant. The expected value of a random variable is often called the mean.

In the special case when all values of y have the same probability of occurrence, then

$$E(y) = \mu_y = (\Sigma\, y_i)/N \qquad \textbf{Eq. 2.2}$$

since the probability of each value of y occurring is 1/N where N is the total number of individuals in the population.

For the 40-acre Douglas-fir stand above, the expected value of the population of heights is obtained by adding all 20,000 heights together and dividing by 20,000.

Linear Transformations of Expectations

It is very common to make linear transformations of random variables and then want to calculate the expected value of the transformed random variable. As a simple example, suppose we add the value of 2 feet to each of the 20,000 heights in the 40-acre slash pine stand. Now, $z = y + 2$ is a new random variable that takes on each of its 20,000 possible values with a probability of 1/20,000. The expected value of the new variable $z = y + 2$ is :

$$\begin{aligned} E(z) &= E(y+2) = [\Sigma\, (y_i + 2)]/20{,}000 \\ &= (\Sigma\, y_i)/20{,}000 + (\Sigma\, 2)/20{,}000 \\ &= E(y) + 2. \end{aligned}$$

Using similar methods, it can be shown (see Problem 2.2) for any constants b and c,

$$E(by + c) = bE(y) + c\,. \qquad \textbf{Eq. 2.3}$$

For any two random variables, x and y,

$$E(bx + cy) = bE(x) + cE(y) . \qquad \textbf{Eq. 2.4}$$

Variance

The variance of any random variable y is denoted Var(y), V(y) or σ^2_y and is defined as:

$$Var(y) = E[y - E(y)]^2 = E(y^2) - [E(y)]^2 . \qquad \textbf{Eq. 2.5}$$

Using the definition of expected value above, this is expands to:

$$\begin{aligned} Var(y) &= \Sigma\ [y_i - E(y)]^2\ f(y_i) \\ &= \Sigma\ [y_i^2\ f(y_i)] - [\Sigma\ y_i\ f(y_i)]^2 . \end{aligned} \qquad \textbf{Eq. 2.6}$$

When $f(y_i) = 1/N$,

$$Var(y) = (\Sigma y_i^2)/N - (\Sigma y_i/N)^2 . \qquad \textbf{Eq. 2.7}$$

For the 40-acre Douglas-fir stand, we would calculate the variance in two steps, substituting 20,000 for N in each step: 1) sum up the squares of all 20,000 height values and divide by 20,000 and 2) subtract from that the square of the expected value ($30^2 = 900$).

Linear Transformations of Variances

We state without proof several important properties of variances.

$$Var(y + c) = Var(y) . \qquad \textbf{Eq. 2.8}$$

$$Var(by) = b^2 Var(y) . \qquad \textbf{Eq. 2.9}$$

$$Var(by + c) = b^2 Var(y) . \qquad \textbf{Eq. 2.10}$$

$$Var(x + y) = Var(x) + Var(y) + 2Cov(x,y) . \qquad \textbf{Eq. 2.11}$$

$$\text{Var}(bx + cy) = b^2\text{Var}(x) + c^2\text{Var}(y) + 2bc\text{Cov}(x,y)\ . \qquad \textbf{Eq. 2.12}$$

For the last two equations, see the next section for definition of Cov(x,y). If x and y are independent random variables, then the covariance is zero and

$$\text{Var}(bx + cy) = b^2\text{Var}(x) + c^2\text{Var}(y)\ . \qquad \textbf{Eq. 2.13}$$

Covariance and Correlation

For any two random variables x and y, two measures of association between x and y are the covariance and correlation. The covariance between x and y is denoted as Cov(x,y) or σ_{xy} and is defined as:

$$\begin{aligned}\text{Cov}(x,y) &= E\{[x - E(x)][y - E(y)]\} \\ &= E(xy) - E(x)E(y)\ . \qquad \textbf{Eq. 2.14}\end{aligned}$$

When f(x,y) is a constant 1/N, the covariance becomes

$$\text{Cov}(x,y) = (\Sigma\ x_iy_i)/N - [(\Sigma\ x_i)(\Sigma\ y_i)]/N^2\ .$$

The correlation between two random variables is denoted Corr(x,y) or r_{xy} and is defined as:

$$\text{Corr}(x,y) = \text{Cov}(x,y)/[\text{Var}(x)\text{Var}(y)]^{½}\ . \qquad \textbf{Eq. 2.15}$$

The correlation is a standardized covariance in the sense that the maximum and minimum values for the correlation are 1 and -1. This occurs when x and y are perfectly correlated (i.e. the same trait).

Linear Transformations of Covariances and Correlations

Several algebraic properties associated with covariances and correlations are stated below without proof (b,c,d, and e are constants; x, y, w and z are random variables).

$$\text{Cov}(y,y) = \text{Var}(y)\ . \qquad \textbf{Eq. 2.16}$$

$$\text{Corr}(y,y) = 1 . \qquad \textbf{Eq. 2.17}$$

$$\text{Cov}(bx,cy) = bc\text{Cov}(x,y) . \qquad \textbf{Eq. 2.18}$$

$$\text{Corr}(bx,cy) = \text{Corr}(x,y) . \qquad \textbf{Eq. 2.19}$$

$$\text{Cov}(bx+d,cy+e) = bc\text{Cov}(x,y) . \qquad \textbf{Eq. 2.20}$$

$$\text{Corr}(bx+d,cy+e) = \text{Corr}(x,y) . \qquad \textbf{Eq. 2.21}$$

$$\text{Cov}(x+y,w+z) = \text{Cov}(x,w) + \text{Cov}(x,z) + \text{Cov}(y,w) + \text{Cov}(y,z) \qquad \textbf{Eq. 2.22}$$

CONCEPTS OF LINEAR STATISTICAL MODELS

Linear statistical models are so widely used in the biological sciences that it is easy to overlook the underlying assumptions involved in their use. A linear **statistical** model is distinguished from a linear **deterministic** model by the presence of at least one random variable (often a random error term). For example in a completely randomized field test of 9 different provenances of Douglas-fir, a possible representation of the data (e.g. 10-year volumes) is:

$$y_{ij} = \mu + p_i + w_{ij}$$

where

μ is the overall mean (an unknown constant, not a random variable),

p_i represents 9 different provenance effects (i=1,2,...,9) and each of these provenance effects is an unknown constant, and

w_{ij} represents many random effects associated with both genetic and environmental differences among trees within the same provenance (as random variables, the w_{ij}'s have probability distributions and means and variances associated with them).

The w_{ij}'s are the only random variables in this model and without them the model is deterministic. That is, the deterministic model implies that each provenance mean can be modeled as just the sum of the overall mean and the provenance effects and that each tree within a provenance will have the exact response (no variability among trees within a provenance). While this is an over-simplified example which seems very unrealistic, the distinction between constants and random variables is not always easy to make and yet the exact specification of the model greatly influences the appropriate analytical approach. We discuss these concepts further below.

Fixed and Random Effects

To specify a linear statistical model (hereafter called linear model) for an experiment, the investigator must decide whether each factor (e.g. location, block, provenance, family, etc) is best represented by **fixed effects** (i.e. constants) or by **random effects** (i.e. random variables). Sometimes this is easy, sometimes it is debatable, and sometimes the decisions may change depending on the purpose of the analysis (see Searle 1971 for a thorough discussion).

In classical agricultural and forestry experiments, treatments such as fertilizers, irrigation levels, and plant spacings are always treated as fixed effects. This is because we are only interested in the exact treatments that have been applied. Also, we expect that in repeated runs of the experiment, the treatments could be repeated exactly and would have the exact same effects (Searle 1971).

In genetic experiments such as progeny tests, the actual genetic entries (such as families) in a specific experiment (such as a test location) may represent only a small sample of all the entries of interest, For example, we might test the frost hardiness of 25 families in an experiment. Those particular families may not be of interest except in providing information regarding genetic variability. Thus, we treat the 25 as a random sample from a much larger population of families about which we wish to draw conclusions regarding frost hardiness.

In deciding whether a set of effects is to be considered fixed or random, the important concerns are the inferences to be drawn from the experiment and hence the sampling process. If interest centers on the exact treatment levels in the experiment, then the treatments should be considered fixed effects (i.e. constants). In this case the levels of the treatments represent the entire population of treatments to which the inferences will apply. On the other hand, if the conclusions are to apply to a broader population of which the treatment levels of a certain factor in the experiment are a sample, then the factor is random. As an example, in a classical analysis of a randomized complete block design, blocks are considered random factors because the exact blocks chosen are envisioned as only a sample of a larger population of blocks to which we want to apply the inferences from the experiment.

A given linear model can contain both random and fixed effects (called a mixed model) and some effects may be considered fixed for one analysis and then random in the next. Year effects, site effects and block effects are good examples of factors that can be considered either fixed or random depending upon the purpose of the analysis. For example, if a progeny test was planted in 3 locations, the locations may or may not be a truly a random sample of all the locations about which conclusions are to be drawn. However, by considering locations as a random effect, the inferences about the variability among genotypes are broadened to some larger (perhaps even poorly defined) population of locations. In terms of the sampling process, the three locations are considered only a sample drawn from some larger population of locations.

On the other hand, if the purpose of the analysis is to rank the genotypes in terms of their performance in these particular 3 tests, then it may be reasonable to consider locations as fixed. Then the sampling process is envisioned as having drawn the entire population of locations on which these genotypes have been tested and across which they will be ranked. The purpose of the analysis and its effect on the choice of fixed *vs*. random is discussed in the next section.

Goals of Analyses of Linear Models

The most difficult conceptual part of any statistical analysis can be specifying the linear model: 1) what are the factors, 2) are the factors nested, cross-classified, etc, 3) are the factors fixed or random, 4) what assumptions are made about the variances and covariances among model effects, and 5) what are the goals of the analysis? After these decisions are made, the rest of the analysis is essentially computational (though not necessarily easy!). In terms of the last consideration, there are three very distinct goals of analysis.

1. Estimating fixed effects: In a simplified sense, the goal of this analysis aims to determine whether there are differences among a given set of fixed effects (such as 3 different fertilizer treatments) and then to develop estimates of the effect for each treatment level (such as the mean responses for the three fertilizer treatments). This type of analysis includes regression and most applications of ANOVA taught in a one-year statistical methods course. The linear model employed may include both fixed and random factors, but the goal of the analysis focuses on the fixed effects with the random effects being viewed as "nuisance" factors.

An example of this type of analysis is the mixed model usually employed for a Randomized Complete Block design. Blocks are considered random, treatments fixed. But, the goals are to make inferences about the fixed set of treatment effects. Blocks are not of any particular interest in themselves, and are considered random to allow the inferences about treatments to apply to a broader population of blocks than the treatments were tested in (Anderson and McClean 1974 p.57).

2. Estimating variance components for random effects: The goal of this type of analysis is to estimate the amount of variability associated with one or more of the random factors in a linear model. This analysis is quite common in forest genetics. For example, considering genotypic effects as a random factor, we often estimate the variability (i.e. variance) among genotypes for several traits. If the level of genetic variability for a particular trait is large compared to that due to the environment, we conclude that there is good opportunity to make gains from selection. In this case, the genotypes in the experiment are nearly always considered a sample of the overall population of genotypes about which inferences are made regarding genetic variability; they are therefore considered random factors. In fact, if they are not considered random variables, it does not even make sense to estimate variance components because constants (i.e. fixed effects) do not vary (i.e. do not have variances).

In variance component analysis in forestry, blocks, locations and years are "nuisance" factors and are usually considered as random effects in order that the variance component estimates for genotypes apply to a larger population of blocks, years and locations than those specific ones in the experiments. However, when heritability estimates are calculated (see formula of Namkoong et al. 1966 and Shelbourne 1969), variances due to these "nuisance" factors are omitted from the phenotypic variance (i.e. the denominator of heritability). This implies that selections will only be made after adjustment of the phenotypic means for the **fixed** set of "nuisance" effects in which they occur. This means that for the purpose of estimating variance components the "nuisance" factors are considered random, but for heritability estimation and subsequent selection they are considered fixed. While this may seem contradictory at first, it emphasizes that the same factor can be considered fixed for one kind of analysis and random for another.

Cotterill (1987) gives a thorough discussion of the adjustment of phenotypic means for blocks and location effects prior to selection, but briefly consider two sets of 25 different families. Set one is planted on a good location and after 5 years the test mean is 20 feet; the 25 families at the poorer second location average only 10 feet. There is no reason to believe that the 25 families in set one are all genetically superior to those in set 2; and yet without adjustment for the specific location effects, the best say 10 families selected might all come from site 1. Thus, it seems logical in this instance (and increases the selection efficiency) to adjust the phenotypic family means to reflect the for the **fixed** set of location effects. This entails expressing the phenotypic family means as deviations from the location means.

3. Predicting random effects: The first two types of analyses, estimating fixed effects and estimating variance components of random effects are both called "estimating" because the parameters of interest are constants: fixed effects in the first case and the variances (and covariances) of random variables in the second case. Remember that while the variances are unknown, they are constants associated with the distribution of the underlying random variable. Together, these first two types of analyses occupy the totality of what is covered in nearly all statistical texts on the analysis of linear models.

However, the most important analysis in any breeding program involves a type of analysis with a different objective: predicting a random variable from observed data. The problem is then one of predicting random effects, not estimating fixed effects (see Searle 1974 and Henderson 1984 for discussions of the concept of prediction *vs* estimation).

As an example consider the objective of most progeny tests: to predict the breeding values for a set of parents based on the performance of their offspring (Allard 1960). This is clearly one of prediction as the breeder is not interested in the performance of the offspring in the specific tests, but of future offspring in other environments. The breeding values of the parents are best thought of as unobservable random variables whose values can never be known exactly, but only predicted. Also, in most cases only a small sample of the total number of parents are included in any given single progeny test. So, there may be 1000 parents for which breeding values are being predicted, but offspring from only 50 of those parents might be included in any given field location. Thus, these 50 are logically regarded as a sample from some larger population about which we wish to make inferences (and hence are considered random, not fixed).

Much of this book is concerned with the prediction of unobservable random variables. For this purpose, we employ a mixed model with years, locations and blocks considered as fixed effects and the genetic effects considered random. The rationale for treating the "nuisance" factors as fixed is the same as presented in the previous section: we wish to adjust each observed phenotypic mean for its specific set of fixed effects example prior to ranking the genotypes.

Defining a Linear Model

Throughout much of this book, we use open-pollinated (considered half-sib) progeny test data as examples to illustrate the concepts of prediction and here we set up a linear model for such data and then (in following sections) illustrate many of the concepts of expected values, variances and covariances previously described in this chapter.

Consider the hypothetical (though realistic) situation in which many hundred mass selections are made as part of the initial stages of a tree improvement program. Open-pollinated seed from these selections are then collected either from the ortets growing in the natural stands in which they were selected or from their clones growing in grafted seed orchards. The open-pollinated seed are then planted in field tests and any given test location may contain offspring from some or all of the parents (i.e. may contain many or few open-pollinated families). If each test location is planted in a Randomized Complete Block design, a linear model to represent the data for a given trait measured at a given age is:

$$y_{ijkl} = \mu + E_i + B_{ij} + f_k + fe_{ik} + p_{ijk} + w_{ijkl} \, . \qquad \textbf{Eq. 2.23}$$

where

μ = a fixed general mean,

E_i = fixed effect of ith test environment, $i = 1, 2, \ldots t$,

B_{ij} = fixed effect of jth block in i^{th} test, $j = 1, 2, \ldots b_i$,

f_k = random effect of kth family, $k = 1, 2, \ldots s_i$,
$E(f_k) = 0$, $Var(f_k) = \sigma^2_f$,

fe_{ik} = random effect of kth family in i^{th} test,
$E(fe_{ik}) = 0$, $Var(fe_{ik}) = \sigma^2_{fe}$,

p_{ijk} = random plot error of kth family in j^{th} block of ith test,
$E(p_{ijk}) = 0$, $Var(p_{ijk}) = \sigma^2_p$,

w_{ijkl} = random tree error of lth tree in ijk^{th} plot,
$E(w_{ijkl}) = 0$, $Var(w_{ijkl}) = \sigma^2_w$, and
the covariances between all pairs of factors are assumed zero.

There are t E_i effects to estimate (one for each of the t test locations) and each is expressed as a deviation from the overall mean μ. Thus, over all test environments, the E_i effects sum to zero. As an example, test location 3 may have a test mean that is 5 units below the overall average mean of 25 units. The estimate of E_3 is -5 units. An estimate of the test mean for location 3 is obtained as $\hat{\mu} + \hat{E}_3$.

The B_{ij} effects are the deviations of the block means at any test site from the test environment mean. Thus, there are b_i of these deviations to be estimated at the ith test site. The estimates ($\hat{B}_{ij}$'s) will sum to zero at each test site. That is, if there are four blocks at the third test site, the four estimates of block effects (i.e. $\hat{B}_{31}$, $\hat{B}_{32}$, $\hat{B}_{33}$ and $\hat{B}_{34}$) are expressed as deviations from the test location mean and will add to zero (assuming balanced data).

The f_k effects are assumed random and are associated with the average genetic effects of the open-pollinated families planted in each test. When open-pollinated families are considered true half-sib families (as done in this book), f_k is the general combining ability of the kth parent or equivalently is equal to ½ of the breeding value of the kth parent.

Note that breeding value and general combining ability are defined as average performances of offspring over the population of all possible test locations. At any particular test location, a given parent's offspring may interact with that environment to produce a response that is higher or lower that the overall f_k. This genotype x environment interaction is modeled as the fe_{ik} effects for each parent (i.e. family). If the kth parent's offspring are in eight tests, then there will be 8 fe_{ik} effects for that parent (fe_{1k}, fe_{2k}, ... , fe_{8k}).

In each block, the performances of the offspring from a given parent can be averaged together to obtain a subclass mean or plot mean. There is a plot mean for every family in each block. The p_{ijk} effects are then estimated as:

$$\hat{p}_{ijk} = (\text{plot mean})_{ijk} - (\hat{\mu} + \hat{E}_i + \hat{B}_{ij} + \hat{f}_k + \hat{fe}_{ik}) \; .$$

Each $\hat{p}_{ijk}$ estimates the effect of the kth family in block ij as a deviation from what would be expected for that plot based on the additive effects of the other factors.

The random tree effects, w_{ijkl}, are specific to an individual tree and are estimated as the deviation of the tree measurement from the estimated plot mean. These effects are influenced both by environmental differences among trees within the same plot and by genetic differences among trees within the same family.

Averages, Variances and Covariances with Linear Models

Averages: Forest geneticists use many types of averages from their experiments: plot means, family means from a given test, family means averaged over many tests, etc. It is often instructive and sometimes imperative to express these averages in terms of the factors (effects) of the underlying linear model. Assuming Eq. 2.23 as the linear model, these concepts are illustrated for a family mean from a given test location. To obtain an average for a particular family, say the kth family, growing in the ith test location, we apply Eq. 2.2 to the linear model and average over all the b blocks and n trees of that family within each block:

$$\bar{y}_{i.k.} = (\sum_{j}^{b} \sum_{l}^{n} y_{ijkl})/bn \; . \qquad \textbf{Eq. 2.24}$$

The two dots for the j and l subscripts indicate that the average has been obtained across all trees and blocks for the kth family in the ith test. For example, if this were family number 3 in test 2 that had 8 trees in each of 5 blocks, then

$$\bar{y}_{2\cdot3\cdot} = (\sum_{j}^{5} \sum_{l}^{8} y_{2j3l})/40 \; .$$

The summation is only for those trees in the third family at the second test location and we have assumed a balanced design with the same number of trees in each block. The assumption of balance is not necessary, but simplifies the notation. Also note that a different family might have a different formula if it were in only 2 of the three blocks.

Now consider a family mean in terms of the linear model Eq. 2.23. This is done to see what effects in the linear model influence the value of a family mean and to assess the magnitude of their impact.

$$\bar{y}_{i.k.} = (\sum_{j}^{b}\sum_{l}^{n} y_{ijkl})/bn$$

$$= [\sum_{j}^{b}\sum_{l}^{n} (\mu + E_i + B_{ij} + f_k + fe_{ik} + p_{ijk} + w_{ijkl})]/bn \; .$$

To evaluate this expression, the summation operators are distributed and thus each term is evaluated individually. For example, the summation of the constant term μ over bn trees results in bn μ terms. Similarly, E_i is a constant with respect to summation over l and j. This is because there is a constant location effect, E_i, common to all trees in this test. Thus,

$$\sum_{j}^{b}\sum_{l}^{n} E_i = bnE_i \; .$$

The summation over the B_{ij} effects is over all of the block effects at the ith test location and since they are expressed as deviations from the test mean will sum to zero. So a particular family mean will not contain any block effects (and hence not be influenced by block effects) as long as the data are balanced. See Chapter 3 for discussion of unbalanced data.

Summations for the last four terms in the linear model are evaluated below.

$$\sum_{j}^{b}\sum_{l}^{n} f_k = bnf_k \qquad \text{(since } f_k \text{ is constant for the summation).}$$

$$\sum_{j}^{b}\sum_{l}^{n} fe_{ik} = bnf_{ik} \qquad \text{(since } fe_{ik} \text{ is constant for the summation).}$$

$$\sum_{j}^{b}\sum_{l}^{n} p_{ijk} = n(\sum_{j}^{b} p_{ijk}) \qquad (p_{ijk} \text{ is constant for summation over l})$$

$$= np_{i \cdot k} \qquad \text{(the dot indicates summation over j).}$$

$$\sum_{j}^{b}\sum_{l}^{n} w_{ijkl} = w_{i \cdot k \cdot} \qquad \text{(dots indicate summation over j and l).}$$

There are bn p_{ijk} terms in each family mean (one for each tree) and these are divided into b groups of n each. Within each block, all n trees have the same plot effect (since they are in the same plot). This is why the summation simplifies.

The final step is to collect all terms and distribute the bn from the denominator of Eq. 2.24. After some algebra, the result is:

$$\bar{y}_{i.k.} = (\sum_{j}^{b} \sum_{l}^{n} y_{ijkl})/bn$$

$$= [\sum_{j}^{b} \sum_{l}^{n} (\mu + E_i + B_{ij} + f_k + fe_{ik} + p_{ijk} + w_{ijkl})]/bn$$

$$= \mu + E_i + f_k + fe_{ik} + (p_{i \cdot k}/b) + (w_{i \cdot k \cdot}/bn) \quad . \qquad \textbf{Eq. 2.25}$$

While the amount of algebra may appear formidable at first, after some practice it is possible to shortcut many of the steps. Note that Eq. 2.25 expresses a family mean in general for the kth family at the ith location. If the data are not balanced, then b and n are specific to the number of blocks and trees for the specific family of interest. So, a different family may have a different number of trees (i.e. a different bn).

The interpretation of Eq. 2.25 indicates that several effects influence the observed family means from any given progeny test. If the test is on a good site, then the E_i effect will be a large positive deviation inflating the family means for all families tested at that location. The f_k is the overall average genetic influence passed onto the kth family from parent k (i.e. ½ the breeding value of parent k for true half-sib families). The fe_{ik} is the specific effect of the kth family on the ith site and is the family by test location interaction. If a particular family (say the kth family) performs much better on the ith test site than it does on average, fe_{ik} is a large positive effect that increases the family mean for that family in that test.

The two terms involving p_{ijk} and w_{ijkl} are averages of random effects that are expected to be zero, but will not necessarily average to zero for a particular family mean. In a particular family mean, there are b plot effects that are averaged and bn tree effects that are averaged. If these random effects are extremely variable, then there is more chance that for a particular family mean they will not average to zero and will influence that family's observed performance. For example, if the planting microsites within a block are extremely variable and just by random chance the trees from a particular family (say the kth family) are planted in better than average microsites, then its family mean will be inflated.

We have presented a specific kind of average (i.e. for an observed family mean) for a specific linear model; however, for other types of averages (e.g. plot means) and other linear models, the method is completely general:

1) Write the desired average as calculated from the observed data (as in Eq. 2.24),
2) Express observations in terms of the assumed linear model, and
3) Carry out the averaging process with the observations expressed in terms of the effects of the linear model (as in Eq. 2.25).

Conceptually, the interesting part of the family mean in Eq. 2.25 is the fact that both genetic and several other effects influence observed family means. If there are only few blocks and few trees per block, then any given family mean will be more influenced by environmental effects due to random placement of its trees in certain plots and certain microsites. So, a family mean far above average can be due to positive genetic effects and/or due to random placement of its trees on good plots and good microsites within plots. If there are many blocks and many trees within a block, this will tend to reduce spurious contributions by the last two terms to the family means, thus rankings of families will be more precise.

Variances: Continuing with the example of open-pollinated data, consider the variance among family means at a given test location. That is, if we were to repeat the experiment many times, on the average how much variability would exist among observed family means? This is the variance of $\bar{y}_{i\cdot k\cdot}$ and is obtained by distributing the variance operator across the terms in Eq. 2.25. Since the terms are independent, there are no covariance terms:

$$\begin{aligned} Var(\bar{y}_{i\cdot k\cdot}) &= Var[\mu + E_i + f_k + fe_{ik} + (p_{i\cdot k}/b) + (w_{i\cdot k\cdot}/bn)] \\ &= Var(\mu) + Var(E_i) + Var(f_k) + Var(fe_{ik}) \\ &\quad + Var(p_{i\cdot k})/b^2 + Var(w_{i\cdot k\cdot})/b^2n^2 \\ &= \sigma^2_f + \sigma^2_{fe} + \sigma^2_p/b + \sigma^2_w/bn \ . \end{aligned}$$ **Eq. 2.26**

This expression is obtained by evaluating each of the variances according to the rules for variance operations given in Eqs. 2.8 - 2.13. $Var(\mu)$ and $Var(E_i)$ are both zero because μ and E_i are both fixed effects (constants). The variances of f_k and fe_{ik} are defined in the linear model (Eq. 2.23) and these are just repeated above using the σ symbol. The last two terms deserve more explanation.

$$
\begin{aligned}
\mathrm{Var}(p_{i\cdot k}/b) &= \mathrm{Var}(p_{i\cdot k})/b^2 && \text{(from Eq. 2.9)} \\
&= \mathrm{Var}(\sum_j^b pijk)/b^2 && \text{(substituting from Eq. 2.25)} \\
&= \mathrm{Var}(p_{i1k} + p_{i2k} + \ldots + p_{ibk})/b^2 && \\
&= [\sum_j^b \mathrm{Var}(pijk)]/b^2 && \text{(Eq. 2.13; } p_{ijk} \text{ are independent)} \\
&= (\sum_j^b \sigma^2{}_p)/b^2 && \text{(by definition in Eq. 2.23)} \\
&= (b\sigma^2{}_p)/b^2 && \text{(since } \sigma^2{}_p \text{ is a constant)} \\
&= \sigma^2{}_p/b \ .
\end{aligned}
$$

The second step above replaces the abbreviated notation ($p_{i.k}$) with the equivalent, but more explanatory summation formula. This expression can be written in even more explicit form as in step three: $\mathrm{Var}(p_{i1k} + p_{i2k} + \ldots + p_{ibk})/b^2$. There are b of these plot terms being summed; the terms represent the plot effects associated with family k at location i. Since these effects are independent, the variance operator can be distributed and the variance of all plot effects are assumed equal to $\sigma^2{}_p$ (steps four and five above). The last two steps follow as shown.

$\mathrm{Var}(w_{i\cdot k\cdot})/b^2n^2$ is evaluated in a similar manner. There are bn of these terms (one for each tree in the family mean) and each one has variance equal to $\sigma^2{}_w$. Thus, $(bn\sigma^2{}_w)/b^2n^2 = \sigma^2{}_w/bn$.

The magnitude of expected variability among observed family means at a given test location is then influenced by more than just genetic differences among the families. If there is substantial plot-to-plot and/or tree-to-tree variation, certain portions of this variability tend to inflate the variance among family means.

The variance in Eq. 2.26 is the **expected** variance among family means at a given location based on the linear model. The observed variance among family means at a given location can be larger or smaller than this expected variance which is that expected on the average. The variance in Eq. 2.26 is a constant, while that observed at any given location is a random variable. If the experiment were repeated many times, we expect the observed variances to average to Eq. 2.26 (if the assumptions of the model were met). As an example, suppose that the particular sample of families planted at a given test site are more similar to each other than expected of a random sample of families. The observed variance among family means in that test might then be smaller than the expected (i.e. true) variance expressed in Eq. 2.26 as a result of sampling error.

Covariances and Correlations: There are many types of covariances and correlations that are sometimes used in the application of linear models. We have occasion to use some of these in subsequent chapters and in this section show the complete derivation of the covariance between two traits when family means from a given test location are used as the basis for the calculation. For example, the five-year measurements for diameter and height from 30 open-pollinated families planted in a single test location could be used by first obtaining the 30 family means for height and the 30 for diameter and then obtaining a covariance between the two traits using the 30 pairs of family means. The derivation of the expected value of this covariance in terms of the linear model effects is given below using two traits, s and t, assuming balanced data :

$$\begin{aligned} \mathrm{Cov}(\bar{y}^s_{i\cdot k\cdot},\bar{y}^t_{i\cdot k\cdot}) &= \mathrm{Cov}(\mu^s + E^s_i + f^s_k + fe^s_{ik} + p^s_{i\cdot k}/b + w^s_{i\cdot k\cdot}/bn, && \textbf{Step 1} \\ &\qquad \mu^t + E^t_i + f^t_k + fe^t_{ik} + p^t_{i\cdot k}/b + w^t_{i\cdot k\cdot}/bn) \\ &= \mathrm{Cov}(\mu^s,\mu^t) + \mathrm{Cov}(E^s_i,E^t_i) + && \textbf{Step 2} \\ &\quad \mathrm{Cov}(f^s_k,f^t_k) + \mathrm{Cov}(fe^s_{ik},fe^t_{ik}) + \\ &\quad \mathrm{Cov}(p^s_{i\cdot k},p^t_{i\cdot k})/b2 + \mathrm{Cov}(w^s_{i\cdot k\cdot},w^t_{i\cdot k\cdot})/b^2n^2 \\ &= \sigma_f^{st} + \sigma_{fe}^{st} + \mathrm{Cov}(p^s_{i\cdot k},p^t_{i\cdot k})/b^2 + && \textbf{Step 3} \\ &\quad \mathrm{Cov}(w^s_{i\cdot k\cdot},w^t_{i\cdot k\cdot})/b^2n^2 \\ &= \sigma_f^{st} + \sigma_{fe}^{st} + (\sigma_p^{st})/b + (\sigma_w^{st})/bn && \textbf{Eq. 2.27} \end{aligned}$$

where σ_f^{st} is the covariance between true family effects for traits s and t and the other three covariance terms are defined similarly.

For step 1, the family means for traits s and t are written in terms of their linear models. Step 2 results by applying Eq. 2.22 to create covariances between all possible linear model effects and then realizing that covariances between non-matching effects are all zero by definition of the model. For example, $\mathrm{Cov}(f^s_k, fe^t_{ik})$ is the covariance between the average family effect for trait s and the family by location interaction term for trait t; this has a covariance of zero. Step 2 also uses Eq. 2.18 to pull the constants out from within the covariance operation. Step 3 substitutes covariance notation for the first two model terms; this is analogous to the variance notation used in Eq. 2.26.

For the last step, evaluate the last two terms in a manner similar to that done for variances in the preceding section. For the summed plot effects, write out $p^s_{i.k}$ and $p^t_{i.k}$ in detail as in the previous section. There are b^2 possible covariances between the b plot effects for trait s and the b plot effects for trait t. Even though there are b^2 such terms only b are non-zero because plot effects only have a non-zero correlation if they are in the same block (i.e. when j is the same for both s and t). So for example, if the two traits are height and diameter, being on a fertile plot within a block might increase the family means for both traits and therefore increase the covariance between family means for the two traits.

An important interpretation of Eq. 2.27 is that random interaction, random plot effects and random tree effects all influence the covariance and correlation between two traits based on observed family means on a test site. These are called type A covariances and correlations (Burdon 1977) and are widely used in forestry to make genetic inferences even though their interpretation is quite complicated. Type B covariances and correlations employ family means from different sites and are much more straightforward to interpret in terms of the genetic relationships between two traits (See Chapters 3 and 7 for further discussion).

Variances of Deviations and Finite Sample Size Corrections

Often in application of best linear prediction and selection indices it is necessary to express observations as deviations from their means. This attempts to adjust observations for unwanted influences of spurious fixed effects (such as effects of block and test location). As an example, consider the family mean in Eq. 2.25 expressed as a deviation from its expected value. The expected value of $\bar{y}_{i \cdot k \cdot} = \mu + E_i$. These are the fixed effects associated with a specific test location. By expressing family means as deviations from these fixed effects (i.e. $\bar{y}_{i \cdot k \cdot} - \mu - E_i$), family means tested in different test locations can be compared on equal footing. For example, families tested on a very good site might all average taller than those tested on a poorer site; however, when expressed as deviations from their respective site means, both sets of families would average zero. This would allow ranking across both tests to occur without all families from the better test site being ranked too high.

If the actual observations in prediction methods are expressed as deviations, then it is necessary to calculate the variances and covariances of these deviations. If we assume that the fixed effects are known constants (and therefore do not have to be estimated from the data), variances of deviations are identical to that of unadjusted means. For the family mean from an open-pollinated progeny test, $Var(\bar{y}_{i \cdot k \cdot} - \mu - E_i) = Var(\bar{y}_{i \cdot k \cdot})$ which is given in Eq. 2.26. This is because μ and E_i are assumed known constants and therefore have no variance.

In actual practice, fixed effects are not known without error but rather are estimated from the data. These estimated fixed effects are random variables that have associated variances. The variances and covariances of deviations from estimated fixed effects are not the same as those of the unadjusted observations, but rather are altered by finite sample size corrections associated with the precision on the estimated fixed effects. In the general case, consider an observation, y_i, that is expressed as a deviation from its estimated fixed effects, $\hat{\alpha}_i$. Then $Var(y_i - \hat{\alpha}_i) = Var(y_i) + Var(\hat{\alpha}_i) - 2Cov(y_i, \hat{\alpha}_i)$. Thus, the variance of y_i (i.e. the unadjusted observation) is altered by the last two terms of the expression. These last two terms are the finite sample size correction and are associated with the precision of the estimate of a_i. If large amounts of data result in low error variance of $\hat{\alpha}_i$, then $\hat{\alpha}_i$ can be considered a constant and its variance and covariance can be treated as zero. This formula is completely general and can always be used to calculate variances of deviations.

If the fixed effects are based on large sample sizes (e.g. large numbers of families, blocks and test locations), then variances of deviations approach those of the unadjusted observations. If, on the other hand, the fixed effects are estimated poorly (i.e. small sample sizes), then the variances and covariances of the deviations will be more affected. In practice, if there are more than 20 families in a given progeny test, then finite sample size corrections will be minimal. Throughout most of this book, we assume that fixed effects are known constants and therefore we do not account for finite sample size corrections in the calculations of variances and covariances. This greatly facilitates description of other concepts. For more discussion of finite sample size corrections see Chapter 10 and Falconer (1981), Burdon (1982), Bulmer (1985), Harvey and Townsend (1985) and Land et al. (1986).

Example 2.1: To illustrate the process of calculating variances of deviations, consider a family mean from an open-pollinated progeny test (Eq. 2.25) expressed as a deviation from the overall test location mean. The overall test location mean, $\bar{y}_{i\cdots}$, has an expected value of $\mu + E_i$ and thus can be used to express family means from each test location as deviations from their estimated fixed effects. Then $Var(\bar{y}_{i\cdot k\cdot} - \bar{y}_{i\cdots}) = Var(\bar{y}_{i\cdot k\cdot}) + Var(\bar{y}_{i\cdots}) - 2Cov(\bar{y}_{i\cdot k\cdot}, \bar{y}_{i\cdots})$. The first term, $Var(\bar{y}_{i\cdot k\cdot})$ is given in Eq. 2.26. The second term can be evaluated in the same manner by averaging family means across all s families at the ith location and calculating the variance in terms of the linear model effects:

$$Var(\bar{y}_{i\cdots}) = \sigma^2_f/s + \sigma^2_{fe}/s + \sigma^2_p/sb + \sigma^2_w/sbn.$$

The covariance term requires more explanation and arises because the kth family mean contributes to the overall test location mean. If there are only a few families in the test, then each family (say the kth family) will have a noticeable impact on the overall test location mean. If the kth family has a large positive effect (i.e. is a very good performer), then this will tend to increase the estimate of the overall test location mean. Thus, this covariance is always positive.

To evaluate the covariance, express each mean in terms of the linear model effects, collect like effects and then evaluate the covariance of each set of effects. Constant terms μ, B and E are omitted as they are constants and do not contribute to variances and covariances.

$$Cov(\bar{y}_{i\cdot k\cdot}, \bar{y}_{i\cdots}) = Cov\{[f_k + fe_{ik} + p_{i\cdot k}/b + w_{i\cdot k\cdot}/bn], [(f_{\cdot} + fe_{i\cdot})/s + p_{i\cdot\cdot}/sb + w_{i\cdots}/sbn]\}.$$

Remember that a dot indicates summation over that subscript; so, f./s indicates that s family effects have been averaged (i.e. summed and divided by s). Collecting like effects and noting that unlike effects are assumed to have zero correlation, the covariance becomes

$$\begin{aligned}\mathrm{Cov}(\bar{y}_{i\cdot k\cdot},\bar{y}_{i\cdots}) &= \mathrm{Cov}[f_k,(f./s)] + \mathrm{Cov}[fe_{ik},(fe_{i\cdot}/s)] + \\ &\quad \mathrm{Cov}[(p_{i\cdot k}/b),(p_{i\cdots}/sb)] + \\ &\quad \mathrm{Cov}[(w_{i\cdot k\cdot}/bn),(w_{i\cdots}/sbn)].\end{aligned}$$

Each of these covariances is then evaluated as shown in detail for the first term.

$$\begin{aligned}\mathrm{Cov}[f_k,(f./s)] &= (1/s)\ \mathrm{Cov}[f_k,(f_1 + f_2 + \ldots + f_k + \ldots + f_s)] \\ &= (1/s)\ \mathrm{Cov}(f_k,f_k) \\ &= (1/s)\ \mathrm{Var}(f_k) \\ &= \sigma^2_f/s.\end{aligned}$$

The first step above makes use of Eq. 2.18 to bring the constant, 1/s, out in front of the covariance term and then expands the summation of f.. For the second step, note that of all s terms in the summation, f_k only is correlated with one of them, f_k. That is, there are s family effects in the overall test location mean, but the kth family mean has only its own family effect in common with those s family effects. The third step makes use of Eq. 2.16 and the last step simply substitutes from the definition of the linear model, Eq. 2.23. When each of the above covariances are evaluated in this manner, $\mathrm{Cov}(\bar{y}_{i\cdot k\cdot},\bar{y}_{\cdots})$ is shown to equal $\mathrm{Var}(\bar{y}_{i\cdots})$.

The last step in evaluating $\mathrm{Var}(\bar{y}_{i\cdot k\cdot} - \bar{y}_{i\cdots})$ is to collect terms from each step.

$$\begin{aligned}\mathrm{Var}(\bar{y}_{i\cdot k\cdot} - \bar{y}_{i\cdots}) &= \mathrm{Var}(\bar{y}_{i\cdot k\cdot}) + \mathrm{Var}(\bar{y}_{i\cdots}) - 2\mathrm{Cov}(\bar{y}_{i\cdot k\cdot},\bar{y}_{i\cdots}) \\ &= [(\sigma^2_f + \sigma^2_{fe} + \sigma^2_p/b + \sigma^2_w/bn) \\ &\quad + (\sigma^2_f/s + \sigma^2_{fe}/s + \sigma^2_p/sb + \sigma^2_w/sbn) \\ &\quad - 2(\sigma^2_f/s + \sigma^2_{fe}/s + \sigma^2_p/sb + \sigma^2_w/sbn)] \\ &= [(\sigma^2_f + \sigma^2_{fe} + \sigma^2_p/b + \sigma^2_w/bn) \\ &\quad - (1/s)(\sigma^2_f + \sigma^2_{fe} + \sigma^2_p/b + \sigma^2_w/bn)] \\ &= [(s-1)/s]\ (\sigma^2_f + \sigma^2_{fe} + \sigma^2_p/b + \sigma^2_w/bn).\end{aligned}$$

The variance of a family mean expressed as a deviation from an estimated location mean is the variance of a family mean reduced by a factor, (s-1)/s. This finite sample size correction factor corrects the variance of a family mean to account for the fact that the test location mean is estimated by averaging over a finite number of families.

RANDOM VECTORS

Definitions

Random vector: A random vector is a vector containing elements that are random variables. Suppose we have a 3 x 1 column vector whose elements are the open-pollinated family means for a given parent from three different test locations (such as the first three elements of **y** in Example 1.2). Then since each element is a random variable, the vector is a random vector.

Each element of a random vector has an expected value which is found by applying the techniques discussed previously in this chapter. If a new vector is created of the same dimension whose elements are the expected values of each of the elements of **y**, this is written as E(**y**). For the 3 x 1 **y** vector above,

$$E(\mathbf{y}) = \begin{bmatrix} E(y_1) \\ E(y_2) \\ E(y_3) \end{bmatrix}$$

So, in general the expectation of a random vector, **y**, is $E(\mathbf{y}) = \{E(y_i)\}$; that is, the expected values of the elements of **y**. While the random vector contains elements that are random variables, the vector of expected values contains only constants.

Variance of a random vector: Each of the random variables in a random vector also has a variance, $Var(y_i)$, and each pair of elements has a covariance, $Cov(y_i,y_j)$. It is often useful to arrange these together in a matrix called the variance-covariance matrix of **y**, written Var(**y**). For the 3 x 1 **y** vector above,

$$\mathrm{Var}(\mathbf{y}) = \begin{bmatrix} \mathrm{Var}(y_1) & \mathrm{Cov}(y_1,y_2) & \mathrm{Cov}(y_1,y_3) \\ \mathrm{Cov}(y_1,y_2) & \mathrm{Var}(y_2) & \mathrm{Cov}(y_2,y_3) \\ \mathrm{Cov}(y_1,y_3) & \mathrm{Cov}(y_2,y_3) & \mathrm{Var}(y_3) \end{bmatrix}$$

The variances of the three elements of **y** are on the diagonal of Var(**y**). The covariances are the off-diagonal elements and $\mathrm{Cov}(y_i,y_j) = \mathrm{Cov}(y_j,y_i)$ is found two places: in the ith row, j^{th} column and in the jth row, i^{th} column. That is, Var(**y**) is a symmetric matrix.

To generalize, Var(**y**) is formally defined as E{[**y** - E(**y**)] [**y** - E(**y**)]'}. If **y** has dimension n x 1, this multiplication is (n x 1) (1 x n) and the result is n x n. Thus, the variance-covariance matrix of an n x 1 random vector, **y**, is an n x n symmetric matrix with the variances of the elements of **y** on the diagonal and covariances for the off-diagonal elements.

While **y** is a random vector containing elements that are random variables, Var(**y**) is a matrix of **constants**. That is, each element of the variance-covariance matrix is a constant (usually an unknown constant) parameter associated with the distribution of a random variable or the joint distribution of two random variables. In other words, each y_i in **y** is a random variable that has experimental error associated with it; however, $\mathrm{Var}(y_i)$ is a constant parameter describing the distribution of y_i and $\mathrm{Cov}(y_i,y_j)$ is a constant parameter associated with the joint distribution of y_i and y_j.

Covariance of two random vectors: In an analogous fashion as Var(**y**) was defined in the preceding section, Cov(**y**,**x**') = E{[**y** - E(**y**)] [**x** - E(**x**)]'}. If **y** and **x** have dimensions n x 1 and m x 1, respectively, the covariance matrix will have dimension n x m. Each element in the covariance matrix is a covariance between an element of **y** and an element of **x**. Thus element 1 is $\mathrm{Cov}(y_1,x_1)$. Cov(**y**,**x**') is often written as Cov(**y**,**x**) omitting the transpose operator on **x**, and this convention is used in this text. Note that Cov(**y**,**x**') = [Cov(**x**,**y**')]'.

Theorems

Just as it is common to make linear transformations on scalar random variables, so too this is done with random vectors. Let **w** be the n x 1 random vector obtained by premultiplying the n x 1 random vector **y** by **A**, an n x n matrix of constants. That is, let **w** = **Ay**. Note that **A** is composed of many different elements, but each a_{ij} is a constant, not a random variable. Each element in **w** is the sum of products in which each product is a constant times a random variable. For example if **A** is 3 x 3 and **y** is 3 x 1, then the first element of **w** is $w_1 = a_{11}y_1 + a_{12}y_2 + a_{13}y_3$.

We will sometimes use the following theorems concerning linear transformations of random vectors.

$$E(\mathbf{A}) = \mathbf{A}\ . \qquad \textbf{Eq. 2.28}$$

$$E(\mathbf{w}) = E(\mathbf{Ay}) = \mathbf{A}E(\mathbf{y})\ . \qquad \textbf{Eq. 2.29}$$

$$Var(\mathbf{w}) = Var(\mathbf{Ay}) = \mathbf{A}Var(\mathbf{y})\mathbf{A}'\ . \qquad \textbf{Eq. 2.30}$$

$$Cov[\mathbf{Ay},(\mathbf{Bx})'] = Cov(\mathbf{Ay}, \mathbf{x}'\mathbf{B}') = \mathbf{A}Cov(\mathbf{y},\mathbf{x}')\mathbf{B}' \qquad \textbf{Eq. 2.31}$$

PROBLEMS

Problem 2.1: A binomial random variable, z, takes on the value 1 with probability of p and the value of 0 with probability of q. Use the definitions of expected value (Eq. 2.1) and variance (Eq. 2.5) to derive the expected value and variance for z. (Hint: $p + q = 1$).

Problem 2.2: Using the definition of expected value (Eq. 2.2), prove that $E(by + c) = bE(y) + c$ where b and c are constants and y is a random variable.

Problem 2.3: Show algebraically that $E[y - E(y)]^2 = E(y^2) - [E(y)]^2$.

Problem 2.4: Consider a Douglas-fir plantation in which mensurationists have measured the heights (x) and diameters (y) of a sample of 10-year old trees. Evaluate the following linear transformations: 1) in terms of the Var(x), Var(y) and Cov(x,y) (briefly describe the meaning of each) and 2) numerically assuming E(x) = 30 ft, E(y) = 10 in, Var(x) = 400, Var(y) = 100, Cov(x,y) = 160 (give the units of measurement with all answers).

a. Var(2x + 5).
b. Cov(2x,3y).
c. Corr(2x,3y).
d. Var(2x + 3y).

Problem 2.5: Using Eq. 2.14, show algebraically that Cov(bx,bx) = b^2Var(x).

Problem 2.6: Use the basic definitions to show Eq. 2.17. State this property in words.

Problem 2.7: "Standardization" is a commonly used transformation in tree improvement (See Chapter 3 for further details). Consider a random variable, y, for which we know the true expected value, μ, and variance, σ^2. Derive the mean and variance of the standardized variable, z, where $z = (y - \mu)/\sigma$.

Problem 2.8: Consider a balanced open-pollinated progeny test with 30 families in a Randomized Complete Block design with 4 blocks and 10 trees of each family in each block. For each of the factors in Eq. 2.23 (i.e. μ, E_i, etc), state how many effects (i.e. levels) there are for that factor.

Problem 2.9: Express the following in terms of the model factors in Eq. 2.23:

a. A plot mean of the kth family in the jth block at the ith location.
b. A family mean for the kth family averaged over three locations.
c. The variance of the family mean in (b) above.

Problem 2.10: Assume the linear model in Eq. 2.23 with b=3 blocks and n=10 trees per family in each block. Consider the following family means all for the kth family (Eq. 2.25 expresses these family means in terms of the linear model factors):

y_1 = family mean from test location 1 for 5-yr height (h5),
y_2 = family mean from test location 1 for 5-yr DBH (d5),
y_3 = family mean from test location 2 for 10-yr height (h10), and
y_4 = family mean from test location 2 for 10-yr DBH (d10).

a. Let **y**' = [y_1 y_2 y_3 y_4] and write **V** = Var(**y**) in words. For example, v_{11} is the variance of family means for 5-yr height from location 1.
b. Write the elements of **V** in terms of the linear model effects.

Problem 2.11: Let **y** be the data vector from Problem 2.11. Suppose an investigator wanted to create a 2 x 1 random vector of the family means averaged across test locations (one mean for each trait) that placed twice the weight on 10-year data as on 5-year data.

a. Find a matrix **A** so that **Ay** = **w** where **w**' = [$y_1 + 2y_3$, $y_2 + 2y_4$].
b. Find **S** = Var(**Ay**).

Problem 2.12: Referring to Example 2.1 in the section on finite sample size corrections for variances of deviations, show that $\text{Cov}[(p_{i\cdot k}/b),(p_{i\cdot\cdot}/sb)] = \sigma^2_p/sb$. This should entail expanding each of these terms and conducting the algebra.

Chapter 3
Concepts of Progeny Test Analysis

INTRODUCTION

Forest tree breeders establish progeny tests primarily to predict breeding values of parents based on the performance of their offspring. The breeder then ranks parents according to their predicted breeding values (or some function of their predicted breeding values) to make selections. Progeny tests are generally replicated over a number of years and planting locations with the objective of sampling a wide range of environmental conditions. Because of large numbers of parents being evaluated, economic incentives to establish tests quickly, differential survival, etc. there is often a great deal of unbalance across all progeny tests in forest tree breeding programs. For example, all families are not always planted in the same year or location, and some families may be represented at only a few sites (Hatcher et al. 1981, Cotterill et al. 1983). This can make the goal of accurately and precisely ranking parents quite complex as data from a large number of progeny tests with diverse characteristics must be combined into one value representing the relative genetic worth of the parent. It is common in forestry to calculate a family mean of all offspring of a particular parent in a given test, and then somehow average those means across all tests in which the parent is represented. Different approaches to calculating the average, and different transformations of the data are sometimes utilized to deal with the problems associated with 'messy' progeny test data. The types of 'messy data' problems common in tree breeding programs include parents tested in different series of tests that may differ in mating design, field design, precision, mean, variances, age, and degree of correlation with the trait of interest. Also, parents are often in different numbers of tests. In this chapter we discuss a number of factors that make progeny test data difficult to analyze, the effects these may have on selection, and some common approaches to ranking parents used in tree breeding. In this chapter, we assume that open-pollinated progeny from several parents are planted in randomized complete block designs at one or more physical site locations. This facilitates discussion, but most concepts presented do not depend on this choice of design.

ESTIMATING FAMILY PERFORMANCE IN A SINGLE TEST

Before discussing multiple progeny tests, we first consider a number of methods of calculating a family mean in a single test. The intent is to develop an unbiased, precise estimate of the performance of a given family at a particular location. The simplest approach is the **arithmetic average** across all trees in a given family. If a family is not represented equally (i.e. by the same number of trees) in all blocks, however, an arithmetic average will be a biased estimator of the family effect. This is because the fixed block effects will not sum to zero; blocks with more trees from a family will contribute more block effects to the family mean. For example, if a family has 10 trees in a very fast growing block but only two trees in a slow growing block, the simple arithmetic family mean calculated across the 12 trees will be upwardly biased by having 10 "good" block effects and only 2 trees having "bad" block effects.

To circumvent this problem, family means are often calculated as the **mean of the plot means** by first averaging all trees from a given family in a given block, and then averaging across all blocks. Using this approach there is one plot mean for each family in each block (assuming no missing plots), block effects will sum to zero and the estimate of the family mean will be unbiased (i.e. unaffected by block effects) if a family is represented in each block. When there are no missing plots, the "mean of the plot means" for a given family is equivalent to what is called a least squares adjusted mean (LSM) for that family (Freund and Littell 1981; Milliken and Johnson 1984). While raw (i.e. arithmetic) means are a function of the number of observations in each plot and hence affected by block effects, LSM's are not.

When families do not appear in all blocks, least squares means can still be used to estimate family means. LSM's are free of bias due to unequal representation across blocks, and are essentially the family marginal means that would be expected if all families were in all blocks (Freund and Littell 1981). Finally, **generalized or weighted least squares** (GLS) can be utilized to estimate family means taking into account differences in precision among blocks.

If there are large differences among families in number of trees per block (or missing plots), or among blocks in number of trees (e.g. due to survival differences) or if there are differences among blocks for within block environmental variability, simple arithmetic averages using raw phenotypic values can cause either biased and/or imprecise family rankings. If families are represented in all blocks and blocks have approximately the same error variance, the "mean of the plot means" approach to estimation of family effects will generally be satisfactory, although GLS may be marginally more efficient.

FAMILIES TESTED IN DIFFERENT SERIES OF PROGENY TESTS

In most forest tree breeding programs all parents are not included in all progeny tests, so data from two or more different series of progeny tests must be used to develop comparable breeding values to rank the parents across all series of tests simultaneously. There will always be differences between any two progeny tests, and it is obviously desirable to put more emphasis or weight on data from 'good' progeny tests, and calculate the rankings accordingly. How 'good' a progeny test is can, at least in part, be determined by answering two questions:

1) Are the observations from the test reliable in a genetic sense (or in more quantitative terms, is the trait measured heritable)?

2) Does the information from the test provide information about the trait of real interest, i.e. what is the genetic correlation between the trait measured and the trait being predicted?

For example, assume we wish to rank loblolly pine parents on the basis of their breeding value for height growth at an age of 25 years. We have ten-year-old progeny test data on two traits, height growth and needle color. Under criteria 1, ten-year height growth is a moderately heritable trait, while needle color is highly heritable. Under criteria 2 however, information on ten-year height growth may provide some information about 25-year height growth, but information on needle color will likely be of no value.

Tests of Different Precision

Often progeny tests at two different locations differ in experimental error, i.e. have different levels of environmental and experimental variability associated with observations. Higher experimental error may be caused by inherently high microsite variability, poor site preparation, poor layout of blocks, poor quality of the seedlings, poor planting technique, etc. Another difference could be in the experimental designs, whether intentionally through the use of different numbers of blocks, families, or planting configuration, or unintentionally through differences in survival on the two sites. In either case measurements from any two progeny tests are quite likely to be of different levels of precision. More experimental error implies measurements of lower precision that have lower heritability and higher variance of family means.

Example 3.1: To examine the effects of different numbers of replication on precision and variances, suppose you know the following parameters apply to half-sib progeny tests at two locations (A and B): $\sigma^2_f=10$, $\sigma^2_{fe}=8$, $\sigma^2_p=25$, $\sigma^2_w=90$. For progeny test A with 10 blocks and 10 trees per plot, the variance of family means ($Var((\bar{y}_{i\cdot k\cdot}))$ can be calculated from Eq. 2.26 as 21.4 = (10 + 8 + 25/10 + 90/(10•10)). The heritability of family means (h^2_f) equals $\sigma^2_f/Var(\bar{y}_{i\cdot k\cdot})$ = 10/21.4 = 0.47. Compare those values to test B with 4 blocks and 5 trees per plot: $Var(\bar{y}_{i\cdot k\cdot})$ = 28.75 = (10 + 8 + 25/4 + 90/(4•5)), h^2_f = 0.35 = 10/28.75. Because of different designs, test B has lower heritability and higher variance of family means than test A.

Different levels of precision might be encountered in a third test (location C) that has the same design as B (4 blocks and 5 trees per block) but higher values of environmental or experimental variability. Suppose σ^2_p = 50 and σ^2_w = 200, while the two genetic variances are the same as tests A and B. The variance among family means in test C is 40.5 (largest of all three tests) and the family heritability for test C is 0.25 (smallest of all three tests).

Tests With Different Means and Variances

Estimating site means: As discussed in Chapter 2, different tests often have different site means. For example, average height of all trees in one test may be 15 feet, while in another test (on a better site) the average height may be 18 feet. When comparing families in the two tests, family means should be adjusted for the fixed effects of the specific location in which they occur by expressing the family means as deviations from the overall test or site mean. Site means are often estimated through the use of genetic checks, which are planted in many progeny tests to provide a baseline for comparison across tests which do not have identical family representation. Checklots should be genetically broad-based so that performance is not markedly affected by genotype-environment interaction (Lowe et al. 1983). If each test includes a large random sample of families, the overall test mean will probably be a better estimate of the site or location effect than the checklot mean, as the overall test mean will generally be more precisely estimated and may be less biased by genotype-environment interaction.

Effect of different variances on selection: Another effect often observed when tests have different means is that they have different variance structures as well. As discussed in Example 3.1, different test designs or relative amounts of experimental error can make family means in one test more variable than in another test. It is also possible for different variances to occur because of scale effects, i.e. measurements with larger absolute values tend to have larger variances. Variances for many biological traits increase as the mean of the trait increases (Snedecor and Cochran 1967), and this can have a significant effect on the ranking and selection of parents. A common example in forestry is that as tests age, growth measurements increase and variances also increase.

Consider the case where all families that have family means above a certain truncation point are selected. If there are two distributions of family means (e.g. from two different test locations) that have a common mean but different variances, and if a single truncation point is applied to both distributions, a higher percentage of families will be selected from the distribution with the higher variance (assuming selection of less than half of the population) (Figure 3.1). Depending on differences in variances and intensity of selection, the difference between the percentage selected from the two distributions can be substantial.

Example 3.2: Consider two populations of family means (A and B) each having 1000 families measured for the same trait y that is normally distributed. Assume $\bar{y}_A = \bar{y}_B = 100$, $Var(y_A)=100$, and $Var(y_B)=121$ where $\bar{y}_A$ and $\bar{y}_B$ are the means over all 1000 families in each population and y_A and y_B are individual family means from population A and B. Suppose we select all family means that are 20% above the mean (i.e. above $y_i=120$ where i can be A or B). The standard normal deviate (Z_i) equivalent to the y=120 truncation point for each population is:

$$Z_i = (120 - 100)/ [Var(y_i)]^{½}.$$

Thus, $Z_A = 2.00$ and $Z_B = 1.82$. From a normal table (Appendix 1), the cumulative percentage of the distribution above the truncation point can be determined for populations A and B, and are 0.0228 and 0.0344, respectively. Therefore, we would expect to select approximately 23 (i.e. 1000 • 0.0228) families from population A and 34 from population B. This is 50% more families

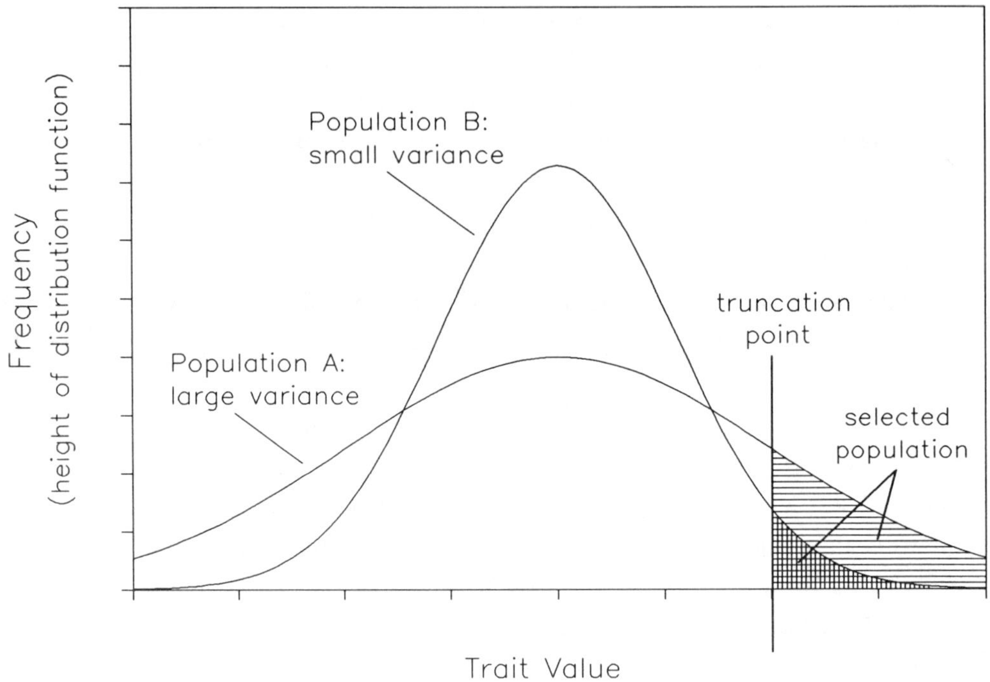

Figure 3.1. The effect of different variances on the percentage chosen when using truncation selection.

selected from the population with larger variance. If this larger variance results from reduced precision, it seems highly undesirable to select more families from the population with the most experimental error (and lowest heritability).

A similar effect of different variances is observed when calculating a grand family mean by averaging family means for a given parent across several tests having different variances. The grand family mean for the parent is most affected by family means from tests with larger variances. The effects of different variances on selection and family rankings are generally considered undesirable in forest tree breeding and procedures have been developed to homogenize variances of observations from different progeny tests prior to averaging across tests or selection (Hatcher et al. 1981, Cotterill et al. 1983). Some of these approaches are described in "Common Approaches to Ranking Parents" at the end of this chapter.

Tests With Different Correlations To The Trait Of Interest

One of the criteria used in evaluating the quality of progeny test data is its degree of relationship to the trait of interest or economic importance. Different progeny tests in which the same trait is measured may have different degrees of correlation with the genetic value of the trait being predicted. This is intuitively apparent when discussing growth traits measured in tests of varying age. In many cases, a trait of high economic importance is volume yield at rotation age; thus, the trait to predict is genetic value for volume yield at economic maturity. If progeny from a given parent are grown in a number of different tests of varying ages, data on volume yield from younger tests will likely be less correlated with mature volume than data from older tests. In predicting the genetic value for mature volume yield for that family, we would like to place more weight on older data. In other words, volume data from different aged tests are considered different traits and some traits (i.e. from older tests) have higher correlation with the trait being predicted (mature volume).

Even growth traits measured in different progeny tests of the same age may be differently correlated to the genetic value being predicted, mature growth at harvest.

Example 3.3: Suppose offspring from many different Douglas-fir parents are planted in two progeny tests: elevation 1000 and 2000 feet. Both tests are measured for volume growth at age 20 and are equally precise. We wish to rank (and subsequently select) parents which produce best growth for a breeding region at elevation 2000 feet; therefore we want to predict breeding values for volume growth at age 60 at 2000 feet. Although the same trait, volume growth at age 20, is measured in both tests, the data from the test at 2000 feet will likely be more correlated with the breeding value being predicted. While data from both tests provide information about growth at 2000 feet, we can treat the data from the two tests as if they are two distinct traits that have different correlations with growth at age 60 at 2000 feet.

Genetic correlations: Recall that we want to use the performance of offspring in some specific tests to predict breeding values of the parent trees, or in other words, the performance of future offspring from the parents. To do this it is useful to know the correlation between observations on offspring in different environments. Genetic correlations have been generally estimated when both traits are measured on the same individuals (Lambeth 1980, 1983). This type of correlation is designated a type A correlation.

Correlations can also be calculated using measurements on different individuals within a genetic group, and these correlations are designated type B correlations (Burdon 1977). Type B correlations arise from treating a trait (e.g., volume growth) expressed in two different environments as two different traits. This is a reasonable approach as the genes affecting growth in one environment are not necessarily the same genes affecting growth in another environment (Falconer 1981). The type B genetic correlation (r_{Bg}) is of the form:

$$r_{Bg} = Cov_{f(x,y)} / (\sigma_{f,x}\, \sigma_{f,y})$$

where $Cov_{f(x,y)}$ = genetic covariance among families for the trait as it is expressed in environments x and y, and $\sigma^2_{f,x}$ and $\sigma^2_{f,y}$ are variances among true (i.e. genetic) family means expressed in environments x and y.

A type B genetic correlation equal to one indicates that there is no genotype-environment interaction between a trait expressed in two environments. Conversely, type B correlations will be less than one whenever there is genotype-environment interaction, and as the relative amount of genotype-environment interaction increases, type B correlations decrease.

When combining data across multiple progeny tests to rank parents according to their breeding value, a common approach is to assume that all measurements of the same trait (e.g. height at age 5) are equally correlated with the breeding value of interest (e.g. volume at age 30). An alternative, and perhaps more desirable, approach would be to consider traits measured in different locations as different traits and somehow weight the observations from different tests according to their correlation with the breeding value being predicted.

Tests of Different Age

Often tree improvement programs will establish progeny tests over a long period of time, and therefore it is common that progeny test information on different parents will vary widely in age. To summarize what has been alluded to throughout the above discussion, for any kind of growth trait, different aged tests are affected by nearly all possible 'messy' data problems. Growth observations in older tests will generally be of higher precision and heritability, have larger means and variances, and be more correlated with the growth trait at harvest age than growth observations in younger tests.

FAMILIES TESTED IN DIFFERENT NUMBERS OF TESTS

If families are represented in different numbers of tests and are ranked by the grand family mean across all tests, families with the highest (and lowest) rankings will tend to be those tested in the fewest numbers of tests. This problem occurs because of the increasing variance of grand family means with decreasing numbers of tests, and the subsequent effects of different variance structures on truncation selection.

Consider grand family means from open-pollinated progeny tests in which different parents are in different numbers of tests. There may be some sort of standardization or transformation of all family means in each test prior to averaging across tests, but this does not alter the arguments presented below. The grand family mean for the kth parent is obtained by averaging across all individuals, blocks and locations in which offspring from the kth parent occur. Using the linear model in Eq. 2.23 and assuming balanced data:

$$(\bar{y}_{\cdot\cdot k\cdot}) = [(\sum_{i}^{t} \sum_{j}^{b} \sum_{l}^{n} y_{ijkl})/tbn]$$

$$= [\sum_i^t \sum_j^b \sum_l^n (\mu + E_i + B_{ij} + f_k + fe_{ik} + p_{ijk} + w_{ijkl})/tbn]$$

$$= \mu + f_k + (fe_{\cdot k})/t + (p_{\cdot\cdot k}/tb) + (w_{\cdot\cdot k\cdot})/tbn \qquad \textbf{Eq. 3.1}$$

and the variance among grand family means is

$$Var(\bar{y}_{\cdot\cdot k\cdot}) = Var[\mu + f_k + (fe_{\cdot k})/t + (p_{\cdot\cdot k}/tb) + (w_{\cdot\cdot k\cdot})/tbn]$$

$$= \sigma^2_f + \sigma^2_{fe}/t + \sigma^2_p/tb + \sigma^2_w/tbn. \qquad \textbf{Eq. 3.2}$$

Note that $Var(\bar{y}_{\cdot\cdot k\cdot})$ decreases with increasing values of t, i.e. as a family is included in more tests, and this will occur regardless of how the family mean is calculated or transformed. This is not a problem if all families are included in equal numbers of tests, as then all family means will come from distributions with the same variance. But suppose there are large differences in the number of tests in which families are included. Then, parents represented in few tests will have a larger variance of grand family means than families in many tests. If grand family means are compared (i.e. ranked) without accounting for this difference, those families with the highest (or lowest) values will tend to be those which are represented in a small number of tests. Thus, a higher percentage of selected parents will be those that are represented in fewer tests. This effect of different variance structures that results in more likelihood of selecting poorly tested parents is equivalent to that in Example 3.2 where the different variances arose from differences at single test locations. Here, grand family means with higher variance arise because they are in fewer tests.

COMMON APPROACHES TO RANKING PARENTS

A number of different approaches to combining data from multiple progeny tests into one value in order to rank parents have been utilized in forest tree breeding. Most of these approaches are the same ones used in estimating family means in a single test, and nearly all treat the problem of ranking parents as one of estimating fixed family effects.

The simplest approach to combining data is simply taking an **arithmetic average** across all tests to calculate a grand family mean, which is considered an estimate of the general combining ability (GCA), or half of the breeding value. In general, family means are corrected for the environmental mean, i.e. family means are expressed as deviations from a commercial check or the test mean, before averaging. Using family means expressed as deviations can correct for differences in test means (fixed environmental effects), but all other 'messy' data situations discussed previously are potential problems.

An improvement over arithmetic averages is to use **least squares estimates** of family means to account for bias due to unequal representation of families in all tests. Least squares estimates are the family means that would be expected with all families in all tests, and can also be considered estimates of GCA. When using least squares estimates an additive model of family and test effects is assumed, i.e. there is no family-test interaction.

A related approach is **generalized or weighted least squares** which weights observations in different progeny tests by the residual mean squares of the tests. Family means from precise progeny tests have a greater influence on the estimate of the family effect or GCA than those from imprecise tests, and thus generalized least squares accounts for differences in test mean, non-orthogonality across tests, and test precision. (For other presentations of least squares and generalized least squares, see Kmenta 1971, Ch. 11 and Ch. 12; for a forestry application, see Cotterill et al. 1983.)

It is quite common in forestry progeny tests for different tests to have markedly different variances of family means, the tests being of different ages and different precisions. This is particularly true for growth traits where the variance is often correlated with the mean. One approach to homogenizing variances from different tests is through the use of **averaged standard scores.** Observed family means from a given test, are standardized to a distribution with a mean of zero and variance of one (see Problem 2.8), and then averaged across tests (Hatcher et al. 1981). A second approach with the same goal of homogenizing variances is a **log transformation** of the observed values (Cotterill et al. 1983). Both transformations attempt to correct for different variances (which are correlated to the test means) when ranking parents. A problem with any type of transformation, however, is back-transformation, which in this case is interpreting the grand family means of the transformed variable in terms of the real variable of interest. For example, averaged standard scores for height at 5 years may be reasonably correlated with volume at 25 years, however it would be useful to be able to make some kind of conversion of the unitless standard scores into genetic gain in units of volume at maturity.

SUMMARY AND CONCLUSIONS

All the above approaches (including the transformation approaches) treat family effects as fixed effects that can be estimated using some type of averaging approach. If families are present in all tests (or roughly the same number of tests) and if the progeny test data are relatively uniform (same age, nearly equal precision and approximately equal correlation with the breeding value of interest), the above approaches will probably rank parents quite satisfactorily. In this situation, few selection errors will be made; that is, the same parents will be selected as "winners" regardless of which method of analysis is used.

If data are of quite different quantity or quality among the parents being ranked, the prediction methods described in subsequent chapters may give much more precise and more interpretable predictions of genetic values. These prediction methods treat the genetic values being predicted as random variables and account for test data of different quality and quantity.

PROBLEMS

Problem 3.1: Suppose two progeny tests each include a different set of 100 families, and you know the following parameters apply to both tests: $\sigma^2_f=6$, $\sigma^2_{fe}=6$, $\sigma^2_p=10$, $\sigma^2_w=78$. Further suppose Test A has 10 blocks and 10 trees per plot, while Test B has 3 blocks and 5 trees per plot. Use Eq. 2.26 to determine the variances of family means ($\mathrm{Var}(\bar{y}_{i\cdot k\cdot})$) in tests A and B. If only those families that average 5 feet above the mean are selected, how many families would you expect to select from Test A? Test B?

Problem 3.2: Consider Tests C and D that contain different groups of 100 families each and have the same mean for height, 25 feet. The coefficient of variation (CV) calculated on a family mean basis is 0.2 in Test C and 0.5 in Test D.

How many families from Test C are expected to be more than 6 feet above the mean? From D?

Problem 3.3: If Test E (with another group of 100 families) and Test C (from Problem 3.2) have the same variance of family means, but the mean in Test E is 27 feet. If the truncation point for selection is 30 feet, how many families from Test C will we select? From E?

Problem 3.4: Assume that the following parameters apply to all progeny tests: $\bar{y}_{i\cdots}=20$ ft, $\sigma^2_f=6$, $\sigma^2_{fe}=6$, $\sigma^2_p=15$, $\sigma^2_w=73$, b=number of blocks=6, n=number of trees per plot=6. You have a group of 200 parents tested in 1 test, a second group of 200 tested in 4 tests, and a third group of 200 tested in 8 tests.

a. What is the variance of grand family means ($\text{Var}(\bar{y}_{\cdot\cdot k\cdot})$ as in Eq. 3.2) across 1 test (t=1)? 4 tests? 8 tests?

b. If the truncation point is $\bar{y}_{\cdot\cdot k\cdot}=25$ ft, how many families will be selected from the group tested in 1 test? 4 tests? 8 tests?

c. Explain in words why the number selected from thc group in four tests (S_4) is approximately equal to the number selected from 8 tests (S_8).

Problem 3.5: Now assume that the following parameters apply to all progeny tests: $\bar{y}_{i\cdots}=20$ ft, $\sigma^2_f=8$, $\sigma^2_{fe}=4$, $\sigma^2_p=15$, $\sigma^2_w=73$, b=number of blocks=6, n=number of trees per plot=6. You have a group of 200 parents tested in 1 test, a second group of 200 tested in 4 tests, and a third group of 200 tested in 8 tests.

a. What is the variance of grand family means ($\text{Var}(\bar{y}_{\cdot\cdot k\cdot})$) across 1 test (t=1)? 4 tests? 8 tests?

b. If the truncation point is $\bar{y}_{\cdot\cdot k\cdot}=25$ ft, how many families will be selected from the group tested in 1 test? 4 tests? 8 tests?

c. Compare the ratio S_1:S_4 to the same ratio in Problem 3.4. Which is larger? Why?

Chapter 4
Theory of Best Linear Prediction

INTRODUCTION

In general, all selection processes first rank the candidates using some function of the observed data and then choose the selected portion as those with the largest (or smallest) values of that function. More explicitly, a function of **y**, an n x 1 vector of observed data records, is used to predict **g**, a q x 1 vector of genetic values of q genetic entities. The vector of predicted genetic values is calculated from $\hat{\mathbf{g}} = f(\mathbf{y})$ and the genotypes with the best $\hat{g}$ values are selected.

Example 4.1: Suppose we have measured (i.e. observed) five-year heights on open-pollinated families from 2 parents. Family 1 was planted in three test locations but family 2 was missing from location 1. One simple, but not optimal, approach to predicting breeding values g_1 and g_2 for 5-year height of the two parents is to calculate the family means for each parent as follows:

$$\hat{g}_1 = 1/3\ (y_1 + y_2 + y_3) \quad \text{and} \quad \hat{g}_2 = 1/2\ (y_4 + y_5)$$

where y_1, y_2 and y_3 are the family means for parent 1 from tests 1, 2, and 3, and y_4 and y_5 are the two family means observed for parent 2 in tests 2 and 3.

These are linear functions of the data (e.g. we did not square the family means) and are represented in matrix notation as $\hat{\mathbf{g}} = \mathbf{B'y}$ or

$$\underset{\hat{\mathbf{g}}}{\begin{bmatrix} \hat{g}_1 \\ \hat{g}_2 \end{bmatrix}} = \underset{\mathbf{B'}}{\begin{bmatrix} 1/3 & 1/3 & 1/3 & 0 & 0 \\ 0 & 0 & 0 & 1/2 & 1/2 \end{bmatrix}} \underset{\mathbf{y}}{\begin{bmatrix} y_1 \\ y_2 \\ y_3 \\ y_4 \\ y_5 \end{bmatrix}}$$

To make maximum progress from the selection, it is necessary to use a function of the data that results in the predicted genetic values being ranked as closely as possible to the ranking of the true (but always unknown) genetic values. Best linear prediction is a method of finding a function of the data that, when certain assumptions hold, is in some senses optimal. In this chapter, we begin with a discussion of the distinction between estimating fixed effects *vs* predicting random variables. This is followed by the concepts, assumptions and theory involved with best linear prediction and a description of the statistical properties of best linear predictions. The chapter concludes with a series of topics related to various aspects of the theory and application of best linear prediction.

The theory in this chapter is completely general in the sense that it applies to selection processes at all levels from selection of the best individuals to the selection of the best parents, clones or provenances. While the theory is general, we illustrate the concepts with examples dealing with selection of the best parents based on their predicted breeding values developed from progeny test data. It was this problem of predicting parental breeding values from complex and messy progeny test data that motivated our interest in BLP. We believe this is an important application of BLP and is also the subject of Chapters 5-8. Selection of individuals is the subject of Chapters 9 and 10.

PREDICTION VS ESTIMATION

All of the data analysis techniques discussed in Chapter 3 treat genetic values as fixed effects which are estimated from the data (model I of Eisenhart 1947). The estimates then apply specifically to the sample of genotypes actually present in the tests and not to any broader population. When the exact inference population is specified in forestry, we believe it is almost always more appropriate to treat the breeding values as random variables (model II of Eisenhart 1947) rather than fixed effects. This approach has been widely adopted in dairy science (Henderson 1973, 1977, 1984; Searle 1974) where sire breeding values are treated as random variables. In forestry, we have somewhat of a dichotomy in that family effects are treated as random variables for the estimation of variance components and then as fixed effects for the purposes of ranking families (see Chapter 2 for more discussion).

While the distinction between the two approaches (fixed *vs* random) may seem subtle and even semantic, they lead to different linear models and therefore to different functions of the data that are used to rank the genotypes. This results in different properties of the ranking criteria between random and fixed approaches.

Example 4.2: Consider Example 4.1 where we used a fixed effects approach (family averages). The family averages are estimates of family effects which apply to: 1) five-year height, 2) only the sample of offspring tested and 3) only the test environments in which the sample of offspring was planted. They also assume that these two families are the entire population of parents for which we wish to estimate breeding values. On the contrary, in most progeny test situations, the parents represented at a given test location are just a sample of the total population of parents for which we need to predict breeding values. Further, we often want the predicted breeding values to apply to: 1) some other trait than actually measured (e.g. mature volume instead of five-year height), 2) other offspring from the parents, not to those planted in the progeny tests, and 3) to other test environments.

Thus, the problem becomes one of predicting breeding values, not simply estimating family effects. When the data are completely balanced and equally precise from all test locations and for all parents, the two analytical approaches yield the same genotype rankings; however, the predicted breeding values will almost never be similar between the two systems (even though they may rank genotypes similarly). In fact, the fixed effects approach necessarily estimates the family effects for the trait measured and these estimates will therefore be in the units measured (e.g. feet at 5 years). When the genetic effects are assumed random, the predicted values are expressed in the units of that trait being predicted, called the predictand (e.g. cubic feet at rotation age). The trait being predicted may or may not be the same as that measured (e.g. measured 5 year heights used to predict breeding values for volume at rotation age).

In general, the predictions developed when genetic effects are assumed random have more desirable statistical properties than when they are assumed fixed (see "Properties of Best Linear Predictions" later in this chapter). Best linear prediction, discussed in Chapters 4-8, is one such approach to predicting breeding values which is appropriate when means of extraneous environmental and treatment effects are well estimated by simple approaches such as simple averages. When these fixed "nuisance" effects need estimating simultaneously with the prediction of random genetic effects, then the appropriate technique is Best Linear Unbiased Prediction discussed in Chapters 11 and 12.

THEORY OF BEST LINEAR PREDICTION

Specification of the Problem

The objective is to develop a function of the observed n x 1 **y** vector of data that predicts, both accurately and precisely, **g**, a q x 1 non-observable random vector of genetic values. The dimension of **g** reflects both the number of traits and the number of genotypes for which predictions are desired. For example, q = 2s to predict breeding values for 2 traits on each of s parents.

The observed **y** vector must be jointly distributed (correlated) with **g** or else it will not be useful for predictions. However, the exact form of the distribution is almost never known and the derivation of BLP does not rely on knowledge of the exact joint distribution. The derivation of best linear prediction does assume that the following first and second moments of the joint distribution are known (Henderson 1973, 1977, 1984).

E(**g**) = $\boldsymbol{\tau}$ is a q x 1 vector representing the expected values of the genetic values being predicted. In forestry, this can often be assumed **0** so that the predicted genetic values will be expressed as deviations around zero. However, if previous selection has occurred, $\boldsymbol{\tau}$ may in some cases reflect the expected progress from that selection.

E(**y**) = $\boldsymbol{\alpha}$ is the n x 1 vector of expected values of the observed data (the first moments). The exact content of $\boldsymbol{\alpha}$ depends on the underlying linear model used to represent **y**, and these expected values are associated with the fixed effects portion of the linear model for each data point. Each data point may have a different expected value depending upon which set of fixed effects are associated with that observation. Commonly in forestry, the elements of $\boldsymbol{\alpha}$ represent environmental (e.g. test site location) means, but they may also represent block, treatment and other fixed effects. Herd, year and season are common fixed effects in dairy cattle breeding.

Var(**y**) = **V** is the n x n matrix of variances and covariances among the observations (the second moments of the data).

Cov(**y**,**g**') = **C** is the n x q matrix of covariances between the observations and the genetic values being predicted.

Example 4.3: Continuing with the hypothetical data of Example 4.1, the first and second moments are detailed below assuming 1) the observed data are open-pollinated family means of five-year heights (i.e. the **y** vector in Example 4.1), 2) these family means can be expressed in terms of the underlying linear model by Eq. 2.25, 3) we wish to predict breeding values of both 5-year height and 25-year (i.e. rotation age) volume for the two parents and 4) E(**g**) = **0**.

We want to predict

$$\hat{\mathbf{g}} = \begin{bmatrix} \hat{g}_{1v} \\ \hat{g}_{1h} \\ \hat{g}_{2v} \\ \hat{g}_{2h} \end{bmatrix}$$

where $\hat{g}_{1v}$ and $\hat{g}_{1h}$ are the predicted breeding values for 25-year volume and 5-year height for parent 1, etc.

The first and second moments are as follows.

$$E(\mathbf{y}) = \boldsymbol{\alpha} = \begin{bmatrix} E(y_1) \\ E(y_2) \\ E(y_3) \\ E(y_4) \\ E(y_5) \end{bmatrix} = \begin{bmatrix} \mu + E_1 \\ \mu + E_2 \\ \mu + E_3 \\ \mu + E_2 \\ \mu + E_3 \end{bmatrix}$$

where $\mu + E_1$, $\mu + E_2$ and $\mu + E_3$ are location means for test sites 1, 2 and 3, respectively (see Eq. 2.25 for derivation).

$$\mathrm{Var}(\mathbf{y}) = \mathbf{V} = \begin{bmatrix} \mathrm{Var}(y_1) & \mathrm{Cov}(y_1,y_2) & \mathrm{Cov}(y_1,y_3) & \mathrm{Cov}(y_1,y_4) & \mathrm{Cov}(y_1,y_5) \\ \mathrm{Cov}(y_1,y_2) & \mathrm{Var}(y_2) & \mathrm{Cov}(y_2,y_3) & \mathrm{Cov}(y_2,y_4) & \mathrm{Cov}(y_2,y_5) \\ \mathrm{Cov}(y_1,y_3) & \mathrm{Cov}(y_2,y_3) & \mathrm{Var}(y_3) & \mathrm{Cov}(y_3,y_4) & \mathrm{Cov}(y_3,y_5) \\ \mathrm{Cov}(y_1,y_4) & \mathrm{Cov}(y_2,y_4) & \mathrm{Cov}(y_3,y_4) & \mathrm{Var}(y_4) & \mathrm{Cov}(y_4,y_5) \\ \mathrm{Cov}(y_1,y_5) & \mathrm{Cov}(y_2,y_5) & \mathrm{Cov}(y_3,y_5) & \mathrm{Cov}(y_4,y_5) & \mathrm{Var}(y_5) \end{bmatrix}$$

where the diagonal elements are variances of open-pollinated family means at a given test location (Eq. 2.26) and off-diagonal elements are covariances between open-pollinated family means either at the same or at two different site locations (see Chapters 2 and 5 for details). As examples, $\mathrm{Var}(y_2)$ and $\mathrm{Var}(y_4)$ represent the variance among family means at test location 2 and will be equivalent if parents 1 and 2 are equally tested at location 2. $\mathrm{Cov}(y_2,y_4)$ is the covariance between the open-pollinated family means for families 1 and 2 at test location 2; this will be zero if the families are unrelated. Finally, the last type of covariance is exemplified by $\mathrm{Cov}(y_1,y_2)$, the covariance between open-pollinated family means from parent 1 at test locations 1 and 2. This is the type B covariance of Burdon (1977).

$$\mathrm{Cov}(\mathbf{y},\mathbf{g}') = \mathbf{C} = \begin{bmatrix} \mathrm{Cov}(g_{1v},y_1) & \mathrm{Cov}(g_{1h},y_1) & \mathrm{Cov}(g_{2v},y_1) & \mathrm{Cov}(g_{2h},y_1) \\ \mathrm{Cov}(g_{1v},y_2) & \mathrm{Cov}(g_{1h},y_2) & \mathrm{Cov}(g_{2v},y_2) & \mathrm{Cov}(g_{2h},y_2) \\ \mathrm{Cov}(g_{1v},y_3) & \mathrm{Cov}(g_{1h},y_3) & \mathrm{Cov}(g_{2v},y_3) & \mathrm{Cov}(g_{2h},y_3) \\ \mathrm{Cov}(g_{1v},y_4) & \mathrm{Cov}(g_{1h},y_4) & \mathrm{Cov}(g_{2v},y_4) & \mathrm{Cov}(g_{2h},y_4) \\ \mathrm{Cov}(g_{1v},y_5) & \mathrm{Cov}(g_{1h},y_5) & \mathrm{Cov}(g_{2v},y_5) & \mathrm{Cov}(g_{2h},y_5) \end{bmatrix}$$

where the first column elements are the covariances of the observations with the breeding value for 25-year volume for the first parent, the second column covariances are between the observations and the breeding value for 5-year height for the first parent, etc.

Derivation of BLP Formula

The goal of this section is to provide a complete and detailed derivation of the best linear prediction formula. This provides the information necessary to understand how, what and where assumptions are made and perhaps therefore to develop a feel for the importance of violating those assumptions. However, it is not necessary to understand the following derivation in order to use best linear prediction or to appreciate how best linear prediction "handles" messy data. Many readers may want to skim this section until the need for understanding the derivation motivates a more closer reading. A derivation along the same lines is sketched on page 39 of Henderson (1984).

We develop the formula for predicting a single genetic value, g_i, as opposed to the entire vector of breeding values, **g**. But, this is completely general since the ith element could be anywhere in **g** and therefore the formula applies to every element g_1, ... g_q in the **g** vector.

As with linear regression, BLP considers only linear functions of the data and we want to find $\hat{g}_i = a_i + \mathbf{b}_i'\mathbf{y}$ such that the predicted value, $\hat{g}_i$, tends to be close to the true breeding value, g_i. Note that g_i, a_i and $\mathbf{b}_i'\mathbf{y}$ are all scalar values (and hence conform for addition) that pertain to the ith parent. The $\mathbf{b}_i$ and **y** vectors are both n x 1, and $\mathbf{b}_i'\mathbf{y}$ forms a linear combination of the observed values. As an example of a linear function (not BLP), the first row of the **B'** matrix in Example 4.1 is used to form the linear function of the data for estimating the family mean of the first parent, i.e. $\mathbf{b}_1'\mathbf{y}$.

The approach we take is to find that linear combination of the data that minimizes the expected value (over repeated sampling) of the squared difference between the true and predicted breeding values. That is, find a_i and $\mathbf{b}_i$ to minimize $E(\hat{g}_i - g_i)^2$. The solution then possesses the minimum error variance among all possible linear combinations of the data (see next section on Properties of Best Linear Predictions).

To simplify the notation we omit the subscript, i, but g, a, and **b** still apply to the ith parent and **y** is the entire n x 1 vector of data. Thus the problem is to find $\hat{g} = a + \mathbf{b}'\mathbf{y}$ such that $E(\hat{g} - g)^2$ is a minimum.

The derivation proceeds in 5 steps by 1) substituting $a + \mathbf{b}'\mathbf{y}$ for $\hat{g}$ in $E(\hat{g} - g)^2$, 2) completing the square, 3) distributing the expected value, 4) taking the partial derivatives with respect to both a and **b** and setting the equations equal to zero and 5) solving the two equations simultaneously.

$$\begin{aligned} L &= E(\hat{g} - g)^2 = E(a + \mathbf{b'y} - g)^2 && \text{Step 1} \\ &= E(a^2 + \mathbf{b'yy'b} + g^2 + 2a\mathbf{b'y} - 2ag - 2g\mathbf{b'y}) && \text{Step 2} \\ &= a^2 + \mathbf{b'}E(\mathbf{yy'})\mathbf{b} + E(g^2) + 2a\mathbf{b'}E(\mathbf{y}) - 2aE(g) - 2\mathbf{b'}E(\mathbf{y}g) && \text{Step 3a} \\ &= a^2 + \mathbf{b'Vb} + \mathbf{b'}\boldsymbol{\alpha\alpha'}\mathbf{b} + E(g^2) + 2a\mathbf{b'}\boldsymbol{\alpha} - 2a\tau - 2\mathbf{b'c} - 2\tau\mathbf{b'}\boldsymbol{\alpha} && \text{Step 3b} \end{aligned}$$

Step 2 makes use of the following results from matrix algebra (cf Chapter 2 for further details):

a) The transpose of a scalar is that scalar so $\mathbf{b'y} = (\mathbf{b'y})' = \mathbf{y'b}$.

b) For scalars, multiplication is commutative so the order of the terms does not matter.

Step 3a uses the following (cf Chapter 2):

a) The expected value of a constant is just that constant and a^2 is a constant.

b) For any linear function of a random vector (e.g. $\mathbf{b'y}$), $E(\mathbf{b'y}) = \mathbf{b'}E(\mathbf{y})$.

c) $E(\mathbf{b'yy'b}) = \mathbf{b'}E(\mathbf{yy'})\mathbf{b}$ from Graybill (1976) theorem 1.9.3 part 4.

For Step 3b:

a) Recall

$E(\mathbf{y}) = \boldsymbol{\alpha}$ an n x 1 vector of constants,

$E(g) = \tau$ the expected genetic value of the ith genotype,

$Var(\mathbf{y}) = \mathbf{V} = E(\mathbf{yy'}) - \boldsymbol{\alpha\alpha'}$ (cf Chapter 2),

$Cov(\mathbf{y},g) = \mathbf{c}$ an n x 1 vector (i^{th} column of $\mathbf{C}$ in the previous section) and

$\mathbf{c} = E(\mathbf{y}g) - E(\mathbf{y})E(g) = E(\mathbf{y}g) - \boldsymbol{\alpha}\tau$.

b) Also note

$\mathbf{b'Vb} = \mathbf{b'}\{E(\mathbf{yy'}) - \boldsymbol{\alpha\alpha'}\}\mathbf{b} = \mathbf{b'}E(\mathbf{yy'})\mathbf{b} - \mathbf{b'}\boldsymbol{\alpha\alpha'}\mathbf{b}$, and

$2\mathbf{b'c} = 2\mathbf{b'}E(\mathbf{y}g) - 2\tau\mathbf{b'}\boldsymbol{\alpha}$.

c) So, add and subtract off $\mathbf{b'}\boldsymbol{\alpha\alpha'}\mathbf{b}$ and $2\tau\mathbf{b'}\boldsymbol{\alpha}$ from Step 3a and collect terms to yield Step 3b.

Now, take the partial derivatives (see Chapter 2) of 3b and equate to zero.

$$\partial L/\partial a = 2a + 2\mathbf{b'}\boldsymbol{\alpha} - 2\tau \dashrightarrow 0 \quad \text{or}$$

$$a = \tau - \mathbf{b'}\boldsymbol{\alpha}\,. \qquad \text{Step 4a}$$

$$\partial L/\partial \mathbf{b} = 2\mathbf{Vb} + 2\boldsymbol{\alpha\alpha}'\mathbf{b} + 2a\boldsymbol{\alpha} - 2\mathbf{c} - 2\tau\boldsymbol{\alpha} \dashrightarrow 0 \text{ or}$$

$$\mathbf{Vb} + \boldsymbol{\alpha\alpha}'\mathbf{b} + a\boldsymbol{\alpha} = \mathbf{c} + \tau\boldsymbol{\alpha}. \qquad \text{Step 4b}$$

Substituting for a in 4b from 4a,

$$\mathbf{Vb} + \boldsymbol{\alpha\alpha}'\mathbf{b} + \tau\boldsymbol{\alpha} - \boldsymbol{\alpha\alpha}'\mathbf{b} = \mathbf{c} + \tau\boldsymbol{\alpha}, \text{ therefore}$$

$$\mathbf{Vb} = \mathbf{c}, \text{ thus}$$

$$\mathbf{b} = \mathbf{V}^{-1}\mathbf{c} \text{ and } \mathbf{b}' = \mathbf{c}'\mathbf{V}^{-1}. \qquad \text{Step 5}$$

Finally, remember that $\hat{g} = a + \mathbf{b}'\mathbf{y}$ and using the results from 4a and 5,

$$\hat{g} = \tau - \mathbf{b}'\boldsymbol{\alpha} + \mathbf{b}'\mathbf{y} = \tau + \mathbf{b}'(\mathbf{y} - \boldsymbol{\alpha})$$

$$= \tau_i + \mathbf{c}_i'\mathbf{V}^{-1}(\mathbf{y} - \boldsymbol{\alpha}) \qquad \textbf{Eq. 4.1}$$

where the subcript i has been re-inserted in the final step to emphasize that this is a prediction for the ith genotype.

To complete the derivation, remember that Eq. 4.1 applies to the prediction of the ith element of **g** where i = 1, 2 ... q. Thus, Eq. 4.1 is appropriate for every element in **g**. Further, $\mathbf{V}^{-1}(\mathbf{y} - \boldsymbol{\alpha})$ is the same for all elements of **g** and **c** is just the ith column of **C** from the previous section. So we can predict the whole vector of genetic values at one time by writing,

$$\hat{\mathbf{g}} = \boldsymbol{\tau} + \mathbf{C}'\mathbf{V}^{-1}(\mathbf{y} - \boldsymbol{\alpha}). \qquad \textbf{Eq. 4.2}$$

This is the main formula for BLP used in Chapters 5-8 and for selection index in Chapters 9 and 10. Note that because **C** is the covariance between **g** and **y** and **V** is the variance of **y**, $\mathbf{C}'\mathbf{V}^{-1}$ is heuristically Cov(**g**,**y**) "divided by" Var(**y**) and can be thought of as multiple regression coefficients relating **g** to **y** (Thompson 1979). A different set of coefficients is developed for each trait and genotype depending upon the exact trait being predicted and the data observed for that genotype. To see this, consider Eq. 4.1 which applies to the single i^{th} element being predicted within **g**. Then, $\mathbf{c}_i'\mathbf{V}^{-1} = \mathbf{b}_i$ is a 1 x n row vector of coefficients that are used to predict the genetic value of the ith element of **g**. These coefficients in $\mathbf{b}_i$ are the weights applied to the data points to predict the ith genetic value. Data points that do not contribute any information to the ith genetic value (such as data collected on an unrelated family) will have corresponding coefficients (i.e. elements in $\mathbf{b}_i$) that are zero. So in Equation 4.2, $\mathbf{C}'\mathbf{V}^{-1}$ is q x n and represents q sets of coefficients (one for each trait-genotype element in **g**). That is, every different trait and genotype for which a genetic value is being predicted has a different set of coefficients.

BLP is a generalization of classical selection index first used for plants and animals by Smith (1936) and Hazel (1943), respectively; however, the classical approach assumes equal amounts and quality of data available for all genotypes (see Chapters 9 and 10). Hence, a single vector of weights (coefficients) is developed and used for all genotypes. BLP, as seen above, develops a different vector of coefficients for each genotype to reflect the nature of the data available for that genotype. Of course, genotypes with identical quality and quantity of data will still have identical sets of coefficients. The more general BLP approach minimizes the error variance of the predicted genetic values.

Properties of Best Linear Predictions

Throughout the derivation, we assumed that the first and second moments of the joint distribution of **g** and **y** were known. In this section, we outline many desirable statistical properties of best linear predictions (see Henderson 1963, 1973, 1977, 1984). The derivation of these properties also assumes that the first and second moments are known constants. However, in practice this is never the case; they must be estimated. If the estimates are good ones (both precise and accurate), then best linear predictions will nearly possess the properties described below. However, if poor estimates are used, the predictions will not necessarily even come close to having these properties. We discuss the importance of these assumptions and their violation in more depth at the end of this Chapter and in Chapter 7. Here, we assume that $E(\mathbf{g}) = \boldsymbol{\tau}$, $E(\mathbf{y}) = \boldsymbol{\alpha}$, $Var(\mathbf{y}) = \mathbf{V}$, and $Cov(\mathbf{y},\mathbf{g}') = \mathbf{C}$ are all known constants. These properties are all derived and discussed in Henderson (1963, 1973, 1977, 1984).

1. The predictions are unbiased: This means that in repeated sampling, the many calculations of predicted genetic values for each and every genotype would average to their mean genetic values.

Proof:

$$\begin{aligned} E(\hat{\mathbf{g}}) &= E\{\boldsymbol{\tau} + \mathbf{C'V}^{-1}(\mathbf{y} - \boldsymbol{\alpha})\} & \\ &= E(\boldsymbol{\tau}) + \mathbf{C'V}^{-1}\{E(\mathbf{y}) - E(\boldsymbol{\alpha})\} & \text{(since } \mathbf{C'V}^{-1} \text{ is a constant)} \\ &= \boldsymbol{\tau} + \mathbf{C'V}^{-1}(\boldsymbol{\alpha} - \boldsymbol{\alpha}) & \text{(since } \boldsymbol{\tau} \text{ and } \boldsymbol{\alpha} \text{ are constants)} \\ &= E(\mathbf{g}) & \text{(since } E(\mathbf{g}) = \boldsymbol{\tau} \text{ and } \boldsymbol{\alpha} - \boldsymbol{\alpha} = \mathbf{0}). \end{aligned}$$

2. The error variance of the predictions is minimized: Within the class of all possible linear functions of the observed data, Eq. 4.2 minimizes the error variance of the predictions. This is just a restatement of the derivation of Eq 4.2 above.

3. The correlation between the predicted and true genetic values is maximized: Within the class of all possible linear functions of the data, predictions from Eq. 4.2 have the highest correlation with the true unknown genetic values (over repeated sampling). This is proven on page 40 of Henderson (1984) and also in Chapter 9.

4. A multivariate normal joint distribution results in three other properties: If the joint probability distribution between **g** and **y** is multivariate normal, then (see Henderson 1984, p 40):

a) BLP is the best (minimum error variance) predictor among all possible functions and transformations (not just linear functions) of the data,

b) BLP maximizes the probability of selecting the better of two candidates (see also Searle 1974),

c) BLP maximizes expected genetic progress (gain) for a fixed number of selections made (see also Bulmer 1980).

5. BLP of a linear transformation of genetic values is that function of the predictions: That is, any linear function of the predicted genetic values is the best linear prediction of that same linear function of the underlying true genetic values. This is extremely important when economic weights are used to combine multiple traits into an aggregate genetic value for each genotype (for example in selection indices). It means that the economic weights can be applied after the predictions and still result in BLP of the aggregate genetic value. The topic is further discussed in Chapters 7, 9 and 10.

Proof:

Let **M** be any s x q matrix of constants where s is the number of genotypes for which aggregate genetic values are desired. For example, if 2 traits were predicted for each of s genotypes, then q = 2s. Then, **Mg** = **h** is that s x 1 vector of desired aggregate genetic values to be predicted. Cov(**Mg**,**y'**) = **B** is a s x n matrix of covariances between the aggregate genetic values and the observations. Then BLP of **h** is found from Eq. 4.2 as

$$\begin{aligned} \hat{\mathbf{h}} &= E(\mathbf{h}) + \mathbf{B}\mathbf{V}^{-1}(\mathbf{y} - \boldsymbol{\alpha}) && \text{(where } \mathbf{B} \text{ simply replaces } \mathbf{C}'\text{)} \\ &= E(\mathbf{Mg}) + \mathbf{MC'V}^{-1}(\mathbf{y} - \boldsymbol{\alpha}) && \text{(since Cov}(\mathbf{Mg},\mathbf{y}')=\mathbf{M}\text{Cov}(\mathbf{g},\mathbf{y}')=\mathbf{MC}'\text{)} \\ &= \mathbf{M}\{E(\mathbf{g}) + \mathbf{C'V}^{-1}(\mathbf{y} - \boldsymbol{\alpha})\} \\ &= \mathbf{M}\hat{\mathbf{g}}. && \textbf{Eq. 4.3} \end{aligned}$$

6. The variance among the predictions can be calculated: For each genetic value predicted, it is possible to calculate the estimated spread among the predictions that will occur with repeated sampling. It turns out that predictions will tend to spread out more (have larger variance) when they are based on "better" (larger quantity, higher quality) data. As the data become better and better, the predicted genetic values approach the true genetic values and the variances of the predictions approach the genetic variances. This is discussed in depth in Chapter 5. Note that the expression in Eq. 4.4 is a q x q matrix with estimated variances among the predictions on the diagonal and covariances between the predictions on the off-diagonal.

Derivation:

$$\begin{aligned} \text{Var}(\hat{\mathbf{g}}) &= \text{Var}\{\boldsymbol{\tau} + \mathbf{C'V}^{-1}(\mathbf{y} - \boldsymbol{\alpha})\} && \text{(from Eq. 4.2)} \\ &= \text{Var}(\mathbf{C'V}^{-1}\mathbf{y}) && \text{(since } \boldsymbol{\tau} \text{ and } \boldsymbol{\alpha} \text{ are constants)} \\ &= \mathbf{C'V}^{-1}\{\text{Var}(\mathbf{y})\}\mathbf{V}^{-1}\mathbf{C} && \text{(from Eq. 2.30 letting } \mathbf{C'V}^{-1} \text{ be } \mathbf{A}\text{)} \\ &= \mathbf{C'V}^{-1}\mathbf{C} && \textbf{Eq. 4.4} \end{aligned}$$

7. The covariance between predicted and true genetic values can be calculated: $Cov(\hat{\mathbf{g}},\mathbf{g}')$ is a q x q matrix of covariances with covariances between the true and predicted genetic values for a given element of **g** (a specific trait of a given genotype) on the diagonal. Off the diagonal are covariances between the predicted value of one element of **g** and the true value of another. This covariance matrix is useful for deriving the result in 8 below and also for calculating estimates of the correlation between predicted and true genetic values (See Property 9 below). Note at the bottom of the derivation that the covariance between true and predicted breeding values is equivalent to the variance among the predictions (from Property 6 above).

Derivation:

$$Cov(\hat{\mathbf{g}},\mathbf{g}') = Cov(\boldsymbol{\tau} + \mathbf{C}'\mathbf{V}^{-1}\mathbf{y} - \mathbf{C}'\mathbf{V}^{-1}\boldsymbol{\alpha}, \mathbf{g}') \quad \text{(from Eq. 4.2)}$$

$$= Cov(\mathbf{C}'\mathbf{V}^{-1}\mathbf{y},\mathbf{g}') \quad \text{(since } \boldsymbol{\tau} \text{ and } \boldsymbol{\alpha} \text{ are constants)}$$

$$= \mathbf{C}'\mathbf{V}^{-1}Cov(\mathbf{y},\mathbf{g}') \quad \text{(since } \mathbf{C} \text{ and } \mathbf{V} \text{ are constants)}$$

$$= \mathbf{C}'\mathbf{V}^{-1}\mathbf{C} \quad \textbf{Eq. 4.5}$$

$$= Var(\hat{\mathbf{g}}) \quad \text{(from Eq. 4.4).}$$

8. The error variance of the predictions can be calculated: The estimated error variance of the predictions that would occur with repeated sampling can be calculated and provides information about the precision of the predictions.

Derivation:

$$Var(\hat{\mathbf{g}} - \mathbf{g}) = Var(\hat{\mathbf{g}}) + Var(\mathbf{g}) - 2Cov(\hat{\mathbf{g}},\mathbf{g}')$$

$$= Var(\hat{\mathbf{g}}) + Var(\mathbf{g}) - 2Var(\hat{\mathbf{g}}) \quad \text{(from Property 7)}$$

$$= Var(\mathbf{g}) - Var(\hat{\mathbf{g}})$$

$$= \mathbf{G} - \mathbf{C}'\mathbf{V}^{-1}\mathbf{C} \quad \textbf{Eq. 4.6}$$

where **G** is a q x q matrix of known genetic variances and covariances.

9. The correlation between true and predicted genetic values can be calculated: Another measure of precision of the predictions is to see how closely they correlate with the true genetic values. If more and better data are used the correlations will be higher. To avoid confusion between elemental *vs* matrix operations, the derivation below develops the estimated correlation between the true and predicted for a single trait of a single genotype (i.e. the ith element of **g**).

Derivation:

$$\begin{aligned} \text{Corr}(\hat{g},g) &= \text{Cov}(\hat{g},g)/\{\text{Var}(\hat{g})\ \text{Var}(g)\}^{½} \\ &= \text{Var}(\hat{g})/\{\text{Var}(\hat{g})\ \text{Var}(g)\}^{½} \quad \text{(from Property 7 above)} \\ &= \{\text{Var}(\hat{g})/\text{Var}(g)\}^{½} \end{aligned}$$

Eq. 4.7

TOPICS IN BEST LINEAR PREDICTION

Applying the theory presented in previous sections of this chapter is not always straightforward and the purpose of the following sections is to introduce some topics relating to the application of best linear prediction. Chapters 5-8 deal with most of these topics in more depth and also provide numerical examples of BLP. Introduction of the topics here is meant to clarify some aspects of the theory and to highlight some general aspects of applying prediction methods (BLP, selection index and BLUP). Detailed discussions of some other important aspects of prediction methods are deferred to the chapters on selection index and BLUP theory (Chapters 9 and 11). On a first reading, the following sections may be skimmed until the reader has worked through some numerical examples of BLP.

Specifying the Data Vector: y

To apply any prediction method, the vector containing the data must be written and a linear model specified for each observation. The data vector has dimension n x 1 and therefore contains n observations. In many ways, it is the specification of the data vector (and subsequent linear models) that determines the computational complexity of the problem. If n is large, computing the inverse of the n x n **V** matrix may be extremely difficult, costly, time consuming or even intractable. Thus, whenever subclass or other types of means can be used as data instead of individual measurements, computations can be greatly simplified.

As an example consider the prediction of parental breeding values from progeny test data which is the subject of Chapters 5-8. The raw progeny test measurements are observations on individual offspring from the parents being tested. These individuals are usually planted in replicated, randomized experimental designs (for example randomized complete block designs replicated across several test site locations). Alternative possibilities for use as data in the **y** vector include: 1) individual measurements, 2) plot means (each mean is the average of all individuals of a given family in a given block), 3) family means at a location (each mean is the average of all individuals of a given family across all blocks at that location), and 4) family means averaged across all locations.

Choice of increasingly higher level means as the unit of observation to use in the **y** vector will reduce the computational complexity of the problem, but may also result in loss of information (and hence the results will not be BLP). Taking the example of the previous paragraph, suppose that all parents being tested are equally represented at a given test location. That is, all families at a given test location are represented in all blocks and by the same number of trees in all blocks. Then no information would be lost by using family means at a test location as the data in the **y** vector. If this meant averaging over say 5 blocks and 8 individuals per block, then the length of the **y** vector (i.e. n) would be reduced by a factor of 40. However, if tests were quite imbalanced, then some family means might be represented by only 10 trees (from say only 2 of the five blocks) and others by 40. Use of family means could then result in considerable loss of information. Still, if 1) least square means were used to estimate family means and 2) the fact that family means based on different numbers of trees have different variances and covariances was accounted for in specifying **V**, then use of family means may still be suitable.

The choice of what to use as the unit of observation in the **y** vector is quite important and is linked closely to the linear model specified for each observation. A linear model must be written for each element of **y** and complete specification includes assumptions about 1) the variance of each element, covariances among elements, covariances with the genetic values being predicted. There is not just a single linear model for all of the elements of **y**. Each observation may not only have different effects in a linear model, but may also have different variances and covariances. This topic is discussed further below and in Chapter 7 after the principles of BLP have been illustrated with several examples. Also for selection of individuals (i.e. for predicting the breeding values of individuals), the topic is discussed in Chapters 9 and 10.

First Moments of the Data: E(y) = $\boldsymbol{\alpha}$

Each of the n observations in the data vector, **y**, has an expected value. The vector of these n expected values is $\boldsymbol{\alpha}$. The theoretical expected values are always determined by writing the linear model for each observation and then calculating its expectation. In Example 4.3 in which family means are the observations in **y**, the expected values are shown to be the $\mu + E_i$ for each observation; thus data (i.e. family means) from the same location have the same expectation. This results because the linear model written for each family mean (Eq. 2.25) contains only two fixed effects (an overall mean and the test location effect). If individual measurements were the units of observation with Eq. 2.23 as the assumed linear model, then the expected value would also contain a block effect. Other linear models that include other fixed effects (such as treatments, provenances, etc) would result in different expected values; thus the specification of the linear model is a critical step.

In the prediction of genetic values (e.g. the BLP formula of Eq. 4.2), the expected values are used to express the data as deviations from their fixed effects. These deviations, **y** - $\boldsymbol{\alpha}$, can be thought of as the data adjusted for "nuisance" effects. Continuing with Example 4.3, the 5 x 1 vector of expected values contains site location means for test sites 1, 2 and 3. Then **y** - $\boldsymbol{\alpha}$ expresses each family mean as a deviation from the test location mean. The resultant deviations "adjust" family means for the spurious effects of different environmental effects at each site. In Example 4.3, family one is tested on location one, but family two is not. Suppose that location one was an excellent site (high rate of growth due to climatic and edaphic factors). Expressing the family means as deviations in Equation 4.2 ensures that the predicted breeding value for parent one is not upwardly biased because it was tested at the location with a higher mean.

Depending upon the data and the linear model, treatment effects, block effects, year effects and other fixed effects may be contained in the expected value of each observation. In all cases, the deviations (i.e. **y** - $\boldsymbol{\alpha}$) "adjust" the data for the fact that some observations may have come from blocks, treatments, years, etc. with higher means.

Best linear prediction (Eqs. 4.2 - 4.7) assumes that the fixed effects, $\boldsymbol{\alpha}$, are known constants when in practice they are never known, but rather must be estimated from the data. The estimates can be obtained in several ways: 1) the same genetic check lots (sometimes called genetic standards) can be grown in all blocks, test locations, years, etc., 2) simple averages can be calculated using data in each block, test location, year, etc., 3) generalized least squares (GLS) estimates can be calculated (See Chapters 3, 11 and 12). The objectives, of course, are to obtain estimates of the fixed effects that are both accurate (unbiased) and precise as well as computationally tractable. In some applications, check lots and/or simple averages can meet these objectives (Chapters 5 - 10), while in other instances these approaches will not be nearly as good as GLS. For convenience, when we use simple averages and check lots to estimate fixed effects, we use Eq. 4.2 to predict genetic values and call it best linear prediction. When GLS estimates of fixed effects are used, the prediction method is called best linear unbiased prediction (See Chapters 11 and 12).

The important point is that applications of BLP and selection index are really only approximations to BLP because the fixed effects are assumed known constants when in fact they are estimated with some error (the following sections dealing with second moments show that these also must be estimated but are assumed known). When the estimates are precise and accurate, then the predictions will be nearly BLP. However, any biases or imprecision in estimating the fixed effects will be reflected in the predictions. These errors are **not**, however, reflected in formulas 4.4 - 4.7 which relate to the precision of the predictions. For example Eq. 4.6, which estimates the error variance of the predictions, assumes that the fixed effects are known constants. Thus, the error variance estimate will be too low because it does not reflect the error in estimating the fixed effects.

The examples of fixed effects used in this section have mostly been spurious effects associated with blocks, locations, years, treatments, etc. However, some genetic effects, such as selection generation, breed, variety, provenance, and seed source, may also be assumed fixed. While these are discussed in the next section in more detail, we note here that these fixed effects also are reflected in $\boldsymbol{\alpha}$. For example, consider an experiment predicting parental breeding values for several parents within each of several different seed sources (or breeds or varieties). If seed source effects are treated as fixed, then, $\mathbf{y} - \boldsymbol{\alpha}$ expresses the data as deviations from all fixed effects including seed source effects. In general, this will be clear by specifying the linear model for each observation in $\mathbf{y}$ and calculating its expected value.

First Moments of the Genetic Values: $\mathbf{E(g)} = \boldsymbol{\tau}$

In many applications of BLP in this book, we assume that the expected value of the genetic values being predicted is zero (i.e. $\boldsymbol{\tau} = \mathbf{0}$) and hence that all genetic values being predicted have the same expectation (i.e. zero). In many forestry applications, this assumption will be valid and it has the effect of centering the predictions around zero; but as noted in the previous paragraph, there will be instances when this assumption is not appropriate. If the genetic values being predicted have different fixed effects (say because they come from different seed sources which are assumed fixed), then these different fixed genetic effects are reflected in $\boldsymbol{\tau}$. Remember that these fixed genetic effects are also reflected in $\boldsymbol{\alpha}$, but that $\boldsymbol{\alpha}$ also contains other fixed effects. Each element in $\boldsymbol{\alpha}$, say α_i, is the expected value of the corresponding observation (i.e. $E(y_i)$). When expressed in terms of the linear model for the ith observation, α_i contains "nuisance" fixed effects (such as block, location, treatments, etc.) as well as genetic fixed effects.

To see how BLP treats fixed genetic effects, examine Equation 4.2 and remember that they appear twice: once in $\boldsymbol{\alpha}$ with the other fixed effects and once in $\boldsymbol{\tau}$. Heuristically, 1) the fixed effects get subtracted off from the data, 2) the deviations, i.e. $\mathbf{y} - \boldsymbol{\alpha}$, get multiplied by $\mathbf{C'V}^{-1}$, and then 3) the genetic fixed effects get added back as $\boldsymbol{\tau}$. While this may seem redundant to first subtract them off and then add them back on, we shall see in great detail in later chapters, that multiplication by $\mathbf{C'V}^{-1}$ "shrinks" the deviations back toward zero. So, the effect is to shrink only that portion of the data associated with random effects (e.g. breeding values of parents within a seed source), not fixed genetic effects (e.g. seed sources). Two examples are presented below.

Consider in more detail an experiment in which many parents within each of several seed sources are being progeny tested and parental breeding values are considered random and seed source effects fixed. The deviations, $\mathbf{y} - \boldsymbol{\alpha}$, express the data as deviations from their expectations; thus if family means at a given test location are used as the observations in $\mathbf{y}$, $\boldsymbol{\alpha}$ will include both test location and seed source effects. Each family mean is expressed as a deviation above or below the specific location and seed sources it represents. Since the deviations average to zero over repeated sampling (see Property 1 earlier in this Chapter), the predicted breeding values will be shrunken to center around seed source means. That is, within each seed source, predicted parental breeding values average to zero. To predict the breeding values of parents across all seed sources, the fixed seed source effects are added back in (by adding $\boldsymbol{\tau}$ in Eq. 4.2). Then, the predicted parental breeding values for all parents within a seed source will average (across repeated sampling) to the seed source effect (instead of to zero).

Another example is detailed in Chapter 8 where two groups of parents have undergone different amounts of previous selection for a specific trait (selected *vs* unselected) and therefore belong to two different selection groups. Many parents are being progeny tested from each group. The groups are assumed fixed with the difference between the means of the two groups, (τ_1 - τ_2), representing the gain from selection. Data are expressed as deviations from fixed effects which include group means and therefore $\mathbf{C'V^{-1}(y - \alpha)}$ first predicts parental breeding values which center around group means. Then, τ (which in this case means τ_1 or τ_2) is added back onto each predicted parental breeding value so the predicted breeding value for each parent will reflect which group the parent is in.

Since the fixed genetic effects are never known in practice but rather are estimated, this is another aspect that makes predicted breeding values only approximately BLP. Also as with α, the errors associated with estimating τ are not accounted for in Equations 4.3 - 4.7; hence, the predictions will appear more precise than they really are (biased). If the genetic effects are well estimated, the estimated precision of the predictions will be nearly unbiased. This topic is discussed more in Chapters 11 and 12.

Second Moments of the Data: Var(y) = V

Each observation in **y** has a variance associated with it and a covariance (which may be zero) with all other elements in **y**. These variances and covariances are in the units of measurement and form the elements of **V**. Example 4.3 provides a simple example, and details of how to specify **V** in more complicated situations are in nearly all subsequent chapters. The process is always the same 1) specify a linear model for each element of **y**, 2) calculate the theoretical values for the variance and covariances (i.e. the elements of **V**) based on the linear model, and 3) develop a suitable estimate for each element.

Because specifying **V** is considered in much detail later, only two points are important here: 1) **V** is completely general and 2) **V** is assumed known in development of the theory while in practice it must be estimated. First, the theory of BLP allows for complete generality of **V**. Each element in **y** may have a different variance and set of covariances with other elements. For example, suppose family means are used as data for predicting parental breeding values and that each family mean (i.e. each element of **y**) is based on different numbers of individual measurements, then each family mean would have a different variance (covariances would also be affected). Different variances (and covariances) may also arise from different quality (i.e. precision) of data or data of different ages: family means from tests with higher variances should have higher variances in **V**. In many ways this complete generality of **V** is part of the power of BLP (and also BLUP) because the second moments of the data can be completely reflected. However, sometimes it may greatly simplify computations to assume some degree of homogeneity of variances (e.g. to assume that all family means from the same test have the same variance (and pairwise covariances)). This assumption makes the computations only approximately BLP, but may be suitable. The linear model for each observation reflects the assumptions made about the second moments. Are all observations assumed to have the same genetic and error variances, different genetic variances, and/or different error variances? What about covariances between observations?

As with the first moments, **V** is assumed to be a matrix of known constants for all BLP formula derived in this chapter. In practice, these must be estimated from the data and are not known exactly. Thus, predictions made using the BLP formula (Eq. 4.2) only approximate BLP and the estimated precision of the predicted genetic values from Equations 4.3 - 4.7 will not reflect this additional source of error. The magnitude of errors associated with estimating the second moments is discussed more fully in Chapters 7 and 11, but in general more research is needed on this topic.

Second Moments of the Genetic Values: Var(g) = G

The elements of the q x q **G** matrix are variances and covariances among the genetic values being predicted. Since these are variances among the true (but unobservable) genetic values, they are genetic variances. These variances and covariances have units which correspond to the entities being predicted and not necessarily those measured. For example if breeding values for 25-year-old volume and specific gravity were being predicted for each of 100 parents, the 200 x 1 **g** vector would consist of the unknown, true breeding values being predicted. The diagonal elements of the 200 x 200 **G** matrix would be variances of breeding values which are additive genetic variances (Bulmer 1980; Falconer 1981). If the 100 parents are not related, non-zero off-diagonal elements would be additive covariances between the two traits (i.e. the numerator of additive genetic correlation between the two traits).

If total genetic (clonal) values or other types of genetic values were being predicted, the variances and covariances would correspond to those predictands. In the case of clonal values, the variances would be total genetic variance.

The **G** matrix is not used in Eq. 4.2 to predict breeding values; rather, it is only needed for estimating the precision of the predictions (Eqs. 4.6 and 4.7). As with the other first and second moments, **G** is assumed known, but in practice must be estimated (usually from variance components). The error in estimating **G** adds further to the approximate nature of estimated precisions from Eqs. 4.6 and 4.7.

Second Moments of the Joint Distribution of y and g: Cov(y,g') = C

The n x q **C** matrix contains covariances between the data, **y**, and the genetic values being predicted, **g**. These elements are determined by 1) specifying a linear model for each element in **y**, 2) specifying the exact genetic values being predicted, 3) calculating the theoretical covariance between each element of **y** and each element of **g** and 4) estimating these theoretical covariances in some fashion. This topic is discussed at length in later chapters, but in general only the genetic factors in the linear model will correlate with the genetic values being predicted. Thus, elements of **C** are genetic covariances.

If all elements of **y** have zero covariance with all elements of **g**, then all elements of **C** are zero; in this case, the data provide no information about the genetic values being predicted. In this case, the predicted breeding values are τ or all 0 if τ is assumed **0** (see Eq. 4.2). If some of the elements of **C** are non-zero, then the data provide information about the predictands and the predictions of genetic values will not be all zero. Note that information from many different observations in **y** (different traits and genotypes) can provide information about a given genetic value being predicted. For example if height is measured at a juvenile age and has a non-zero additive genetic covariance with both traits being predicted (say parental breeding values for mature volume and wood specific gravity), then the measurement will provide information to the prediction of both traits. If juvenile percent summerwood is also measured and has non-zero covariance with both predictands, then it too provides information about both traits being predicted. In a sense, best linear predictions can be indirect predictions (analogous to indirect selection in Falconer 1981) and this is part of the power of BLP: all measured data are used for all predictions to increase the precision of each prediction. This topic is discussed more fully in Chapter 7.

As with the other first and second moments, **C** is assumed known in the derivation of all BLP formulas; in practice, it must be estimated either from the data or from external estimates of genetic covariances. The error involved in estimating the elements of **C** further contributes to the approximate nature of Equations 4.1 - 4.7. This topic is discussed in Chapter 7.

SUMMARY AND CONCLUSIONS

None of the prediction methodologies described in this book are perfect for all situations. In particular, use of these methods does not obviate the need to obtain the best data possible (garbage in, garbage out); if either the first or second moments are poorly estimated, then the predicted genetic values may be quite imprecise (and even biased). Still, in many situations good estimates of the first and second moments will be available (nearly "known") and when they are considered known, BLP will make "best" (minimum error variance) use of the data compared to all other linear combinations (e.g. compared to ordinary or weighted least squares). If the joint distribution between **y** and **g** is multivariate normal, then the predictions are "best" among all possible functions (linear and non-linear) of the data. If the second moments are well-estimated, but simple (ordinary least squares) estimates of the first moments are imprecise or biased, then best linear unbiased prediction (Chapters 11 and 12) should be used.

In subsequent chapters, properties, applications, strengths and weaknesses of BLP, selection index and BLUP are discussed. These all are based on the common assumption that the genetic values being predicted are random variables not fixed effects. All three methods also require that a linear statistical model (and all of its inherent assumptions about the factors in the model) be specified for the data and that the second moments be known (or at least estimated). While these may be viewed as weaknesses of the techniques, we believe they are strengths. The process of specifying all of the assumptions and estimating **C** and **V** allows the importance of those assumptions and estimates to be tested by sensitivity analysis (see Chapter 7). It may turn out that predictions are much more sensitive to some assumptions and estimates than others. Thus, it would be possible to calculate the predictions several times using a variety of model assumptions and moment estimates and to see which genotypes rank high for all or many of the model formulations. Also, this directly points out what types of research and data are most urgently needed. Once the most critical parameters have been re-estimated or assumptions relaxed (e.g. to allow heterogeneous variances in **V**), the model can be reformulated and genetic values recalculated.

BLP also allows data from a variety of sources (tests of different ages, precision and experimental designs) and relatives (e.g. ancestral and other relatives) to all be used to increase the precision of the genetic predictions. This may reduce testing costs by making maximum use of available data. Finally, BLP can be used to help design efficient testing programs. Using Equation 4.7 to specify a desired degree of precision of the genetic predictions, prediction precisions can be simulated for different mating designs, experimental designs and measurement ages.

PROBLEMS

Problem 4.1: Specify the units (e.g. ft, ft^3, etc.) for all elements in **y**, $\boldsymbol{\alpha}$, **g**, **C** and **V** in Example 4.3.

Problem 4.2: For Example 4.3, find the dimensions of the following: a) Var($\hat{\mathbf{g}}$) and b) Var($\hat{\mathbf{g}}$ - **g**). Tell in words what all elements in each of these matrices means and identify units for each element.

Chapter 5
Best Linear Prediction with Half-sib Progeny Test Data

INTRODUCTION

Although the many desirable statistical properties of BLP can be demonstrated using matrix algebra, it is difficult to develop a sense of how BLP works without actually applying it to some data. The purpose of this chapter is to illustrate how BLP handles the kinds of messy data problems discussed in Chapter 3 through the use of a number of small examples. Using hypothetical half-sib progeny test data and the linear model defined in Chapter 2 (Eq. 2.23), we look at how BLP works on a balanced data set, and then systematically change and delete data to examine how BLP combines data and makes predictions with unbalanced and messy data. Throughout this chapter, open-pollinated families are assumed to be half-sib families and the terms half-sib and open-pollinated (OP) are used interchangeably. Much of this chapter is based on White et al. (1987).

PREDICTING BREEDING VALUES FROM HALF-SIB TEST DATA

We wish to predict the breeding values of parents for trait g from a set of half-sib progeny test data. The trait (or traits) measured in the progeny tests, g^*, may or may not be equivalent to trait g. In this chapter, the breeding value for only one trait is predicted for each parent; predictions of multiple traits and aggregate traits for each genotype are discussed in later chapters. To predict breeding values, the $\boldsymbol{\tau}$, $\boldsymbol{\alpha}$, and **y** vectors and the **C**, **G**, and **V** matrices must either be known, estimated directly from the data, or estimated indirectly from other information sources. Throughout this chapter, we predict relative parental breeding values as deviations from a single population mean. This implies $E(\mathbf{g})=\boldsymbol{\tau}=\mathbf{0}$ and thus it is assumed there are no differences among the parents in fixed genetic effects.

It is also assumed that the observations in **y** are family means averaged over the individuals of a family in a single test; that is, there is one observation for every test a parent is in. Use of family means simplifies computations as the size of the matrices and the number of kinds of covariances are both reduced. Beginning with the linear model defined in Eq. 2.23, each of the n elements in the n x 1 **y** vector is a family mean as shown in Eq. 2.25 and repeated below for convenience. The observation on the kth parent in the ith test is:

$$\bar{y}_{i\cdot k\cdot} = \mu + E_i + f_k + fe_{ik} + (p_{i\cdot k}/b) + (w_{i\cdot k\cdot}/bn) \qquad \textbf{Eq. 5.1}$$

To determine the n elements for $E(\mathbf{y}) = \boldsymbol{\alpha}$, take the expected value for each element in **y**

$$
\begin{aligned}
E(\bar{y}_{i\cdot k\cdot}) &= E(\mu + E_i + f_k + fe_{ik} + (p_{i\cdot k}/b) + (w_{i\cdot k\cdot}/bn)) \\
&= E(\mu) + E(E_i) + E(f_k) + E(fe_{ik}) + E(p_{i\cdot k}/b) + E(w_{i\cdot k\cdot}/bn) \\
&= \mu + E_i \,. \qquad \textbf{Eq. 5.2}
\end{aligned}
$$

Thus, the expected value of a half-sib family mean from a single test reduces to $\mu + E_i$ since all but the first two terms are random variables with expectations of zero. This means that each of the n family means in **y** has an expected value of $\mu + E_i$ which is the overall site mean of the test in which the family mean is observed. If there are t different progeny tests, there are t site means. These t site means are never known exactly (as is assumed by BLP theory) but can be estimated either by growing genetic checklots on each site or by averaging all family means on given site. In this chapter, we assume that precise, unbiased estimates of $\mu + E_i$ are available (see Chapter 7 for further details).

Since it is assumed in this chapter that all parents come from the same population, the fixed genetic effects for all parents are the same and arbitrarily set to zero (i.e. $E(\mathbf{g}) = \boldsymbol{\tau} = \mathbf{0}$). With this assumption, the BLP formula (Eq. 4.2) reduces to

$$
\hat{\mathbf{g}} = \mathbf{C}'\mathbf{V}^{-1}(\mathbf{y} - \boldsymbol{\alpha}) \qquad \textbf{Eq. 5.3}
$$

In complete generality (as described in Chapter 4), **C** and **g** have dimensions n x q and q x 1 where q is the number of genetic values being predicted; q incorporates both the number of genotypes and the number of traits for which values are predicted. In this chapter, there are s parents but only one trait predicted for each so that q = s. Thus, the n x s **C** matrix defines the genetic relationships between the n observed family means in **y**, and the s true breeding values in **g**. There is one column of **C** for each parental breeding value being predicted. The elements of **C** are calculated by genetic theory because the true breeding values are non-observable.

$$
\mathbf{C} = \begin{bmatrix} c_{11} & \cdots & c_{1s} \\ c_{21} & \cdots & c_{2s} \\ \vdots & & \vdots \\ c_{n1} & \cdots & c_{ns} \end{bmatrix} = \begin{bmatrix} Cov(y_1,g_1) & \cdots & Cov(y_1,g_s) \\ Cov(y_2,g_1) & \cdots & Cov(y_2,g_s) \\ \vdots & & \vdots \\ Cov(y_n,g_1) & \cdots & Cov(y_n,g_s) \end{bmatrix}
$$

The first element, c_{11}, is the covariance between the first family mean and the breeding value of the first parent. For example, if the first observation in the **y** vector (y_1) is the family mean for parent 1 observed in the first open-pollinated progeny test location ($\bar{y}_{1\cdot1\cdot}$), then (from Eq. 5.1)

$$
\begin{aligned}
c_{11} &= Cov(y_1,g_1) \\
&= Cov(\bar{y}_{1\cdot1\cdot},g_1) \\
&= Cov(f_1 + fe_{11} + p_{1\cdot1}/b + w_{1\cdot1\cdot}/bn,\ g_1)
\end{aligned} \qquad \textbf{Eq. 5.4}
$$

Assuming that the family effect in the linear model for open-pollinated family means for a given trait represents half the breeding value of the female parent for that trait, then for the first observation, $f_1 = \frac{1}{2}g^*_1$ (where g^*_1 is used to denote that the trait measured may or may not be that being predicted). Also assuming that all covariances of cross products are zero, then

$$
\begin{aligned}
c_{11} &= Cov(\tfrac{1}{2}g^*_1,g_1) \\
&= \tfrac{1}{2}Cov(g^*_1,g_1) = \tfrac{1}{2}Cov(g^*,g).
\end{aligned} \qquad \textbf{Eq. 5.5}
$$

Thus, an open-pollinated family mean for the trait measured has covariance equal to $\frac{1}{2}Cov(g^*,g)$ with the breeding value of its female parent and covariance equal to 0 with the breeding values of all other parents (assuming parents are unrelated, and that open-pollinated families are truly half-sib families). Thus, the element c_{11} is 1/2 of the additive genetic covariance between the first observation and the first breeding value being predicted. In situations where the trait measured is the same as the trait for which breeding values are being predicted, $g^*=g$ and Eq. 5.5 reduces to

$$
c_{11} = \tfrac{1}{2}Cov(g^*,g) = \tfrac{1}{2}Cov(g,g) = \tfrac{1}{2}\sigma^2_A \text{ for trait g.} \qquad \textbf{Eq. 5.6}
$$

Whatever the situation, the method of calculating the **C** matrix elements remains the same. First, express the observation in the **y** vector in terms of its linear statistical model. Superimpose the genetic model on the statistical model ($f_k = \frac{1}{2} g_k$ in this case). And finally, calculate through the covariances.

Heuristically, the **C** matrix 1) weights the observations so that data having a closer genetic relationship with any given breeding value are given more weight in predicting that breeding value and 2) scales the observations from the units of measurement to the units of the breeding values being predicted. For the latter point, note that if breeding values for volume (in cm^3) are being predicted from family means of measured heights (in cm), then the units of the covariances in **C** are cm * cm^3.

The n x n **V** matrix contains variances and covariances between observed phenotypic values, i.e. between the n family means in **y**. **V** is a symmetric matrix which for most applications is non-singular.

$$\mathbf{V} = \begin{bmatrix} v_{11} & \cdots & v_{1n} \\ v_{21} & \cdots & v_{2n} \\ \vdots & \ddots & \vdots \\ v_{n1} & \cdots & v_{nn} \end{bmatrix} = \begin{bmatrix} \mathrm{Var}(y_1) & \cdots & \mathrm{Cov}(y_1, y_n) \\ \mathrm{Cov}(y_2, y_1) & \cdots & \mathrm{Cov}(y_2, y_n) \\ \vdots & \ddots & \vdots \\ \mathrm{Cov}(y_n, y_1) & \cdots & \mathrm{Var}(y_n) \end{bmatrix}$$

For measured data that are open-pollinated family means, each diagonal element is the variance of family means at a given planting location. For example, if y_{11} is the family mean for parent 1 in the first test, then $v_{11} = \mathrm{Var}(y_{11}) = \mathrm{Var}(\bar{y}_{1\cdot 1\cdot})$ and from Eq. 2.26

$$v_{11} = \mathrm{Var}(\bar{y}_{1\cdot 1\cdot}) = \sigma^2_f + \sigma^2_{fe} + \sigma^2_p/b + \sigma^2_w/bn \qquad \textbf{Eq. 5.7}$$

The variances of the $\bar{y}_{i\cdot k\cdot}$ can be estimated directly from the data at a given site or calculated from variance component estimates from other data (see Chapters 7 and 8). In the beginning examples below, it is assumed that all family means have the same variances; that is, $\mathrm{Var}(\bar{y}_{i\cdot k\cdot})$ is the same for all sites (i = 1, ... t) and all families (k = 1, ... s). This is not a requirement of BLP and the assumption is relaxed in later examples to allow heterogeneous second moments.

Assuming there are no repeated measurements made at different ages in the same tests, the off diagonal elements of **V** are covariances between either 1) two different families at different locations, 2) two different families at the same location, or 3) the same family at different locations. Assuming only genetic effects from the same family have non-zero correlations (since the parents are unrelated and a given open-pollinated family therefore does not correlate with one from a different parent), all covariances of the first and second kind are zero. For two family means from the same parent but different planting locations, the covariances can sometimes be estimated directly from the data (Burdon 1979). More often for unbalanced data, they can be calculated indirectly using knowledge of the relationships and estimates of the variance components from other data sets. For example, for the kth family measured for the same trait in tests 1 and 2,

$$\begin{aligned} Cov(\bar{y}_{1 \cdot k \cdot}, \bar{y}_{2 \cdot k \cdot}) &= Cov(f_k + fe_{1k} + p_{1 \cdot k}/b + w_{1 \cdot k \cdot}/bn,\ f_k + fe_{2k} + p_{2 \cdot k}/b + w_{2 \cdot k \cdot}/bn) \\ &= Cov(f_k, f_k) \\ &= Cov(\tfrac{1}{2}g_k, \tfrac{1}{2}g_k) = \tfrac{1}{4}\sigma^2_A. \end{aligned} \qquad \textbf{Eq. 5.8}$$

If the traits measured at locations 1 and 2 are not the same, then

$$Cov(\tfrac{1}{2}g^*_k, \tfrac{1}{2}g_k) = \tfrac{1}{4}Cov(g^*, g) \qquad \textbf{Eq. 5.9}$$

In predicting breeding values using Eq. 5.3, $\mathbf{C'V^{-1}}$ roughly serves to weight observations in proportion to the ratio of additive to phenotypic variance. Thus, for families observed with high phenotypic variances relative to the amount of additive genetic variance (such as families measured in imprecise tests or in only a subset of the blocks or at only a few test locations), the predicted breeding values will be "shrunken" or "regressed" towards 0. Heuristically, family means whose values primarily result from environmental, as opposed to genetic, variation, will be discounted back towards the overall population mean of zero.

Finally, the matrix of genetic variances and covariances, **G**, is needed to estimate the error variance of the predictions (EVP) from Eq. 4.6 and the correlation between predicted and true breeding values from Eq. 4.7. **G** is always a symmetric matrix with variances of the g_k on the diagonals and covariances, i.e. $Cov(g_k, g_{k'})$, on the off-diagonal. Since only one trait is predicted for each of the s parents, **G** is an s x s matrix and

$$\mathbf{G} = \begin{bmatrix} G_{11} & \cdots & G_{1s} \\ G_{21} & \cdots & G_{2s} \\ \vdots & \ddots & \vdots \\ G_{s1} & \cdots & G_{ss} \end{bmatrix} = \begin{bmatrix} Var(g_1) & \cdots & Cov(g_1,g_s) \\ Cov(g_2,g_1) & \cdots & Cov(g_2,g_s) \\ \vdots & \ddots & \vdots \\ Cov(g_s,g_1) & \cdots & Var(g_s) \end{bmatrix}$$

The diagonal elements are $Var(g_k)=\sigma^2_A$ and the off diagonal elements are all 0 for unrelated parents. Thus, for unrelated parents, $\mathbf{G} = \sigma^2_A \cdot \mathbf{I}$, where **I** is an s x s identity matrix.

ASSUMPTIONS FOR BLP CASE STUDIES

BLP assumes that the first and second moments are known. For the purposes of the examples in this chapter, we use the somewhat hypothetical values for loblolly pine presented in Table 5.1. The trait is height measured in meters. Values for σ^2_A and h^2 are taken from Lambeth et al. (1983). The additive genetic correlation (r_g) was estimated from Lambeth (1980). Additive covariances between heights at 5 and 20, 10 and 20, and 5 and 10 years could then be calculated. Total phenotypic variance was partitioned into family, family*environment, plot, and within plot error components. All variance components increased with age as the trees are taller and more diverse. We assumed that σ^2_f increased faster than σ^2_{fe} with age and that σ^2_p became relatively less important with age.

Using these variance components and assuming that each progeny test has 6 blocks (b=6) and 10 tree row plots of each family in each block (n=10), the variance of family means ($Var(\bar{y}_{i\cdot k\cdot})$) and family heritability (h^2_f) were calculated for each age.

Given the assumed parameters (Table 5.1) and some hypothetical data (Table 5.2), we now examine how BLP combines data and makes predictions in 5 illustrative cases.

Table 5.1 - Hypothetical data for best linear prediction case studies. Values are representative of *P. taeda* height measured in meters. Variance components are as defined in Eq. 2.23, and $\sigma^2_A = 4\sigma^2_{f\cdot}$

Age	$\bar{y}_{....}$ (m)	σ^2_f	σ^2_{fe}	σ^2_p	σ^2_w	σ^2_A	$Var(\bar{y}_{i\cdot k\cdot})$[a]	$Cov(g_x,g_{20})$[b]	h^2
5	5	0.0075	0.0075	0.0150	0.2700	0.03	0.0220	0.0818	0.10
10	12	0.0400	0.0200	0.0320	0.7080	0.16	0.0771	0.2541	0.20
20	20	0.1500	0.0500	0.0600	1.7400	0.60	0.2390	0.6000	0.30

[a]The variances of family means are calculated from Eq. 2.26 assuming t=1, b=6, n=10.

[b]The values in this column are additive genetic covariances for height at pairs of measurement ages: 5 and 20, 10 and 20 and 20 and 20. Also note that $Cov(g_5,g_{10}) = 0.0568$.

CASE STUDIES

Case 1: Three 5 Year-Old Tests Of Equal Precision, BV5

For Case 1, assume the data in Table 5.2 come from three 5-year-old tests of equal precision, and that all four families are represented in each test. To predict the breeding value for height at age 5 for each of the four parents by applying Eq. 5.3, it is necessary to first construct **V** (the variance-covariance matrix of the observed data), and **C** (the matrix of covariances between the observed data and the trait to be predicted).

The 12 x 1 data vector **y** is a vector of observed family phenotypic means and the 12 x 1 vector $\boldsymbol{\alpha}$ contains expected values of each the elements in **y**. For any family mean, $\bar{y}_{i\cdot k\cdot}$, the expected value is $\mu + E_i$ (from Eq. 5.2); while these expected values are never known, they can be estimated from the data by either the site mean or by checklots. Data in Table 5.2 are already expressed as deviations from their estimated expected values, $y_i - \hat{\alpha}_i$, and the transposed data vector of deviations with elements ordered by family is:

$$\mathbf{y}' - \hat{\boldsymbol{\alpha}}' = (\mathbf{y} - \hat{\boldsymbol{\alpha}})'$$
$$= [0.6\ \ 0.2\ \ 1.0\ \ 1.0\ \ 1.0\ \ 0.0\ \ -0.3\ \ -0.3\ \ 0.6\ \ -0.3\ \ 0.0\ \ -1.3]$$

Next, from Eq. 5.7, diagonal elements of the 12 x 12 **V** matrix are variances of family means, Var($\bar{y}_{i \cdot k \cdot}$), which all are 0.022 at age 5 (Table 5.1). From Eq. 5.8, the off-diagonal elements for the same family in different tests are ¼ σ^2_A (0.0075 at age 5, Table 5.1). All other elements are zero. Thus, the **V** matrix is:

Table 5.2 - Family phenotypic means for height (in meters) expressed as deviations from their estimated expected values for four *P. taeda* families in three open-pollinated progeny tests. These are the data used in BLP case studies 1, 2, 3, and 4.[a]

Family	Test 1	Test 2	Test 3
A	0.6	0.2	1.0
B	1.0	1.0	0.0
C	-0.3	-0.3	0.6
D	-0.3	0.0	-1.3

[a]Each deviation is a family mean minus the appropriate test site mean (which is the average of all family means at that test). For example, the first value, 0.6m, is $\bar{y}_{1 \cdot A \cdot} - \bar{y}_{1 \cdots}$. Averages across the four family deviations in a given test are not zero because these four families are assumed to be a small subset of all families in each test. That is, $\bar{y}_{i \cdots}$ is calculated using many families in each test, but individual data from only four of those families are shown here and used in the case studies.

$$
\mathbf{V} = \begin{bmatrix}
.0220 & .0075 & .0075 & 0 & 0 & 0 & 0 & 0 & 0 & 0 & 0 & 0 \\
.0075 & .0220 & .0075 & 0 & 0 & 0 & 0 & 0 & 0 & 0 & 0 & 0 \\
.0075 & .0075 & .0220 & 0 & 0 & 0 & 0 & 0 & 0 & 0 & 0 & 0 \\
0 & 0 & 0 & .0220 & .0075 & .0075 & 0 & 0 & 0 & 0 & 0 & 0 \\
0 & 0 & 0 & .0075 & .0220 & .0075 & 0 & 0 & 0 & 0 & 0 & 0 \\
0 & 0 & 0 & .0075 & .0075 & .0220 & 0 & 0 & 0 & 0 & 0 & 0 \\
0 & 0 & 0 & 0 & 0 & 0 & .0220 & .0075 & .0075 & 0 & 0 & 0 \\
0 & 0 & 0 & 0 & 0 & 0 & .0075 & .0220 & .0075 & 0 & 0 & 0 \\
0 & 0 & 0 & 0 & 0 & 0 & .0075 & .0075 & .0220 & 0 & 0 & 0 \\
0 & 0 & 0 & 0 & 0 & 0 & 0 & 0 & 0 & .0220 & .0075 & .0075 \\
0 & 0 & 0 & 0 & 0 & 0 & 0 & 0 & 0 & .0075 & .0220 & .0075 \\
0 & 0 & 0 & 0 & 0 & 0 & 0 & 0 & 0 & .0075 & .0075 & .0220
\end{bmatrix}
$$

Finally, the **C** matrix is a 12 x 4 matrix of covariances of 12 observed family means with the 4 breeding values being predicted. From Eq. 5.6 the elements are $\frac{1}{2}\text{Cov}(g_k,g_k)$. In this case, we are measuring height at 5, and trying to predict height at 5. Then $\frac{1}{2}\text{Cov}(g_k,g_k) = \frac{1}{2}\sigma^2_A = 0.015$ at age 5 (Table 5.1).

$$
\mathbf{C} = \begin{bmatrix}
.015 & 0 & 0 & 0 \\
.015 & 0 & 0 & 0 \\
.015 & 0 & 0 & 0 \\
0 & .015 & 0 & 0 \\
0 & .015 & 0 & 0 \\
0 & .015 & 0 & 0 \\
0 & 0 & .015 & 0 \\
0 & 0 & .015 & 0 \\
0 & 0 & .015 & 0 \\
0 & 0 & 0 & .015 \\
0 & 0 & 0 & .015 \\
0 & 0 & 0 & .015
\end{bmatrix}
$$

Breeding values for height at 5 years (BV5) are predicted for the four parents using Eq. 5.3 which yields a 4 x 1 vector with one BV prediction for each parent. Comparing these BLP predicted BV5 values to those estimated as twice the average of each parent's family deviations across the three tests (Table 5.3), note that the four families rank in the same order, because the data are balanced, the means are equally precise, and equally correlated with the trait being predicted. But the BLP predicted BV5's are regressed or shrunken back towards the mean of zero compared to BV's

Table 5.3 - Marginal family deviations (i.e. family deviations from Table 5.2 averaged across three tests) and best linear predictions of 5 and 20 year breeding values (BV5 and BV20, respectively) for Cases 1 and 2. All values are heights in meters.

Family	Averaged family deviations		Best Linear Prediction	
	marginal deviation	BV5[a] (2 x marg.dev.)	BV5	BV20
A	0.60	1.20	0.73	1.99
B	0.67	1.34	0.81	2.21
C	0.00	0.00	0.00	0.00
D	-0.53	-1.06	-0.65	-1.77

[a]An estimated BV5 is obtained by multiplying the marginal deviations by 2 when the open-pollinated families are considered to be half-sibs (i.e., the female parent passes on ½ of its breeding value to its half-sib offspring).

estimated as twice the marginal deviations. Future offspring from these four families (as predicted by BLP BV5 values) are not expected to perform quite as well or poorly as the observed offspring because part of the variation in the observed means is due to sampling and environmental variances.

Predicted BVs are predicted genetic gains: An attractive feature of BLP in this format is that the predicted BV's directly predict genetic gains. Consider that the gain from parental selection based on progeny test data is $\Delta G = 2\ S_D\ h^2_f$ where S_D = selection differential (Namkoong et al. 1966; Shelbourne 1969). The S_D is multiplied by two, because selection takes place on both sexes. The family heritability across three five year old tests is 0.6083 ($h^2_f = \sigma^2_f/(\sigma^2_f + \sigma^2_{fe}/3 + \sigma^2_p/18 + \sigma^2_w/180)$. For parent A, the marginal mean equals 0.6 (Table 5.3), which equals S_D for that parent since it is the deviation of that parent's offspring above the mean. Selecting family A,

$$\Delta G = 2 \times (0.6) \times (.6083) = 0.73$$

which is also the predicted BV5. So if parent A is crossed with other parents of equal breeding value, the offspring are predicted to be 0.73m taller than average when 5 years old. Similarly, offspring from a cross between parents A and B are predicted to be (0.73 + 0.81)/2 = 0.77m above average at 5 years.

Gains from a 1.5 generation orchard or from roguing a 1.0 seed orchard can be calculated by averaging predicted breeding values of the parents established or retained in the orchard because the breeding value of parents in the orchard is, by definition, the gain passed on to the offspring from the orchard. This assumes all parents in the orchard contribute equally as both male and female parents.

Case 2: Three 5-Year-Old Tests Of Equal Precision, BV20

The only change from Case 1 is that we predict breeding values at age 20 (BV20) instead of BV5. The phenotypic observations are the same (Table 5.2) and still considered as measured at age 5, therefore there are no changes in the **y** or **α** vector or **V** matrix. From Eq. 5.5, the non-zero elements of the **C** matrix are $\frac{1}{2}$ Cov(g_k,g^*_k) which equals $\frac{1}{2}$ Cov(g_5,g_{20}) = .0409 (Table 5.1).

The relative ranks of the four families are the same as in Case 1 (Table 5.3). Note that the magnitude of the BV20 values is larger than the BV5 values. This is reasonable; the trees will be taller, therefore family effects and predicted breeding values are larger in absolute value. The BLP formula effected this change of scale from 5-year-old measurements to 20-year-old predicted breeding values by the covariances in **C**. The covariance elements are 0.015 for predicting 5-year breeding values in Case 1 and 0.0409 for predicting 20-year breeding values in Case 2. While correlations are unitless statistics that are independent of the units of measurement, covariances depend on scale. Twenty-year breeding values have larger variances than do 5-year breeding values and this is reflected in the size of the covariances.

Although the absolute values of the BV20's are greater than the absolute values of the BV5's, the predictions have in fact been regressed toward a mean of zero due to the fact that the genetic correlation between 5 and 20 year heights is less than one. For Family A, predicted breeding values at age 5 and age 20 expressed as a percent gain are 15% (0.73m/5m = BV5/$\bar{y}$.... at age 5) and 10% (1.99m/20m = BV20/$\bar{y}$.... at age 20).

Conversion of BV5 into BV20: The predicted BV5 values from Case 1 can be converted into predicted BV20 values manually by determining the ratio between $\Delta G_{20,5}$ and $\Delta G_{5,5}$ where $\Delta G_{i,j}$ = genetic gain in trait i from selection on trait j (Baradat 1976).

$$\frac{\Delta G_{20,5}}{\Delta G_{5,5}} = \frac{1/2\ S_5\ \mathrm{Cov}(g_5,g_{20})/\mathrm{Var}(\bar{y}_{\cdot\cdot k\cdot,5})}{1/2\ S_5\ \mathrm{Cov}(g_5,g_5)/\mathrm{Var}(\bar{y}_{\cdot\cdot k\cdot,5})} = \frac{\mathrm{Cov}(g_5,g_{20})}{\sigma^2_{A5}} = 2.727$$

where S_5 = selection differential at age 5. Therefore, BV20 = 2.727 × BV5. If the parents are represented by unequal amounts or qualities of data then this conversion is difficult to calculate manually. If BV20 is the trait of interest, BLP predicts it in one step rather than two by directly incorporating genetic relationships between the measured and predicted traits (5 and 20 year height in this case) in the **C** matrix. This is indirect prediction of one trait using measurements on a different trait and is completely analogous to indirect selection as discussed in Falconer (1981).

Case 3: Tests of Different Age

In this case we still predict BV20, but Test 3 is assumed to be 10 years old, while Tests 1 and 2 are still 5 years old. The data, $\mathbf{y} - \hat{\boldsymbol{\alpha}}$, are assumed to stay the same (Table 5.2), but because some elements are now assumed to be 10 year data, changes occur in both the **V** and **C** matrices.

The diagonal elements in the **V** matrix corresponding to the third test now change from 0.022 ($\mathrm{Var}(\bar{y}_{i\cdot k\cdot})$ at 5 years from Table 5.1) to 0.0771 ($\mathrm{Var}(\bar{y}_{i\cdot k\cdot})$ at 10 years from Table 5.1). In addition, the off-diagonal elements corresponding to covariances between Test 1 and 3 or Test 2 and 3 measurements change from 0.0075 ($\frac{1}{4}$ $\mathrm{Cov}(g_5,g_5)$) to 0.0142 ($\frac{1}{4}$ $\mathrm{Cov}(g_5,g_{10})$).

$$\mathbf{V} = \begin{bmatrix} .0220 & .0075 & .0142 & 0 & 0 & 0 & 0 & 0 & 0 & 0 & 0 & 0 \\ .0075 & .0220 & .0142 & 0 & 0 & 0 & 0 & 0 & 0 & 0 & 0 & 0 \\ .0142 & .0142 & .0771 & 0 & 0 & 0 & 0 & 0 & 0 & 0 & 0 & 0 \\ 0 & 0 & 0 & .0220 & .0075 & .0142 & 0 & 0 & 0 & 0 & 0 & 0 \\ 0 & 0 & 0 & .0075 & .0220 & .0142 & 0 & 0 & 0 & 0 & 0 & 0 \\ 0 & 0 & 0 & .0142 & .0142 & .0771 & 0 & 0 & 0 & 0 & 0 & 0 \\ 0 & 0 & 0 & 0 & 0 & 0 & .0220 & .0075 & .0142 & 0 & 0 & 0 \\ 0 & 0 & 0 & 0 & 0 & 0 & .0075 & .0220 & .0142 & 0 & 0 & 0 \\ 0 & 0 & 0 & 0 & 0 & 0 & .0142 & .0142 & .0771 & 0 & 0 & 0 \\ 0 & 0 & 0 & 0 & 0 & 0 & 0 & 0 & 0 & .0220 & .0075 & .0142 \\ 0 & 0 & 0 & 0 & 0 & 0 & 0 & 0 & 0 & .0075 & .0220 & .0142 \\ 0 & 0 & 0 & 0 & 0 & 0 & 0 & 0 & 0 & .0142 & .0142 & .0771 \end{bmatrix}$$

The **C** matrix elements corresponding to Test 3 data change from 0.0409 ($\frac{1}{2}$ $Cov(g_5,g_{20})$) to 0.1271 ($\frac{1}{2}$ $Cov(g_{10},g_{20})$).

$$\mathbf{C} = \begin{bmatrix} .0409 & 0 & 0 & 0 \\ .0409 & 0 & 0 & 0 \\ .1271 & 0 & 0 & 0 \\ 0 & .0409 & 0 & 0 \\ 0 & .0409 & 0 & 0 \\ 0 & .1271 & 0 & 0 \\ 0 & 0 & .0409 & 0 \\ 0 & 0 & .0409 & 0 \\ 0 & 0 & .1271 & 0 \\ 0 & 0 & 0 & .0409 \\ 0 & 0 & 0 & .0409 \\ 0 & 0 & 0 & .1271 \end{bmatrix}$$

Given this situation Family A and Family B have switched ranks (Table 5.4), because Family A had its largest and Family B its smallest phenotypic value in Test 3 (see data in Table 5.2). Test 3 measurements are weighted more heavily in determining BV20 than Tests 1 and 2 because the heritability at 10 years and genetic correlation between 10 and 20 year heights are higher. Although rank changes were not involved, the predicted BV20 for Families C and D also changed due to heavier weight on Test 3. The BV20 for Family D became more negative because its lowest value occurred in Test 3, while the BV20 for Family C increased as its highest value got more weight.

Table 5.4 - Predicted breeding values for height at 20 years, variance among the predictions, error variance of prediction, and correlation between predicted and true breeding values for the four case studies as described in the text[a].

Family	Case 2	Case 3	Case 4	Case 5
	Predicted BV20's (m)			
A	1.99	1.96	1.66	1.66
B	2.21	1.44	2.42	2.21
C	0.00	0.40	-0.31	1.12
D	-1.77	-2.01	-1.27	0.00
	Variance among predictors (m^2)			
A	0.136	0.235	0.127	0.113
B	0.136	0.235	0.127	0.136
C	0.136	0.235	0.127	0.076
D	0.136	0.235	0.127	0.076
	Error variance of predictions (m^2)			
A	0.464	0.365	0.472	0.487
B	0.464	0.365	0.472	0.464
C	0.464	0.365	0.472	0.524
D	0.464	0.365	0.472	0.524
	Corr. of predicted and true BV			
A	0.476	0.626	0.460	0.434
B	0.476	0.626	0.460	0.476
C	0.476	0.626	0.460	0.356
D	0.476	0.626	0.460	0.356

[a]Case 2: Tests 5 years old, equal precision.
Case 3: Tests 1 and 2 are 5 years old, Test 3 is 10 years old.
Case 4: Test 3 is low precision, all tests are 5 years old.
Case 5: Families in different numbers of tests, all tests are 5 years old.

Note that in Case 2, predicted BV20 for parent A is 1.99, while in Case 3 it is 1.96. This occurs despite the fact that parent A has its highest value in Test 3. Predicted BV20 for A did not increase because a deviation of one meter (Table 5.2) at 5 years in Case 2 corresponds to 20% of the mean, while a difference of one meter at 10 years is only 8% of the mean. The parent A value in Test 3 did receive more weight, but it does not indicate the same degree of genetic superiority when the data are treated as 10 years old as it did for 5 year data. In this case the net result was that BV20 did not change significantly for parent A.

Case 4: Tests of Different Precision

For this case all tests are assumed 5 years old, but Test 3 is assumed to be an imprecise test. The idea is to demonstrate how to incorporate data with different variance structures. Specifically, the individual tree heritability was arbitrarily decreased from 0.10 to 0.08. Total phenotypic variance, Var(y_{ijkl}), in Test 3 was increased from 0.3 to 0.375, and the increase was partitioned into the σ^2_w and σ^2_p variance components in a 9:1 ratio. The genetic parameters σ^2_f, σ^2_{fe}, and Cov(g_5,g_{20}) are assumed to be population parameters that do not change. In addition, Test 3 was assumed to have only 4 blocks (instead of 6 as in the previous cases) and poor survival leaving 7 trees/family-block (instead of 10). Thus, the variance of family means (Var($\bar{y}_{i\cdot k\cdot}$)) in Test 3 is now 0.0327, while it is still 0.0220 in Tests 1 and 2.

For Case 4,

$$
\mathbf{V} = \begin{bmatrix}
.0220 & .0075 & .0075 & 0 & 0 & 0 & 0 & 0 & 0 & 0 & 0 & 0 \\
.0075 & .0220 & .0075 & 0 & 0 & 0 & 0 & 0 & 0 & 0 & 0 & 0 \\
.0075 & .0075 & .0327 & 0 & 0 & 0 & 0 & 0 & 0 & 0 & 0 & 0 \\
0 & 0 & 0 & .0220 & .0075 & .0075 & 0 & 0 & 0 & 0 & 0 & 0 \\
0 & 0 & 0 & .0075 & .0220 & .0075 & 0 & 0 & 0 & 0 & 0 & 0 \\
0 & 0 & 0 & .0075 & .0075 & .0327 & 0 & 0 & 0 & 0 & 0 & 0 \\
0 & 0 & 0 & 0 & 0 & 0 & .0220 & .0075 & .0075 & 0 & 0 & 0 \\
0 & 0 & 0 & 0 & 0 & 0 & .0075 & .0220 & .0075 & 0 & 0 & 0 \\
0 & 0 & 0 & 0 & 0 & 0 & .0075 & .0075 & .0327 & 0 & 0 & 0 \\
0 & 0 & 0 & 0 & 0 & 0 & 0 & 0 & 0 & .0220 & .0075 & .0075 \\
0 & 0 & 0 & 0 & 0 & 0 & 0 & 0 & 0 & .0075 & .0220 & .0075 \\
0 & 0 & 0 & 0 & 0 & 0 & 0 & 0 & 0 & .0075 & .0075 & .0327
\end{bmatrix}
$$

To predict BV20 as in Case 2, the non-zero elements of the **C** matrix are ½Cov(g_5,g_{20}) = .0409 (Table 5.1). The decreased precision of the observations from Test 3 do not affect the **C** matrix as these elements are the genetic covariances between the observed means and the true breeding values of the trait of interest. Therefore, **C** for Case 4 is equivalent to **C** in Case 2.

When the breeding values are calculated, measurements in Test 3 receive less weight due to the low precision of family means in that test (that is, more of the observed variation in Test 3 is due to spurious environmental as opposed to genetic differences). This is reflected in the changes observed in predicted BV20's (Table 5.4). The largest positive deviation for parent A is in Test 3. Its BV20 dropped from 1.99 in Case 2, where all tests were of equal and better precision, to 1.66 since less weight is placed on the "good" deviation in Test 3. Parent B's lowest value is in Test 3, so its predicted BV20 increased from 2.21 in Case 2 to 2.42. The changes in predicted BV20 for parents C and D are also in the expected direction.

Case 5: Families In Unequal Numbers Of Tests

Often, parent trees are not tested in equal numbers of tests. In Case 5 a number of observations have been arbitrarily deleted to illustrate how BLP deals with this situation. Case 5 is identical to Case 2 (all tests of equal precision and assumed to be 5 years old) with the exception that 5 of the 12 observations have been deleted so that families are represented in unequal numbers of tests (Table 5.5).

Table 5.5 - Family phenotypic means for height (in meters) expressed as deviations from their estimated expected values for four *P. taeda* families. These are the data used in BLP case study 5[a].

Family	Test 1	Test 2	Test 3
A	x	0.2	1.0
B	1.0	1.0	0.0
C	x	x	0.6
D	x	0.0	x

[a]Test marginal deviations do not equal zero as the four families are assumed to be a subset of all families in tests. See Table 5.2 for further explanation.

Since all tests are considered 5 years old and equally precise, the elements of the **C** and **V** matrices are identical to the Case 2 situation except that certain rows and columns have been deleted to reflect the missing data points.

$$(\mathbf{y} - \hat{\boldsymbol{\alpha}})' = [\ 0.2 \quad 1.0 \quad 1.0 \quad 1.0 \quad 0.0 \quad 0.6 \quad 0.0\ ]$$

$$\mathbf{C} = \begin{bmatrix} .0409 & 0 & 0 & 0 \\ .0409 & 0 & 0 & 0 \\ 0 & .0409 & 0 & 0 \\ 0 & .0409 & 0 & 0 \\ 0 & .0409 & 0 & 0 \\ 0 & 0 & .0409 & 0 \\ 0 & 0 & 0 & .0409 \end{bmatrix}$$

$$\mathbf{V} = \begin{bmatrix} .0220 & .0075 & 0 & 0 & 0 & 0 & 0 \\ .0075 & .0220 & 0 & 0 & 0 & 0 & 0 \\ 0 & 0 & .0220 & .0075 & .0075 & 0 & 0 \\ 0 & 0 & .0075 & .0220 & .0075 & 0 & 0 \\ 0 & 0 & .0075 & .0075 & .0220 & 0 & 0 \\ 0 & 0 & 0 & 0 & 0 & .0220 & 0 \\ 0 & 0 & 0 & 0 & 0 & 0 & .0220 \end{bmatrix}$$

No observations on family B were deleted, and the BV20 in Case 2 and Case 5 are the same for Parent B (Table 5.4). The fact that all other families had different amounts of information did not change the prediction for B. The way BLP deals with the problem of different numbers of tests is illustrated in Table 5.6.

For all three family-case combinations presented in Table 5.6 the quality of information from each test is exactly the same. Also all three families are 0.6m above average so would have the same estimated breeding value based on marginal means, however, the families are in different numbers of tests. Note that predicted BV20 increases with the number of tests. This is reasonable, as maintaining a level of performance across an increasing number of tests gives confidence that future offspring are likely to perform well across additional planting sites. Also note that the increase in predicted BV20 from 1 to 2 tests is 0.54, while the increase from 2 to 3 tests is only 0.33. The value of each additional test decreases as the number of tests increases.

Table 5.6 - Predicted BV20's for *P. taeda* families with equal marginal deviations (family deviations averaged across all tests), but in different numbers of tests.

Family, Case	marginal dev.	Tests	BV20
A, Case 2	0.6	3	1.99
A, Case 5	0.6	2	1.66
C, Case 5	0.6	1	1.12

VARIANCE AMONG PREDICTIONS AND ERROR VARIANCE OF PREDICTIONS

BLP vs Least Squares

For BLP, the variance among predictions, Var($\hat{\mathbf{g}}$), and the error variance of the predictions, EVP or Var($\hat{\mathbf{g}}$-$\mathbf{g}$), are both q x q matrices of variances and covariances (see Eqs. 4.4 and 4.6). In this chapter, breeding values are being predicted for only a single trait, so q = s (i.e. the number of parents). For the single trait case when both parents are unrelated, both matrices are diagonal (i.e. have zeros for all off-diagonal elements) because the prediction for a given parent does not depend on information from other parents. For any given parent, the variance among predictions can be interpreted as the expected spread among predicted breeding values associated with a large number of parents with equal information. As information gets better (lower EVP), predictions tend to spread out, and not be "shrunk" back as far towards the mean, because the effective heritabilities are larger. Thus, Var($\hat{\mathbf{g}}$) increases as EVP decreases. The opposite occurs using least squares estimates such as marginal means (Henderson 1973), and so least squares estimates have the largest variance (i.e. most spread among the estimated values) when they are least precisely estimated. Therefore, as illustrated in the Chapter 3 (Figure 3.1), when using least squares estimates there is a tendency to select (i.e. rank as better) those parents which are included in fewer tests and (have the highest EVP).

We demonstrate these results for the case of balanced open-pollinated progeny tests at any given age. The least squares estimates of the breeding values are twice the marginal means. Assuming model 2.23, i.e. still treating f_k as a random variable, the variances of the least squares estimates are below:

$$\begin{aligned} Var(\hat{g})_L &= Var(2\bar{y}_{\cdot\cdot k\cdot}) = 4\ Var(\bar{y}_{\cdot\cdot k\cdot}) \\ &= 4\ (\sigma^2_f + \sigma^2_{fe}/t + \sigma^2_p/tb + \sigma^2_w/tbn). \end{aligned} \qquad \textbf{Eq. 5.10}$$

$$\begin{aligned} Var(\hat{g}\text{-}g)_L &= Var(\hat{g}) + Var(g) - 2Cov(\hat{g},g) \\ &= Var(\hat{g}) - Var(g) \\ &= EVP_L \end{aligned} \qquad \textbf{Eq. 5.11}$$

because, as shown below, $Var(g) = Cov(\hat{g},g)$.

$$\begin{aligned} Cov(\hat{g},g) &= Cov(2y_{\cdot\cdot k\cdot},g) \\ &= 2\ Cov(f_k + fe_{\cdot k}/t + p_{\cdot\cdot k}/tb + w_{\cdot\cdot k\cdot}/tbn,\ 2f_k) \\ &= 4\ Cov(f_k,f_k) \\ &= \sigma^2_A = Var(g) \end{aligned}$$

since $g_k = 2\ f_k$.

So, when twice the marginal means are used to estimate breeding values, these estimates have the above expressions for the variance among the estimates (Eq. 5.10) and the error variance of the estimates (Eq. 5.11). For BLP, the variances of predictors ($Var(\hat{g})$), and error variance of prediction (EVP) are obtained from Eq. 4.4 and 4.6. (Also the correlation between the predicted and true breeding value ($Corr(g,\hat{g})$) can be estimated from Eq. 4.7).

Using the data from Table 5.1 in Eq. 4.4, 4.6, 5.10, and 5.11, and plotting the results over different numbers of tests (Fig. 5.1), it is apparent that BLP predictions always have lower error variance than those based on marginal means. Also, when marginal means estimates are least reliable (Fig. 5.1b), they have the largest spread (Fig. 5.1a). This is especially true for parents tested in only one test. For parents in a

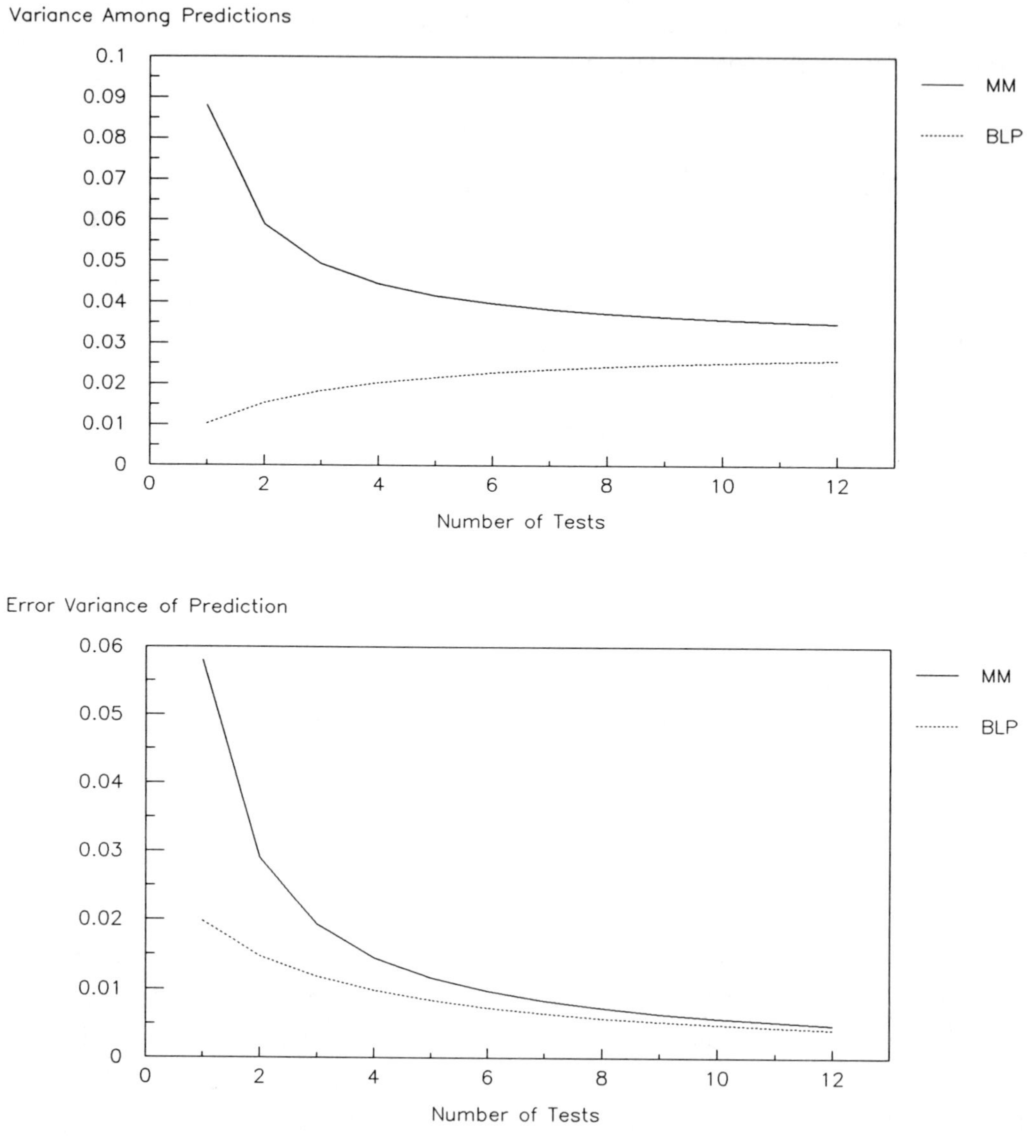

Figure 5.1. a) Variances among predictions (top) and b) error variance of prediction (bottom) for Best Linear Prediction (BLP) and Marginal Means (MM) approaches to predicting breeding values, assuming five-year-old progeny test data is used to predict five-year-old breeding values.

single progeny test, t = 1 in Equation 5.10 and marginal means estimates of breeding values have a large variance (since t occurs in the denominator in Eq. 5.10). This is when the error associated with these estimates is the largest (from Eq. 5.11). If some parents were represented in 1 test and some in 10 tests and if the marginal means estimates were used to rank the parents, the larger variance among the marginal mean estimates for those in a single test would result in the tendency to select as winners those in one test, i.e. the more poorly tested parents (see Chapter 3 and Figure 3.1).

BLP works in the opposite manner. When the quantity of data is less (e.g. parents in only a single test), the predicted breeding values cluster nearer around the population mean of zero; that is, the variance among predictions is small (Figure 5.1a). As the data get more reliable (more tests) and the predictions have smaller error variances (Figure 5.1b), there is a larger spread among predictions (Figure 5.1a). Thus, if some parents were tested in one test and some in 10 and if BLP predicted breeding values were used to rank the parents, there would be a tendency for parents in more tests to rank near the top (and near the bottom) and be selected.

Finally, note that if all parents are tested in more than 4 or 5 tests (given the values in Table 5.1), the variance among predictions and the error variance become relatively constant in both methods and the two methods give similar rankings with similar precision. The variance among predictions asymptotes to σ^2_A for both the marginal means and BLP predictions of breeding values. This is because over an infinite number of tests, all sources of environmental variation are expected to average zero (for example, set $t=\infty$ in Eq. 5.10), and the only source of variability is genetic variation among parental breeding values.

Variances for BLP Case Studies

For Cases 2, 3, and 4 in the above example, all parents are represented equally in all three tests, and the estimated Var($\hat{g}$), EVP, and Corr(g,$\hat{g}$) are the same for all parents (Table 5.4). Compared to Case 2, Var($\hat{g}$) and Corr(g,$\hat{g}$) are larger in Case 3, while EVP is smaller. This occurs because in Case 3, Test 3 is assumed to be 10 years old as opposed to 5 years old in Case 2. Since 10-year data have a larger assumed juvenile-mature correlation and heritability than 5-year data, BV20's are predicted more precisely (lower error) and BLP does not shrink the predictions back to the mean as much as in Case 2 (larger variance among predictions).

Case 4 assumes that Test 3 is 5 years old, but imprecise. Relative to Case 2, the predictions do not spread out as much, i.e. Var($\hat{g}$) is smaller. This is somewhat apparent from the four predicted BV20's (i.e. they are more clustered together than in Case 2), but would be more easily seen if there were more parents. The BV20's in Case 4 are also estimated less precisely than for Case 2 as indicated by their larger EVP and lower correlation.

Finally, in Case 5 the parents are represented in different numbers of tests. In particular, the information for parents C and D, which are in only one test, is relatively imprecise, and the Var($\hat{g}_k$) is estimated as 0.076, and Corr(g,$\hat{g}$) as 0.356. For parents in more tests, the predictions will spread out more; Var($\hat{g}_k$) = 0.113 and 0.136. Also, Corr(g,$\hat{g}$) = 0.434 and 0.476 for parents in two and three tests, respectively. Concomitantly, as parents are represented in more tests, the error variance of the predictions decreases.

SUMMARY AND CONCLUSIONS

The case studies in this chapter are intended to 1) demonstrate how to apply BLP with half-sib progeny test data, 2) illustrate the way BLP predicts breeding values when data on the parents are of unequal quality and quantity, and 3) compare how BLP and least squares estimates of breeding values can result in quite different rankings with unbalanced, messy data.

The case studies here are unrealistic in the sense of very few parents and few observations; also, for each different case only one of the many possible "messy" data problems was examined. In real situations, it is likely that more than one of these problems will be present. For example, in the slash pine data at the University of Florida's Cooperative Forest Genetics Research Program, first-generation parents are represented in from 1 to 22 progeny tests, the tests range from 5 to 15 years old and within any given age the precision of different tests varies drastically. Predictions of breeding values using BLP will account for all of these different problems at the same time as long the linear models are correctly specified and the elements of **C** and **V** are well-estimated. This is discussed in Chapter 8 for the slash pine data.

The discussions in this chapter were also limited to prediction of a single trait. Multiple trait prediction and prediction of aggregate genetic values are discussed in Chapter 7.

PROBLEMS

Problem 5.1: Assume that test 1 (Table 5.2) is 5-years-old, test 2 is 10-years-old, and test 3 is 20-years old. You wish to predict breeding values for height at age 20. Using the values in Table 5.1 to calculate the appropriate second moments, set up the **C**, **V**, and **G** matrices, and using the **y** vector for families A and B, predict the breeding values, and estimate Var($\hat{g}$), EVP, and Corr(g,$\hat{g}$).

Problem 5.2: Repeat Problem 5.1 except that you now wish to predict breeding value for height at age 10.

Problem 5.3: Assume that another family, say family G, is in three 5-year-old OP tests, and that family H is in one 10-year-old OP test, and you wish to predict BV20. Using the values from Table 5.1, determine which predicted breeding value is more precise (has a lower EVP and a higher Var($\hat{g}$) and Corr(g,$\hat{g}$)).

Problem 5.4: Repeat Problem 5.3 with the following exceptions:

1) the 5-year-old tests have 10 blocks (instead of 6), and
2) the 10-year-old tests have 4 blocks (instead of 6), and 6 trees/block (instead of 10 trees/block).

Problem 5.5: You have a group of 500 families which have been tested in one OP progeny test, a second group of 500 families tested in two OP tests, and a third group of 500 families tested in three OP tests. All tests are 5-years-old. You have calculated the Var($\hat{g}$) as 0.076m^2, 0.113m^2, and 0.136m^2 for one, two and three 5-year-old tests, respectively. If you plan to select all parents with breeding values for height greater than 0.5m, what number of the parents tested in one test will you select? What number from the two-test and three-test group will you select? How will these numbers change relative to one another as you move the truncation point higher or lower? Assume that the predicted breeding values are distributed N(0, Var($\hat{g}$)).

Problem 5.6: Most classical approaches to progeny test analysis used in forestry have a tendency to select parents that are less tested (in fewer tests). BLP has the opposite trend; it tends to select those parents represented in more tests (see Problems 3.4 and 3.5). Is either tendency desirable, or should we be selecting equal percentages of parents in large and small numbers of tests?

Problem 5.7: The average BV for volume at age 20 of all the trees in your seed orchard is $1.0m^3$. You have determined that the isolation zone around your orchard is not completely effective and that 30% of the pollen in your orchard comes from unimproved wild stands outside your orchard. What is the expected breeding value of a bulk seed collection from that orchard? State any assumptions which may be necessary.

Chapter 6
BLP with Full-sib and Multiple Sources of Data

INTRODUCTION

To this point BLP has been applied only to half-sib progeny test data, however in theory BLP can utilize data of any type. Many tree improvement programs have extensive full-sib or cross-pollinated (CP) progeny tests where both the male and female parents of every tree in the test are known. Some programs may also have ancestral data available which could be utilized to better predict breeding values of trees in the current generation. Data may also exist from greenhouse or other early sreening tests, and data on the original ortet or several ramets of a clone may also be available. All of these types of data can be to predict genetic values (such as breeding values) by BLP, and any combination of these diverse types of data can be utilized simultaneously in BLP. In every case, one or more linear models must be specified and first and second moments estimated; then BLP can be applied.

The purpose of this chapter is to demonstrate how BLP can be utilized in situations involving full-sib and multiple types of data simultaneously. To expedite description of concepts, discussion is still restricted to predicting breeding values of parents (i.e. selections) for a single trait assuming homogeneous second moments. Predicting breeding values for individuals is considered in Chapters 9 and 10, whilc multiple trait predictions and heterogeneous second moments are discussed in Chapter 7. The theory behind the application (Chapter 4) has not changed, nor have the general procedures required to apply BLP. We begin by considering cross-pollinated progeny tests in a diallel mating design, and examine both balanced and unbalanced situations. Next, data from both half-sib and full-sib tests are used simultaneously to predict breeding values. Once again the unit of observation is taken as family means (i.e. the average of all individuals from a given full- or half- sib family from a specific progeny test) and data for each family from a given test are assumed equal in quality and quantity. Theoretically, other units of observation (e.g. individual measurements, plot means) or family means estimated from unbalanced data could be employed (see Chapter 7 for a complete discussion); however, this would complicate the computations and the description of the concepts.

FULL-SIB PROGENY TESTS

The procedures for using full-sib progeny test data to predict parental breeding values are illustrated in the next several sections by completely detailing such an analysis for the following small example. Assume that four loblolly pine parents (A, B,

C, and D) have been mated in a half-diallel design with no selfs (Griffing 1956). Progeny from the six full-sib families (AxB, AxC, AxD, BxC, BxD, and CxD) are planted in two progeny tests (tests 1 and 2) each of which is a randomized complete block design with six blocks and six trees of each full-sib family planted within each block. The tests have been measured for height at 12 years, and the data from these progeny tests are to be used to predict breeding values of the four parents for 12-year height growth. We assume that accurate and precise variance component estimates exist (Table 6.1) and that the following linear model adequately represents a measurement on each individual phenotype:

$$y_{ijklm} = \mu + E_i + B_{ij} + f_k + fe_{ik} + h_l + he_{il} + fh_{kl} + fhe_{ikl} + p_{ijkl} + w_{ijklm} \qquad \textbf{Eq. 6.1}$$

where

μ	= a fixed general mean,
E_i	= fixed effect of the ith test environment, $i = 1, 2, ...t$,
B_{ij}	= fixed effect of the jth block in i^{th} test, $j = 1, 2, . ..b_i$,
f_k	= random effect of the kth female, $k = 1, 2, ... s_i$, $E(f_k) = 0$, $Var(f_k) = \sigma^2_f$,
fe_{ik}	= random effect of the kth female in i^{th} test, $E(fe_{ik}) = 0$, $Var(fe_{ik}) = \sigma^2_{fe}$,
h_l	= random effect of the lth male, $l = 1, 2, ...r_i$, $E(h_l) = 0$, $Var(h_l) = \sigma^2_f$,
he_{ik}	= random effect of the lth male in i^{th} test, $E(he_{il}) = 0$, $Var(he_{il}) = \sigma^2_{fe}$,
fh_{kl}	= random interaction effect of the kl^{th} full-sib family, $E(fh_{kl}) = 0$, $Var(fh_{kl}) = \sigma^2_d$,
fhe_{ikl}	= random interaction effect of the kl^{th} family in i^{th} test, $E(fhe_{ikl}) = 0$, $Var(fhe_{ikl}) = \sigma^2_{de}$,
p_{ijkl}	= random plot error of kl^{th} family in j^{th} block of ith test, $E(p_{ijkl}) = 0$, $Var(p_{ijkl}) = \sigma^2_p$,
w_{ijklm}	= random tree error of mth tree in $ijkl^{th}$ plot, $E(w_{ijklm}) = 0$, $Var(w_{ijklm}) = \sigma^2_w$, and

the covariances between all pairs of effects are assumed zero.

Also assume the following: $\sigma^2_f = \frac{1}{4}\sigma^2_A$, where σ^2_A is the additive genetic variance; $\sigma^2_{fe} = \frac{1}{4}\sigma^2_{AE}$, where σ^2_{AE} is the additive x environment interaction; $\sigma^2_d = \frac{1}{4}\sigma^2_D$, where σ^2_D is the dominance variance; $\sigma^2_{de} = \frac{1}{4}\sigma^2_{DE}$, where σ^2_{DE} is the dominance x environment interaction. Note that σ^2_p and σ^2_w defined in this model are not equivalent to the same terms defined in the half-sib model of Eq. 2.23. For example, σ^2_w in Eq. 2.23 contains all the dominance variance, while in Eq. 6.1 it contains only 3/4 of the dominance variance.

Specifying the y and $\boldsymbol{\alpha}$ Vectors

We use as observations in the **y** vector, the full-sib family means from each test obtained by averaging over all trees of that family in the test. Any such family mean is represented in terms of the linear model (Eq. 6.1) as:

$$\bar{y}_{i\cdot kl\cdot} = [\sum_{j}^{b}\sum_{m}^{n} (\mu + E_i + B_{ij} + f_k + fe_{ik} + h_l + he_{il} + fh_{kl} + fhe_{ikl} + p_{ijkl} + w_{ijklm})]/bn,$$

$$= \mu + E_i + f_k + fe_{ik} + h_l + he_{il} + fh_{kl} + fhe_{ikl} + p_{i\cdot kl}/b + w_{i\cdot kl\cdot}/bn. \qquad \textbf{Eq. 6.2}$$

Table 6.1 - Hypothetical values for 12-year-old cross-pollinated progeny tests of loblolly pine.[a]

Trait	$\bar{y}_{i\cdots\cdot}$	σ^2_T	σ^2_F	σ^2_f	σ^2_{fe}	σ^2_d	σ^2_{de}	σ^2_p	σ^2_w	$Var(\bar{y}_{i\cdot kl\cdot})$
height (ft)	30	38.7	2.70	1.80	0.90	0.90	0.45	6.30	25.65	8.51

[a] $Var(\bar{y}_{i\cdot kl\cdot})$ is the variance of full-sib family means from Eq. 6.4 assuming six blocks (b=6) and six trees of each family in each block (n=6).
Also note that $\sigma^2_F = \sigma^2_f + \sigma^2_{fe}$.

Each full-sib family mean has an expected value of

$$E(\bar{y}_{i\cdot kl\cdot}) = E(\mu + E_i + f_k + fe_{ik} + h_l + he_{il} + fh_{kl} + fhe_{ikl} + p_{i\cdot kl}/b + w_{i\cdot kl\cdot}/bn)$$

$$= \mu + E_i \qquad \textbf{Eq. 6.3}$$

because expected values of all random effects are zero as specified in the linear model (Eq. 6.1).

The grand test mean across all families, blocks and trees in a given test is:

$$\bar{y}_{i....} = [\sum_{j}^{b}\sum_{k}^{s}\sum_{l}^{r}\sum_{m}^{n} (\mu+E_i+B_{ij}+f_k+fe_{ik}+h_l+he_{il}+fh_{kl}+fhe_{ikl}+p_{ijkl}+w_{ijklm})]/bsrn$$

$$= \mu + E_i + f./s + fe_{i}./s + h./r + he_{i}./r + fh../sr + fhe_{i}../sr + p_{i}.../bsr + w_{i}..../bsrn$$

and the expected value of the grand mean is

$$E(\bar{y}_{i....}) = E(\mu + E_i + f./s + fe_{i}./s + h./r + he_{i}./r + fh../sr + fhe_{i}../sr + p_{i}.../bsr + w_{i}..../bsrn)$$

$$= \mu + E_i$$

So, when the observational elements of **y** are full-sib means, the grand test mean is an estimate of the fixed effects (i.e. the elements of $\boldsymbol{\alpha}$), since it has the same expected value, i.e. the true site mean. The grand mean will only be a "good" estimate of $\mu + E_i$ if there are a large number of unselected female and male parents (i.e. s and r are large). If s and r are small, then even though the expected value of $\bar{y}_{i....} = \mu + E_i$, the actual male and female effects will be less likely to average to zero. In other words, the grand site mean will more likely influenced by the genetic effects of the specific set of males and females included. If the parents are **selected** from a population of second generation parents, then the grand mean estimates $\mu + E_i + \tau$ where τ is the fixed genetic effect associated with the second generation parents (see Chapter 7 for a complete discussion of estimating first moments). Genetic check lots (also called genetic standards) could also be used to estimate the fixed effects associated with test locations, but, for this chapter, we assume that grand means are both unbiased and precise estimate of the elements of $\boldsymbol{\alpha}$.

For this simple diallel example using full-sib family means from the two progeny tests as observations, **y** and $\boldsymbol{\alpha}$ are:

$$\mathbf{y} = \begin{bmatrix} \bar{y}_{1\cdot AB\cdot} \\ \bar{y}_{2\cdot AB\cdot} \\ \bar{y}_{1\cdot AC\cdot} \\ \bar{y}_{2\cdot AC\cdot} \\ \bar{y}_{1\cdot AD\cdot} \\ \bar{y}_{2\cdot AD\cdot} \\ \bar{y}_{1\cdot BC\cdot} \\ \bar{y}_{2\cdot BC\cdot} \\ \bar{y}_{1\cdot BD\cdot} \\ \bar{y}_{2\cdot BD\cdot} \\ \bar{y}_{1\cdot CD\cdot} \\ \bar{y}_{2\cdot CD\cdot} \end{bmatrix} \qquad \boldsymbol{\alpha} = \begin{bmatrix} E(\bar{y}_{1\cdot AB\cdot}) \\ E(\bar{y}_{2\cdot AB\cdot}) \\ E(\bar{y}_{1\cdot AC\cdot}) \\ E(\bar{y}_{2\cdot AC\cdot}) \\ E(\bar{y}_{1\cdot AD\cdot}) \\ E(\bar{y}_{2\cdot AD\cdot}) \\ E(\bar{y}_{1\cdot BC\cdot}) \\ E(\bar{y}_{2\cdot BC\cdot}) \\ E(\bar{y}_{1\cdot BD\cdot}) \\ E(\bar{y}_{2\cdot BD\cdot}) \\ E(\bar{y}_{1\cdot CD\cdot}) \\ E(\bar{y}_{2\cdot CD\cdot}) \end{bmatrix} = \begin{bmatrix} E(\bar{y}_{1\cdots\cdot}) \\ E(\bar{y}_{2\cdots\cdot}) \\ E(\bar{y}_{1\cdots\cdot}) \\ E(\bar{y}_{2\cdots\cdot}) \\ E(\bar{y}_{1\cdots\cdot}) \\ E(\bar{y}_{2\cdots\cdot}) \\ E(\bar{y}_{1\cdots\cdot}) \\ E(\bar{y}_{2\cdots\cdot}) \\ E(\bar{y}_{1\cdots\cdot}) \\ E(\bar{y}_{2\cdots\cdot}) \\ E(\bar{y}_{1\cdots\cdot}) \\ E(\bar{y}_{2\cdots\cdot}) \end{bmatrix} \approx \begin{bmatrix} \bar{y}_{1\cdots\cdot} \\ \bar{y}_{2\cdots\cdot} \\ \bar{y}_{1\cdots\cdot} \\ \bar{y}_{2\cdots\cdot} \\ \bar{y}_{1\cdots\cdot} \\ \bar{y}_{2\cdots\cdot} \\ \bar{y}_{1\cdots\cdot} \\ \bar{y}_{2\cdots\cdot} \\ \bar{y}_{1\cdots\cdot} \\ \bar{y}_{2\cdots\cdot} \\ \bar{y}_{1\cdots\cdot} \\ \bar{y}_{2\cdots\cdot} \end{bmatrix}$$

For this specific example, assume **y** and $\hat{\boldsymbol{\alpha}}$ have been determined and the resulting vector of deviations, $(\mathbf{y} - \hat{\boldsymbol{\alpha}})$ is

$$\mathbf{y} - \hat{\boldsymbol{\alpha}} = \begin{bmatrix} 4.2 \\ 3.8 \\ 1.3 \\ 0.7 \\ 5.6 \\ 6.4 \\ -1.6 \\ -0.4 \\ -4.3 \\ -3.7 \\ 1.1 \\ 0.9 \end{bmatrix}$$

Each element is a family mean (measured in units of feet at 12 years) minus a grand site mean. The elements of $\mathbf{y} - \hat{\boldsymbol{\alpha}}$ do not average to zero because the estimates of the fixed effects, $\bar{y}_{1\cdots\cdot}$ and $\bar{y}_{2\cdots\cdot}$, are assumed to include other full-sib families, say from additional diallels. This might indicate, for example, that "grand" means from several separate (i.e. disconnected) diallels were averaged to obtain the fixed effects estimates. Remember the fixed effects estimate the environmental effects at a given test location and this might best be done by averaging over many disconnected diallels if present at a each location (cf Chapter 7).

In this example we consider only the families in a single diallel and assume that there is no relationship between any of the parents involved in this diallel and the parents involved in any of the other diallels. If there were genetic relationships among parents in different diallels (or the same parent in more than one diallel), then the diallels would not be disconnected (in a genetic sense) and it would be better (i.e. more precise predictions of breeding values) to analyze all diallels at one time (i.e. to include all data from all diallels in one long **y** vector for BLP calculations).

Estimating the Second Moments

Variances: Since all observations are full-sib family means measured at the same age and assumed to have equal variances, the same variance is used for all diagonal elements of **V**. The expected variance of full-sib family means is:

$$\begin{aligned} Var(\bar{y}_{i\cdot k\cdot\cdot}) &= Var(\mu + E_i + f_k + fe_{ik} + h_l + he_{il} + fh_{kl} + fhe_{ikl} + p_{i\cdot kl}/b + w_{i\cdot kl\cdot}/bn) \\ &= \sigma^2_f + \sigma^2_{fe} + \sigma^2_f + \sigma^2_{fe} + \sigma^2_d + \sigma^2_{de} + \sigma^2_p/b + \sigma^2_w/bn \\ &= 2\sigma^2_f + 2\sigma^2_{fe} + \sigma^2_d + \sigma^2_{de} + \sigma^2_p/b + \sigma^2_w/bn. \end{aligned}$$ **Eq. 6.4**

Covariances among the observations: There are three different types of non-zero covariances between observed family means (these are the non-zero off-diagonal elements in **V**) : 1) family means in different tests with two common parents, 2) family means in different tests with one common parent, 3) family means in the same test with one common parent. These are evaluated in Equations 6.5 - 6.7, and in the derivations, fixed effects are ignored as they contribute nothing to the covariances.

For family means in different tests with two common parents,

$$\begin{aligned} Cov(\bar{y}_{i\cdot kl\cdot},\bar{y}_{i'\cdot kl\cdot}) &= Cov(f_k + fe_{ik} + h_l + he_{il} + fh_{kl} + fhe_{ikl} + p_{i\cdot kl}/b + w_{i\cdot kl\cdot}/bn, \\ &\qquad f_k + fe_{i'k} + h_l + he_{i'l} + fh_{kl} + fhe_{i'kl} + p_{i'\cdot kl}/b + w_{i'\cdot kl\cdot}/bn) \\ &= Cov(f_k,f_k) + Cov(h_l,h_l) + Cov(fh_{kl},fh_{kl}) \\ &= 2\sigma^2_f + \sigma^2_d \\ &= \tfrac{1}{2}\sigma^2_A + \tfrac{1}{4}\sigma^2_D. \end{aligned}$$ **Eq. 6.5**

Because both parents are in common, 1/2 the additive and 1/4 of the dominance variance contributes to the covariance and for family means in different tests, plot and tree effects do not contribute to the relationship between family means.

The covariance between family means in different tests with a common female but different male parent is:

$$\begin{aligned}
Cov(\bar{y}_{i\cdot kl\cdot}, \bar{y}_{i'\cdot kl'\cdot}) &= Cov(f_k + fe_{ik} + h_l + he_{il} + fh_{kl} + fhe_{ikl} + p_{i\cdot kl}/b + w_{i\cdot kl\cdot}/bn, \\
&\qquad f_k + fe_{i'k} + h_{l'} + he_{i'l'} + fh_{kl'} + fhe_{i'kl'} + p_{i'\cdot kl'}/b + w_{i'\cdot kl'\cdot}/bn) \\
&= Cov(f_k, f_k) \\
&= \sigma^2_f \\
&= \tfrac{1}{4}\sigma^2_A \qquad \textbf{Eq. 6.6} \\
&= Cov(\bar{y}_{i\cdot kl\cdot}, \bar{y}_{i'\cdot k'l\cdot})
\end{aligned}$$

Note that if the two family means in the different tests have a common male parent rather than female, the derivation proceeds to the same result.

The covariance between full-sib family means in the same test with a common female but different male parent is:

$$\begin{aligned}
Cov(\bar{y}_{i\cdot kl\cdot}, \bar{y}_{i\cdot kl'\cdot}) &= Cov(f_k + fe_{ik} + h_l + he_{il} + fh_{kl} + fhe_{ikl} + p_{i\cdot kl}/b + w_{i\cdot kl\cdot}/bn, \\
&\qquad f_k + fe_{ik} + h_{l'} + he_{il'} + fh_{kl'} + fhe_{ikl'} + p_{i\cdot kl'}/b + w_{i\cdot kl'\cdot}/bn) \\
&= Cov(f_k, f_k) + Cov(fe_{ik}, fe_{ik}) \\
&= \sigma^2_f + \sigma^2_{fe} \\
&= \tfrac{1}{4}\sigma^2_A + \tfrac{1}{4}\sigma^2_{AE} \qquad \textbf{Eq. 6.7} \\
&= Cov(\bar{y}_{i\cdot kl\cdot}, \bar{y}_{i\cdot kl'\cdot})
\end{aligned}$$

Note that if the two family means in the same test have a common male but different female parent the derivation proceeds to the same result. Plot effects do not contribute to the covariance between family means with one common parent in the same test because plot effects are assumed to be random effects associated with a specific plot; the two families are planted in different plots in each and every block and these plot effects are assumed uncorrelated (by specification of the linear model). Intuitively, the plot effect associated with the plot location of family AB in block one should not influence the plot mean of AC in that block.

Covariances between the two families with no parents in common (i.e. kl[th] family mean and k'l'[th] family mean) will be zero whether they are in the same test or different tests, since no effects will be in common,

$$\mathrm{Cov}(\bar{y}_{i \cdot kl \cdot}, \bar{y}_{i \cdot k'l' \cdot}) = \mathrm{Cov}(\bar{y}_{i \cdot kl \cdot}, \bar{y}_{i' \cdot k'l' \cdot}) = 0 \qquad \textbf{Eq. 6.8}$$

Using Eqs. 6.4 - 6.8, and the parameters in Table 6.1, estimates of the second moments necessary to fill in **V** are:

$$\mathrm{Var}(\bar{y}_{i \cdot kl \cdot}) = 8.51,$$
$$\mathrm{Cov}(\bar{y}_{i \cdot kl \cdot}, \bar{y}_{i' \cdot kl \cdot}) = 2\sigma^2_f + \sigma^2_d = 4.50,$$
$$\mathrm{Cov}(\bar{y}_{i \cdot kl \cdot}, \bar{y}_{i' \cdot kl' \cdot}) = \mathrm{Cov}(\bar{y}_{i \cdot kl \cdot}, \bar{y}_{i' \cdot k'l \cdot}) = \sigma^2_f = 1.80,$$
$$\mathrm{Cov}(\bar{y}_{i \cdot kl \cdot}, \bar{y}_{i \cdot kl' \cdot}) = \mathrm{Cov}(\bar{y}_{i \cdot kl \cdot}, \bar{y}_{i . k'l \cdot}) = \sigma^2_f + \sigma^2_{fe} = 2.70.$$

With the data arranged as presented in the **y** vector above, **V** is

$$\mathbf{V} = \begin{bmatrix} 8.51 & 4.50 & 2.70 & 1.80 & 2.70 & 1.80 & 2.70 & 1.80 & 2.70 & 1.80 & 0 & 0 \\ 4.50 & 8.51 & 1.80 & 2.70 & 1.80 & 2.70 & 1.80 & 2.70 & 1.80 & 2.70 & 0 & 0 \\ 2.70 & 1.80 & 8.51 & 4.50 & 2.70 & 1.80 & 2.70 & 1.80 & 0 & 0 & 2.70 & 1.80 \\ 1.80 & 2.70 & 4.50 & 8.51 & 1.80 & 2.70 & 1.80 & 2.70 & 0 & 0 & 1.80 & 2.70 \\ 2.70 & 1.80 & 2.70 & 1.80 & 8.51 & 4.50 & 0 & 0 & 2.70 & 1.80 & 2.70 & 1.80 \\ 1.80 & 2.70 & 1.80 & 2.70 & 4.50 & 8.51 & 0 & 0 & 1.80 & 2.70 & 1.80 & 2.70 \\ 2.70 & 1.80 & 2.70 & 1.80 & 0 & 0 & 8.51 & 4.50 & 2.70 & 1.80 & 2.70 & 1.80 \\ 1.80 & 2.70 & 1.80 & 2.70 & 0 & 0 & 4.50 & 8.51 & 1.80 & 2.70 & 1.80 & 2.70 \\ 2.70 & 1.80 & 0 & 0 & 2.70 & 1.80 & 2.70 & 1.80 & 8.51 & 4.50 & 2.70 & 1.80 \\ 1.80 & 2.70 & 0 & 0 & 1.80 & 2.70 & 1.80 & 2.70 & 4.50 & 8.51 & 1.80 & 2.70 \\ 0 & 0 & 2.70 & 1.80 & 2.70 & 1.80 & 2.70 & 1.80 & 2.70 & 1.80 & 8.51 & 4.50 \\ 0 & 0 & 1.80 & 2.70 & 1.80 & 2.70 & 1.80 & 2.70 & 1.80 & 2.70 & 4.50 & 8.51 \end{bmatrix}$$

Covariances of the observations with breeding values: The final step is to construct the **C** matrix. Assuming that $f_k = \frac{1}{2} g_k$ (i.e. the kth female effect is 1/2 the breeding value of the kth parent), the covariance between an observed full-sib family mean and the breeding value of the parent (g_k) is:

$$
\begin{aligned}
Cov(\bar{y}_{i\cdot kl\cdot}, g_k) &= Cov(f_k + fe_{ik} + h_l + he_{il} + fh_{kl} + fhe_{ikl} + p_{i\cdot kl}/b + w_{i\cdot kl\cdot}/bn,\ g_k) \\
&= Cov(f_k, g_k) \\
&= Cov(f_k, 2f_k) \\
&= 2\sigma^2_f \\
&= \tfrac{1}{2}\sigma^2_A
\end{aligned}
\qquad \textbf{Eq. 6.9}
$$

and, for the parameters in this example, $Cov(\bar{y}_{i\cdot kl\cdot}, g_k) = 3.60$. Since there are a total of 12 observations (six full-sib family means in each of two tests) being used to predict breeding values for four parents, **C** is a 12 x 4 matrix. The first column of **C** contains the covariances between the full-sib means and the breeding value for parent A, the second column of **C** will be covariances of the observations and the breeding value for B, etc. If a given full-sib family mean has a parent in common with the parent whose breeding value is being predicted, then the corresponding element is 3.60. Given the arrangement of the data in **y** above, **C** is

$$
\mathbf{C} = \begin{bmatrix}
3.60 & 3.60 & 0 & 0 \\
3.60 & 3.60 & 0 & 0 \\
3.60 & 0 & 3.60 & 0 \\
3.60 & 0 & 3.60 & 0 \\
3.60 & 0 & 0 & 3.60 \\
3.60 & 0 & 0 & 3.60 \\
0 & 3.60 & 3.60 & 0 \\
0 & 3.60 & 3.60 & 0 \\
0 & 3.60 & 0 & 3.60 \\
0 & 3.60 & 0 & 3.60 \\
0 & 0 & 3.60 & 3.60 \\
0 & 0 & 3.60 & 3.60
\end{bmatrix}.
$$

Constructing G: Variances and Covariances of the Breeding Values

In this example we predict breeding values for height growth at 12 years for four loblolly pine parents. The matrix **G** contains variances and covariances among the breeding values predicted, and for this case **G** is a 4 x 4 matrix. Since $g_k = 2\ f_k$,

$$
\begin{aligned}
Var(g_k) &= Var(2\ f_k) \\
&= 4Var(f_k) \\
&= 4\sigma^2_f \\
&= \sigma^2_A
\end{aligned}
$$

and for this case $Var(g_k) = 7.20$. Since the four parents are assumed unrelated, there are no covariance among the four breeding values. Thus all off-diagonal elements of **G** are zero, and **G** is

$$
\mathbf{G} = \begin{bmatrix} 7.20 & 0 & 0 & 0 \\ 0 & 7.20 & 0 & 0 \\ 0 & 0 & 7.20 & 0 \\ 0 & 0 & 0 & 7.20 \end{bmatrix}
$$

Case Studies with Full-Sib Data

Having specified **y**, $\boldsymbol{\alpha}$, **V**, **C**, and **G**, all formulae described in Chapter 4 can be calculated. To illustrate some principles of how BLP handles full-sib data, consider two cases. In Case 1, the half diallel experiments are assumed balanced (i.e. all 6 full-sib families appear equally at both test sies) and hence the vectors and matrices are as given above. In Case 2, by deleting certain rows and columns from the matrices we examine how missing full-sib families affect the results. For convenience for Case 2, unbalance is only allowed at the family level (as if a cross completely failed so that no trees at all were planted for the deleted full-sib families). The remaining family means are still assumed to have the same variances, etc., as before. Also since the fixed effects estimates are assumed based on many diallels, they are the same for both cases.

For the breeding value predictions, the following equations are utilized from Chapter 4 (note that the parents are assumed to come from the same population so that $\boldsymbol{\tau} = \mathbf{0}$ and is dropped from Eq. 4.2):

a) $\hat{\mathbf{g}} = \mathbf{C'V^{-1}(y - \boldsymbol{\alpha})}$ (Eq. 4.2)

b) $Var(\hat{\mathbf{g}}) = \mathbf{C'V^{-1}C}$ (Eq. 4.4)

c) $Var(\hat{\mathbf{g}} - \mathbf{g}) = \mathbf{G - C'V^{-1}C}$ (Eq. 4.6)

d) $Corr(\hat{g},g) = \{Var(\hat{g})/Var(g)\}^{½}$. (Eq. 4.7)

For comparison with the predicted breeding values, breeding values were also treated as fixed effects and estimated in a separate analysis as twice the general combining ability (GCA's) estimates from the ordinary least squares (OLS) model below. This is another analytical method that requires its own set of assumptions and its own linear model. This assumes that family effects are fixed and for balanced cases, the estimated GCA's are identical to those from Griffing (1956). The model is:

$$\mathbf{y} - \boldsymbol{\alpha} = \mathbf{X}\boldsymbol{\beta} + \mathbf{e}$$

where

- $\mathbf{y}$ is the n x 1 data vector of full-sib family means ($\mathbf{y}$ has dimension 12 x 1 in Case 1),
- $\boldsymbol{\alpha}$ is the n x 1 vector of fixed effects shown above,
- $\boldsymbol{\beta}$ is a 4 x 1 vector of fixed effects to be estimated (i.e. Griffing's GCA for each of the four parents),
- $\mathbf{X}$ is an n x 4 incidence (i.e. design) matrix relating full-sib family means in $\mathbf{y}$ to the fixed effects in $\boldsymbol{\beta}$, and
- $\mathbf{e}$ is an n x 1 vector of errors associated with the full-sib family means ($\mathbf{e}$ has dimension 12 x 1 in Case 1).

The design matrix $\mathbf{X}$ contains only ones and zeros. For example, column 1 applies to parent A and is a 1 in rows 1-6 since these elements in $\mathbf{y}$ have parent A as one of the two parents. The other columns are filled in similarly. The vector of fixed effects, $\boldsymbol{\beta}$, does not contain an overall mean because the data values are expressed as deviations from their fixed effects (i.e. as $\mathbf{y} - \boldsymbol{\alpha}$) and because of the model parameterization. See Chapters 11 and 12 for a more complete discussion of fixed effects, design matrices and model parameterization.

The important point here is that for the balanced case (Case 1), the GCA estimates in $\boldsymbol{\beta}$ are equivalent to Griffing's GCA's calculated using the standard formula presented in Griffing (1956) and Falconer (1981). Equations 6.1 and 6.2 are appropriate for both balanced and unbalanced cases. Because the fixed effects were (hypothetically) estimated from many diallels or from genetic check lots (and not just from these 12 family means), the GCA estimates do not sum to zero.

For the fixed effects model, OLS breeding value estimates are calculated with the ordinary least squares (OLS) formula below (see Graybill 1976); the GCA estimates in $\boldsymbol{\beta}$ are doubled to estimate breeding values.

$$\hat{\boldsymbol{\beta}} = (\mathbf{X'X})^{-1}\mathbf{X'y} \qquad \textbf{Eq. 6.10}$$

Results for case 1: Balanced data--All families in both tests: Using the equations above, the resulting vectors of OLS and BLP breeding values and the matrices of variances are presented in Figure 6.1. Recall that the elements of $\hat{\mathbf{g}}$ are the predicted breeding values for the parents in the units of measurement, therefore parent A would have a predicted breeding value for height growth at 12 years that is 4.96 feet above the mean. For parent A in this example, the OLS estimates of its breeding value is 8.66. The difference between the BLP prediction (4.96) and the OLS estimate (8.66) is another example (similar to those in Chapter 5) of how BLP shrinks the estimated breeding value using an effective family heritability. Heuristically, the BLP breeding value for parent A (4.96 ft) is shrunken from the OLS estimate of 8.66 because the superiority exhibited by the collective offspring of A are due in part to environmental error not expected to be realized in future plantings. Note the BLP breeding values for the other parents are also shrunken back towards zero compared to their OLS estimates.

In predicting a breeding value for a particular parent using full-sib data, BLP predictions incorporate not only information (data) from observed performance of the progeny of that parent, but also from the observed performance of 'connected' progenies. In this example, the predicted breeding value for parent A is affected by the performance of the full-sib families AxB, AxC, and AxD; but in addition, BxC, BxD, and CxD will affect the prediction for A since they are related to some of the progeny of A. In other words, there is a covariance between AxB and BxC, etc., which is reflected in the non-zero off-diagonal elements in **V**. Thus, the predicted breeding values reflect both 1) the family heritability of the trait, and 2) the observations on all related and all connected individuals or families.

Since the data in this example are balanced, i.e. every parent is represented by 3 full-sib families in the each of two tests, all resulting breeding value predictions are equally precise (diagonal elements in Var($\hat{\mathbf{g}}$-$\mathbf{g}$)), have equal spreads (diagonal elements

$$2\hat{\boldsymbol{\beta}} = \begin{bmatrix} 8.66 \\ -3.67 \\ -1.33 \\ 0.66 \end{bmatrix} \quad \hat{\mathbf{g}} = \begin{bmatrix} 4.96 \\ -1.68 \\ -0.57 \\ 0.54 \end{bmatrix} \quad \mathrm{Var}(\hat{\mathbf{g}}) = \begin{bmatrix} 4.24 & 0.26 & 0.26 & 0.26 \\ 0.26 & 4.24 & 0.26 & 0.26 \\ 0.26 & 0.26 & 4.24 & 0.26 \\ 0.26 & 0.26 & 0.26 & 4.24 \end{bmatrix}$$

$$\mathrm{Corr}(\hat{\mathbf{g}},\mathbf{g}) = \begin{bmatrix} 0.77 \\ 0.77 \\ 0.77 \\ 0.77 \end{bmatrix} \quad \mathrm{Var}(\hat{\mathbf{g}}\text{-}\mathbf{g}) = \begin{bmatrix} 2.96 & -0.26 & -0.26 & -0.26 \\ -0.26 & 2.96 & -0.26 & -0.26 \\ -0.26 & -0.26 & 2.96 & -0.26 \\ -0.26 & -0.26 & -0.26 & 2.96 \end{bmatrix}$$

Figure 6.1 -- Comparison of OLS breeding value estimates ($2\hat{\boldsymbol{\beta}}$) and BLP breeding value predictions ($\hat{\mathbf{g}}$) for Case 1. The matrices of variances and covariances of predicted breeding values (Var($\hat{\mathbf{g}}$)), error variance of prediction (Var($\hat{\mathbf{g}}$-$\mathbf{g}$)), and correlation between the predicted and true breeding values (Corr($\hat{\mathbf{g}}$,g)) are also presented.

---

in Var($\hat{\mathbf{g}}$)), and are equally correlated with the true breeding value (elements of Corr($\hat{\mathbf{g}}$,g)). All these values measure efficiency of the predictions in some way, and are interpreted in the same manner as for half-sib progeny test data discussed in Chapter 5.

One significant difference between this case and the examples presented previously is that breeding value predictions for the four <u>unrelated</u> parents are correlated to one another, as can be seen by the non-zero off-diagonal elements of the Var($\hat{\mathbf{g}}$) matrix. The true breeding values for the four unrelated parents are uncorrelated, however, BLP predicted breeding values are not. The element in the first row and second column of Var($\hat{\mathbf{g}}$) equals 0.26, and is the covariance between the predicted breeding values for parent A and parent B. Similarly, there are non-zero off-diagonal elements in Var($\hat{\mathbf{g}}$-$\mathbf{g}$), indicating a covariance between the errors of prediction for two parents. Both of these results occur because, in contrast to the situation with half-sib progeny tests, data on any full-sib family provides information about two parents, and any additional information on one of the parents helps make a better prediction for the other parent.

The positive covariances between BLP breeding values in Var($\hat{\mathbf{g}}$) are because the performance of a given full-sib family (say AxB) affects predicted breeding values of both parents (A and B in this example) in the same direction. If the progeny of family AxB are tall, the predictions for both A and B will tend to be above average, and vice versa. As the number of other parents that A and B are crossed with increases, the covariance (and correlation) of their predicted breeding values will decrease.

The negative covariance between the error of prediction ($\hat{\mathbf{g}}$-g) for A and B indicates that if the predicted breeding value of A is too low, then the predicted breeding value for B will tend to be too high. To further illustrate how predicted breeding values and errors of prediction can be correlated, consider an extreme case: if the full-sib family AxB is the only cross that either parent is involved in, then the predicted breeding values will be perfectly positively correlated. Heuristically, we could not tell if the superiority (or inferiority) of the family AxB was due to parent A or parent B, and therefore will predict equal breeding for the two parents. If the superiority of the family was due primarily to A, and not at all to B, then A's breeding value would be underpredicted (too low), and B's would be overpredicted (too high). Thus, the errors of prediction will be perfectly negatively correlated.

Results for case 2 -- Unbalanced data: Often in tree improvement programs, it is difficult to complete all crosses for a particular mating design because of a lack of pollen or female strobili, incompatibility, weather, or logistical reasons. Case 2 examines how missing crosses affect breeding value predictions in a diallel cross. For Case 2, we arbitrarily assume that cross AxB is not present in either of the two test locations and that cross AxC is present only in test 2. Also assume that the estimates of the fixed effects are based on enough external data that deletion of the data does not affect $\hat{\boldsymbol{\alpha}}$. The new **y** and $\boldsymbol{\alpha}$ are as before but with the first three elements eliminated:

$$\mathbf{y} = \begin{bmatrix} \bar{y}_{2\cdot AC\cdot} \\ \bar{y}_{1\cdot AD\cdot} \\ \bar{y}_{2\cdot AD\cdot} \\ \bar{y}_{1\cdot BC\cdot} \\ \bar{y}_{2\cdot BC\cdot} \\ \bar{y}_{1\cdot BD\cdot} \\ \bar{y}_{2\cdot BD\cdot} \\ \bar{y}_{1\cdot CD\cdot} \\ \bar{y}_{2\cdot CD\cdot} \end{bmatrix} \qquad \boldsymbol{\alpha} = \begin{bmatrix} E(\bar{y}_{2\cdot AC\cdot}) \\ E(\bar{y}_{1\cdot AD\cdot}) \\ E(\bar{y}_{2\cdot AD\cdot}) \\ E(\bar{y}_{1\cdot BC\cdot}) \\ E(\bar{y}_{2\cdot BC\cdot}) \\ E(\bar{y}_{1\cdot BD\cdot}) \\ E(\bar{y}_{2\cdot BD\cdot}) \\ E(\bar{y}_{1\cdot CD\cdot}) \\ E(\bar{y}_{2\cdot CD\cdot}) \end{bmatrix} = \begin{bmatrix} E(\bar{y}_{2\cdots\cdot}) \\ E(\bar{y}_{1\cdots\cdot}) \\ E(\bar{y}_{2\cdots\cdot}) \\ E(\bar{y}_{1\cdots\cdot}) \\ E(\bar{y}_{2\cdots\cdot}) \\ E(\bar{y}_{1\cdots\cdot}) \\ E(\bar{y}_{2\cdots\cdot}) \\ E(\bar{y}_{1\cdots\cdot}) \\ E(\bar{y}_{2\cdots\cdot}) \end{bmatrix} \approx \begin{bmatrix} \bar{y}_{2\cdots\cdot} \\ \bar{y}_{1\cdots\cdot} \\ \bar{y}_{2\cdots\cdot} \\ \bar{y}_{1\cdots\cdot} \\ \bar{y}_{2\cdots\cdot} \\ \bar{y}_{1\cdots\cdot} \\ \bar{y}_{2\cdots\cdot} \\ \bar{y}_{1\cdots\cdot} \\ \bar{y}_{2\cdots\cdot} \end{bmatrix}$$

$$2\hat{\boldsymbol{\beta}} = \begin{bmatrix} 7.36 \\ -6.00 \\ 0.68 \\ 1.23 \end{bmatrix} \qquad \hat{\mathbf{g}} = \begin{bmatrix} 3.67 \\ -3.20 \\ 0.18 \\ 1.06 \end{bmatrix} \qquad \mathrm{Var}(\hat{\mathbf{g}}) = \begin{bmatrix} 3.32 & -0.32 & 0.29 & 0.63 \\ -0.32 & 3.63 & 0.53 & 0.49 \\ 0.29 & 0.53 & 3.99 & 0.24 \\ 0.63 & 0.49 & 0.24 & 4.09 \end{bmatrix}$$

$$\mathrm{Corr}(\hat{g},g) = \begin{bmatrix} 0.68 \\ 0.71 \\ 0.74 \\ 0.75 \end{bmatrix} \qquad \mathrm{Var}(\hat{\mathbf{g}}\text{-}\mathbf{g}) = \begin{bmatrix} 3.88 & 0.32 & -0.29 & -0.63 \\ 0.32 & 3.57 & -0.53 & -0.49 \\ -0.29 & -0.53 & 3.21 & -0.24 \\ -0.63 & -0.49 & -0.24 & 3.11 \end{bmatrix}$$

Figure 6.2 -- Comparison of OLS breeding value estimates ($2\hat{\boldsymbol{\beta}}$) and BLP breeding value predictions ($\hat{\mathbf{g}}$) for Case 2. The matrices of variances and covariances of BLP predicted breeding values (Var($\hat{\mathbf{g}}$)), error variance of prediction (Var($\hat{\mathbf{g}}$-$\mathbf{g}$)), and correlation between the predicted and true breeding values (Corr($\hat{g}$,g)) are also presented.

Thus all matrices above remain unchanged from those in Case 1 except the following rows and columns are eliminated: the first three rows and columns of **V**, and the first three rows of **C**. The matrix **G** remains unchanged as we are still predicting breeding values for four unrelated parents.

Using Equations 6.10, 4.2, 4.4, 4.6, and 4.7, the OLS and BLP breeding value predictions are presented in Figure 6.2. First examine the BLP breeding values and their associated precisions and note that in the **y** vector of observations for Case 2, data from three full-sib crosses are being used to predict the breeding value of parent A, four crosses to predict B, five for C, and six for D. The precision of the resulting predictions depends on the amount of data used to make the prediction; thus Corr($\hat{g}$,g) for A is lowest at 0.68, and Corr($\hat{g}$,g) for D is highest at 0.75. A similar trend is observed for Var($\hat{g}$), where more information leads to a greater spread of predictions, and the opposite trend for Var($\hat{g}$-g), where more information leads to smaller errors of prediction. Note that these trends of increasing precision and larger spread among predictions with increasing data are completely general; however, the rate of increase depends on the second moments specified in **C** and **V**.

In both Cases 1 and 2, six full-sib crosses were used to predict the BLP breeding value for parent D. The predictions, and the associated values for Corr($\hat{g}$,g), Var($\hat{g}$), and Var($\hat{g}$-g), however, are not identical in the two cases. Comparing results from Case 1 to Case 2 for parent D: $\hat{g}$ is 0.54 ft *vs* 1.06 ft, Corr($\hat{g}$,g) is 0.77 *vs* 0.75, Var($\hat{g}$) is 4.24 *vs* 4.09, and Var($\hat{g}$-g) is 2.96 *vs* 3.11. These results again occur because of the connectedness of the data, and the fact that BLP uses all the data to calculate predictions for all the parents. Even though parent D had the same six full-sib families available in both cases, the other parents had less data available. Since these other parents were in the six crosses with parent D, less data on the them decreased the precision of the breeding value prediction for parent D and changed its absolute value.

USING DIFFERENT TYPES OF DATA SIMULTANEOUSLY

In many breeding programs, genetic information may come from a variety of sources. Programs having both half-sib (open-pollinated or polymix) and full-sib (cross-pollinated) tests of the same parents are quite common; further, the cross-pollinated (CP) tests are sometimes established using a variety of mating designs. As programs move into their second and subsequent generations, ancestral data (both half-sib and full-sib) may be available and data on vegetatively propagated relatives may also exist. We illustrate here how to set up **V** and **C** for a simple combination of different data types. Extension to more diverse types of data (e.g. several different mating designs, ancestral data, data from clonal ramets) follow the same general principles:

1) Specify the genetic value being predicted,
2) Specify a linear model for each type of observed data,
3) Estimate the first moments (i.e. fixed effects) for each type of data,
4) Estimate the second moments necessary to estimate **C** and **V**, and
5) Carry out the calculations.

These steps become more complicated as diverse types of data are included in the prediction process. We go through these steps for a very simple example (Case 3) combining half-sib and full-sib data in the same prediction process.

Specifying the Linear Models and Estimating the First Moments

Suppose there are data from a single full-sib progeny test (test 1) whose six full-sib families were created from the four parent loblolly pine half diallel described above for Case 1. Suppose also half-sib data exist from a single open-pollinated test (test 2) on each of the four parents (not the same test 2 as in the previous section). To simplify notation, assume measurements from both tests are for height at 12 years (BLP could also accomodate tests of different ages, see Chapter 7). We use the linear model defined in Eq. 6.1 for the full-sib data, and the model described in Eq. 2.23 for the OP data. The parameter estimates for the CP test are found in Table 6.1. For the OP test, the parameter estimates are assumed the same as for CP tests, except that σ^2_w in the OP test also includes dominance and dominance x environment variance (σ^2_d and σ^2_{de}), thus $\sigma^2_w = 27.0$. Assume that both tests are established in row-plots, with 6 blocks and 6 trees per plot. For purposes of notation, $\bar{y}_{1 \cdot kl \cdot} = \bar{c}_{1 \cdot kl \cdot}$ to indicate that the mean is a full-sib mean from the CP test 1, and $\bar{y}_{2 \cdot k \cdot} = \bar{o}_{2 \cdot k \cdot}$ to indicate that the mean is a half-sib mean from OP test 2. Finally, assume that test means or some other unbiased, precise estimates of fixed effects are available. The **y** and **α** vectors are:

$$\mathbf{y} = \begin{bmatrix} \bar{c}_{1 \cdot AB \cdot} \\ \bar{c}_{1 \cdot AC \cdot} \\ \bar{c}_{1 \cdot AD \cdot} \\ \bar{c}_{1 \cdot BC \cdot} \\ \bar{c}_{1 \cdot BD \cdot} \\ \bar{c}_{1 \cdot CD \cdot} \\ \bar{o}_{2 \cdot A \cdot} \\ \bar{o}_{2 \cdot B \cdot} \\ \bar{o}_{2 \cdot C \cdot} \\ \bar{o}_{2 \cdot D \cdot} \end{bmatrix} \qquad \boldsymbol{\alpha} = \begin{bmatrix} E(\bar{c}_{1 \cdot AB \cdot}) \\ E(\bar{c}_{1 \cdot AC \cdot}) \\ E(\bar{c}_{1 \cdot AD \cdot}) \\ E(\bar{c}_{1 \cdot BC \cdot}) \\ E(\bar{c}_{1 \cdot BD \cdot}) \\ E(\bar{c}_{1 \cdot CD \cdot}) \\ E(\bar{o}_{2 \cdot A \cdot}) \\ E(\bar{o}_{2 \cdot B \cdot}) \\ E(\bar{o}_{2 \cdot C \cdot}) \\ E(\bar{o}_{2 \cdot D \cdot}) \end{bmatrix} = \begin{bmatrix} E(\bar{c}_{1 \cdots \cdot}) \\ E(\bar{c}_{1 \cdots \cdot}) \\ E(\bar{c}_{1 \cdots \cdot}) \\ E(\bar{c}_{1 \cdots \cdot}) \\ E(\bar{c}_{1 \cdots \cdot}) \\ E(\bar{c}_{1 \cdots \cdot}) \\ E(\bar{o}_{2 \cdots}) \\ E(\bar{o}_{2 \cdots}) \\ E(\bar{o}_{2 \cdots}) \\ E(\bar{o}_{2 \cdots}) \end{bmatrix} \approx \begin{bmatrix} \bar{c}_{1 \cdots \cdot} \\ \bar{c}_{1 \cdots \cdot} \\ \bar{c}_{1 \cdots \cdot} \\ \bar{c}_{1 \cdots \cdot} \\ \bar{c}_{1 \cdots \cdot} \\ \bar{c}_{1 \cdots \cdot} \\ \bar{o}_{2 \cdots} \\ \bar{o}_{2 \cdots} \\ \bar{o}_{2 \cdots} \\ \bar{o}_{2 \cdots} \end{bmatrix}$$

Estimating the Second Moments

Variances: There are only two types of observations: full-sib family means from a single CP test and half-sib family means from a single OP test. The variance of a full-sib family mean is calculated according to Eq. 6.4, while the variance of a half-sib family mean in an OP test is calculated according to Eq. 5.7.

Covariances among observations: There are two different types of non-zero covariances among observations for use as off-diagonal elements in **V**: 1) between full-sib family means in the CP test 1 with one common parent calculated according to Eq. 6.7, and 2) between a full-sib family mean in test 1 and a half-sib family mean in OP test 2.

The covariance between a full-sib mean with the female in common with a half-sib mean from a different test is

$$\begin{aligned}\mathrm{Cov}(\bar{c}_{i\cdot kl\cdot},\bar{o}_{i'\cdot k\cdot}) &= \mathrm{Cov}(f_k + fe_{ik} + h_l + he_{il} + fh_{kl} + fhe_{ikl} + p_{i\cdot kl}/b + w_{i\cdot kl\cdot}/bn,\\ &\qquad f_k + fe_{i'}k + p_{i'\cdot k}/b + w_{i'\cdot k\cdot}/bn)\\ &= \mathrm{Cov}(f_k,f_k)\\ &= \sigma^2_f \end{aligned}$$ **Eq. 6.11**

The derivation proceeds to the same result if the male parent of a full-sib family is in common with the female parent of a half-sib family (assuming no reciprocal effects). The only effect causing a covariance is the family effect. All other effects have no covariance because the observed family means are in different tests.

Given the parameters in Table 6.1 (with the modification for σ^2_w from the OP test described above) and Eqs. 5.7, 6.4, 6.7 and 6.11, the four different non-zero elements of **V** are:

$$\begin{aligned}&\mathrm{Var}(\bar{c}_{1\cdot kl\cdot}) = 8.51,\\ &\mathrm{Var}(\bar{o}_{2\cdot k\cdot}) = 4.50,\\ &\mathrm{Cov}(\bar{c}_{1\cdot kl\cdot},\bar{c}_{1\cdot k'l}) = \mathrm{Cov}(\bar{c}_{1\cdot kl\cdot},\bar{c}_{1\cdot kl'}) = 2.70, \text{ and}\\ &\mathrm{Cov}(\bar{c}_{1\cdot kl\cdot},\bar{o}_{2\cdot k\cdot}) = \mathrm{Cov}(\bar{c}_{1\cdot kl\cdot},\bar{o}_{2\cdot l\cdot}) = 1.80.\end{aligned}$$

With the arrangement of data in **y** given above, **V** is:

$$
\mathbf{V} = \begin{bmatrix}
8.51 & 2.70 & 2.70 & 2.70 & 2.70 & 0 & 1.80 & 1.80 & 0 & 0 \\
2.70 & 8.51 & 2.70 & 2.70 & 0 & 2.70 & 1.80 & 0 & 1.80 & 0 \\
2.70 & 2.70 & 8.51 & 0 & 2.70 & 2.70 & 1.80 & 0 & 0 & 1.80 \\
2.70 & 2.70 & 0 & 8.51 & 2.70 & 2.70 & 0 & 1.80 & 1.80 & 0 \\
2.70 & 0 & 2.70 & 2.70 & 8.51 & 2.70 & 0 & 1.80 & 0 & 1.80 \\
0 & 2.70 & 2.70 & 2.70 & 2.70 & 8.51 & 0 & 0 & 1.80 & 1.80 \\
1.80 & 1.80 & 1.80 & 0 & 0 & 0 & 4.50 & 0 & 0 & 0 \\
1.80 & 0 & 0 & 1.80 & 1.80 & 0 & 0 & 4.50 & 0 & 0 \\
0 & 1.80 & 0 & 1.80 & 0 & 1.80 & 0 & 0 & 4.50 & 0 \\
0 & 0 & 1.80 & 0 & 1.80 & 1.80 & 0 & 0 & 0 & 4.50
\end{bmatrix}
$$

Note that the 6 x 6 sub-matrix in the upper left-hand corner of **V** relates to the six observed full-sib family means in test 1 and is identical to the analogous part of the **V** matrix used in Case 1 of this chapter. The lower right-hand 4 x 4 sub-matrix applies solely to the half-sib test data. Finally, the upper right-hand 6 x 4 and lower left-hand 4 x 6 sub-matrices specify the covariances between the full-sib family means observed in test 1 and the half-sib family means observed in test 2.

Covariances of observations with breeding values: To construct **C**, assume that $f_k = ½g_k$, then the covariance between a full-sib family mean, and the breeding value of one of the parents will be $½\sigma^2_A$ (Eq. 6.9). Similarly, the covariance between a half-sib family mean and the breeding value of the parent will be $½\sigma^2_A$ (Eqs. 5.4, 5.6). In this case, all non-zero elements of **C** are 3.60, and **C** is:

$$
\mathbf{C} = \begin{bmatrix}
3.60 & 3.60 & 0 & 0 \\
3.60 & 0 & 3.60 & 0 \\
3.60 & 0 & 0 & 3.60 \\
0 & 3.60 & 3.60 & 0 \\
0 & 3.60 & 0 & 3.60 \\
0 & 0 & 3.60 & 3.60 \\
3.60 & 0 & 0 & 0 \\
0 & 3.60 & 0 & 0 \\
0 & 0 & 3.60 & 0 \\
0 & 0 & 0 & 3.60
\end{bmatrix}
$$

Since we are still predicting breeding values for four unrelated parents, **G** is identical to **G** in Cases 1 and 2 above. While we do not show the results, the calculations proceed as always using the equations from Chapter 4. Interpretation of the results is similar to those previously described in Chapters 5 and 6.

SUMMARY AND CONCLUSIONS

Although Case 3 was an extremely simple example combining different types of data using one full-sib and one half-sib test of the same age, more complex combinations can be incorporated. The use of multiple tests of differing mating designs, ages, precisions and field designs is an extension of what has been presented in this chapter (Cases 1, 2 and 3) and in Chapter 5. Measurements of a single trait, say height, at different ages are handled by treating the different ages as different traits (see Chapters 5 and 7 for more discussion). For example, assume that in Case 3, the full-sib test was 12 years-old, while the half-sib test was 20 years-old, and we wished to predict 20 year-old breeding values. Then the matrices above would change only slightly. The covariance between the full-sib means and the half-sib means ($Cov(\overline{c}_{i\cdot kl\cdot},\overline{o}_{i'\cdot k\cdot})$) in **V** would equal the family covariance for height at ages 12 and 20, not the family variance for height at age 12. Similarly, the covariance between the full-sib family means at age 12 and the breeding value for height at age 20 ($Cov(\overline{c}_{i\cdot kl\cdot},g_k)$ would equal ½ the additive covariance for height at 12 and 20, not ½ the additive variance for height at age 12. Substantially more complex situations can be handled as long as care is taken in specifying linear models and developing estimates of necessary variances and covariances.

One difference between applying BLP to predicting parental breeding values using data just from unrelated, half-sib families (as in Chapter 5) compared to more complicated situations involving either full-sib or multiple types of progeny test data involves computational complexity; that is, inversion of the **V** matrix can become quite complex and time consuming with problems of even moderate size. With half-sib data from unrelated families, data on a particular half-sib family contributes information only about that parent and not any other. As a result, **V** is block diagonal and the prediction of breeding values for traits of a given parent can be calculated separately from those of all other parents (see Chapters 5 and 8). This greatly reduces the size of the matrices because an individual $\mathbf{V}_k$ and $\mathbf{C}_k$ can be set up and used individually for each parent.

With full-sib data and with data from related parents, BLP makes use of all relationships and all "connected" data must be included in a single analysis. This can make **V** large as the number of parents, families and test sites increase. Possible approaches to keeping the computations manageable are 1) find disconnected (or nearly disconnected) sets of data between which the parents and data are unrelated and 2) design tests so that such disconnected sets exist, 3) find more efficient methods of inverting matrices and 4) make simplifying assumptions about **V** (such as is done in dairy science) to allow easier inversion (see Chapter 11). In most present forest tree improvement applications, we believe that approaches 1 and 2 will be sufficient to result in manageable computations.

In many programs, we believe that applications of BLP and BLUP to several, complex types of data can be a very large benefit in terms of combining data from a wide variety of sources (early selection tests, different mating designs, different relatives, different observed traits, different ages, different precisions) into a single predicted breeding value for each trait and genotype. If desired, these can then be aggregated into a single genetic worth for each genotype (using economic weights, see Chapter 7). If the model assumptions hold, then these predicted breeding values have several optimal properties (Chapters 4 and 11).

PROBLEMS

Problem 6.1: You have a four parent diallel as described in Case 1, planted in four 12-year-old progeny tests (instead of only two tests as in Case 1). You wish to predict breeding values (height at 12 years) for the four parents. Assume that precise and accurate fixed effects estimates are available. Given the parameters in Table 6.1,

a) Set up **V**, **C**, **G**.

b) Calculate Var($\hat{\mathbf{g}}$), and Corr($\hat{\mathbf{g}}$,**g**).

c) In Case 1 (described in the text) a four parent diallel was assumed to be planted on two sites. Discuss the results from this problem compared to those from Case 1.

Problem 6.2: Now consider a slightly different mating and testing strategy. Assume a six-parent diallel as diagrammed below:

	A	**B**	**C**	**D**	**E**	**F**
A		x	x	x	x	x
B			x	x	x	x
C				x	x	x
D					x	x
E						x
F						

There are a total of 15 crosses, compared to only 6 when using a four-parent diallel, thus fewer total trees per family can be planted if progeny tests are to require approximately the same amount of land. To accommodate the larger number of crosses, assume that progeny tests will consist of 5 blocks with three trees/family/block (b=5, n=3), and that two tests will be planted (t=2). Also assume that precise and accurate fixed effects estimates are available. You wish to predict breeding values for height at age 12. Given the parameters in Table 6.1,

a) Calculate $\text{Var}(\bar{y}_{i \cdot kl \cdot})$.
b) Set up **V**, **C**, **G**.
c) Calculate $\text{Var}(\hat{\mathbf{g}})$, and $\text{Corr}(\hat{\mathbf{g}},\mathbf{g})$.
d) Compare the results of this problem to those from Case 1 described in the text and to those of Problem 6.1.

Problem 6.3: Consider a factorial mating design involving five parents (A, B, and C are females, D and E are males):

	A	**B**	**C**
D	x	x	x
E	x	x	x

Assume the linear model of Eq. 6.1 and the parameters in Table 6.1, with two tests, 6 blocks, and 6 trees/family/block (t=2, b=6, n=6),

a) Set up **V**, **C**, **G**.
b) Calculate $\text{Var}(\hat{\mathbf{g}})$, and $\text{Corr}(\hat{\mathbf{g}},\mathbf{g})$.
c) Compare the correlation between predicted breeding values of i) two female parents, ii) the two male parents, iii) a male and a female parent.

Problem 6.4: Repeat Problem 6.3 assuming that full-sib family AxD is not present in either of the two tests.

Problem 6.5: You wish to predict the breeding value for trait X for two unrelated parents A and B (g_A and g_B). You have the following measurements available: Trait X on the original ortets (p^x_A and p^x_B), family means for trait X in two OP tests 1 and 2 ($\bar{o}^x_{1\cdot A\cdot}$, $\bar{o}^x_{2\cdot A\cdot}$, $\bar{o}^x_{1\cdot B\cdot}$, $\bar{o}^x_{2\cdot B\cdot}$), and AxB full-sib family means for trait Y in two CP tests 3 and 4 ($\bar{c}^y_{3\cdot AB\cdot}$, $\bar{c}^y_{4\cdot AB\cdot}$). Your breeding program utilized only single pair matings, therefore only these eight observations contribute information about either parent A or B.

a) What are the dimensions of **V**, **C**, and **G**?

b) How many different types of elements (including those equal to zero) will be included in **V**? What are they?

c) How many different types of elements (including those equal to zero will be included in **C**? What are they?

d) How many different types of elements (including those equal to zero will be included in **G**? What are they?

e) What are the relative sizes of 1) $\mathrm{Cov}(\bar{o}^x_{i\cdot k\cdot}, g^x_k)$, 2) $\mathrm{Cov}(\bar{o}^x_{i\cdot k\cdot}, \bar{o}^x_{i'\cdot k\cdot})$, 3) $\mathrm{Cov}(p^x_k, g^x_k)$.

Chapter 7
Best Linear Prediction: Further Topics

INTRODUCTION

Throughout the case studies of BLP in Chapters 5 and 6, a number of simplifying assumptions were made: first moments (E(**g**), E(**y**)) known, second moments (variances and covariances) known, observations in **y** restricted to family means from half- or full-sib progeny tests, all parents unrelated, only one trait measured (although possibly at different ages), and one trait predicted. Also, generally we assumed that the measured traits had homogeneous variance structures. Often in actual applications of BLP in genetic improvement programs, some or all of these assumptions may not be valid. The objective of this chapter is to show how BLP can be applied to broader, more realistic situations.

TOPIC 1: DEFINING THE GENETIC VALUES TO BE PREDICTED

In the general theory of best linear prediction described in Chapter 4, **g** is a q x 1 vector of traits to be predicted. All examples in Chapters 5 and 6 have predicted only a single breeding value for each parent; thus, q = s and **g** has been a s x 1 vector, where s = the number of parents. In theory, however, we could predict a number of breeding values for each parent, or for any combination of parents. For example, if measured data on both juvenile height and diameter in a progeny test were then used to calculate volume from these measured variables, we could predict a breeding value for all three traits (juvenile height, diameter and volume) for each parent, and thus **g** is a 3s x 1 vector (q = 3s). Further we could predict a breeding value for one or more traits not measured (such as mature volume) as long as the new traits were correlated with some of the ones measured. Or, the trait predicted could be a single, aggregate genetic value (genetic worth) that was a linear combination of breeding values and economic weights for a number of traits (see Topic 10). Or, there could be multiple traits predicted that are the same trait, say volume growth, expressed in a number of different target planting environments (see Topic 8).

It is also possible to predict types of genetic values other than breeding values. An example might be total genetic values of clones based on data measured in clonal tests. Finally, it is possible to predict genetic or breeding values of individual offspring in progeny tests as opposed to the parental genetic values predicted in examples so far. This is the subject of Chapters 9 and 10.

The specification of which traits are in **g** affects both the dimensions and the type of second moment estimates needed to construct **C**. The theory presented in

Chapter 4 requires only that the measured data in **y** be correlated with the traits in **g** and that the covariances between the elements of **y** and the elements of **g** (i.e. the elements of **C**) be known. These covariances can be estimated in a variety of ways (see Topics 5, 7, 8 and 9 in this chapter). Once the elements of **C** and **V** are estimated, BLP calculations proceed as described in Chapters 4-6.

TOPIC 2: FIRST MOMENTS OF THE GENETIC VALUES

In most examples in this book, the assumption is made that all genetic values come from a single population. Therefore, the expected values of every genotype is the same and can arbitrarily be taken as 0, i.e. $E(\mathbf{g}) = \mathbf{0}$. This has the effect of centering the predictions around 0. There may be "nuisance" fixed effects associated with the data such as block effects, location effects and treatment effects, but there are no fixed genetic effects.

While this assumption may apply in many forestry applications and is useful for illustrating concepts and applications of BLP, there are instances when the genotypes in **g** come from different genetic populations. In this case, there are fixed genetic effects associated with the elements in **g** that are not always the same and cannot be assumed zero. In this case, BLP assumes that $E(\mathbf{g}) = \boldsymbol{\tau}$ is known. This vector contains the expected value of every genetic value in **g** that is being predicted. Since these are fixed effects, they also appear in $\boldsymbol{\alpha}$ which is the fixed effects for the data (see discussion of $\boldsymbol{\alpha}$ and **g** in Chapter 4). The general procedure is to write a linear model for each data element in **y**. If this includes fixed genetic effects, then these are part of the expected values in $\boldsymbol{\alpha}$ along with the "nuisance" fixed effects. The fixed genetic effects alone are in $\boldsymbol{\tau}$. Other discussions of this subject are in the Topics sections of Chapters 4 and 11.

TOPIC 3: DEFINING THE DATA VECTOR

The first step in defining the data vector **y** is the specification of a linear model (which includes determining if effects are fixed or random and the second moments associated with random effects). Once the model for each individual observation has been determined, the decision can be made whether those individual observations can be consolidated into some kind of mean (such as plot means, family means in a single test, or family means across multiple tests) for use as the elements in the **y** vector.

This depends on the degree of unbalance, if any, in the data, the assumptions made about homogeneity of second moments (see Topic 7), and the genetic values being predicted.

In all examples up to this point, data were assumed balanced with homogeneous second moments within a test. Further, breeding values were predicted for parents, not for individuals in the progeny tests. Thus, family means from either half- or full-sib progeny tests were used in **y** as the units of observation. Given this situation, use of any unit of observation less than a family mean from a single test (such as plot means or individual measurements) would not be of any benefit. For example, if single tree observations are used, the dimensions of **y**, **V**, and **C** (and therefore computational complexity) increases, but the precision with which any parental breeding value is predicted is not affected. The assumptions of balanced data and homogeneous moments within a single test, i.e. moments do not vary from block to block within a test, allows the consolidation of individual observations into family means without loss of information. In some examples, certain types of heterogeneous second moments were assumed across different tests. For example, in Case 3 in Chapter 5, tests of different age had different second moments; and in Case 4, observations from different tests had different levels of precision or variances. In these cases it was not appropriate to use family means across tests in **y**; however, family means from a single test were still appropriate.

The use of family means from a given test instead of individual tree measurements can greatly simplify BLP computations. In cases of nearly balanced data within a test, this will have little effect on final predictions. However, when predicting breeding values for an individual offspring from known parentage in a progeny test as in selection index (Chapters 9 and 10), or when predicting a genetic value in order to select trees for some type of vegetative production, it is necessary to use individual tree observations and plot means in **y**. In many other cases it will be worthwhile to use some sort of mean as the unit of observation. This will depend on the situation.

In some cases, family means from a given test location could still be used as the unit of observation in **y** even when data are unbalanced. Consider a progeny test where there are different numbers of trees per block for different families and thus the precision of family means from the same test differs; family means based on more data have smaller variance. If family means are used as observations, differences in variance are accounted for in **V** by entering the appropriate variances in the diagonal of **V**. Note that family means should be calculated as mean of the plot means or estimated by least squares means (Chapter 3) in order for differential block effects not to affect family means.

There may be situations where the data from a single test are severely unbalanced, say with some families missing in certain blocks, and plot means varying widely in their precision due to either differences in block uniformity or number of trees per plot. In these types of cases it may be advisable to use plot means as the unit of observation. If plot means are recognized as having different precisions, BLP puts more weight on plot means coming from blocks which are more uniform. Families in more blocks, or in the more precise blocks, end up with more precise breeding value predictions.

As a final example, consider the case of progeny tests that have different treatments (such as fertilizer rates) cross classified with the family effects. If the data (such as family means) from different treatments have different variances or are differently correlated with the trait being predicted, it may be advisable to use as elements in **y**, the family means calculated by averaging across data from a single treatment at a single test location. For example, if there are two treatments in each test and family means from the fertilized treatment have different variance and/or are more correlated with the breeding values being predicted, then family means for the fertilized and unfertilized treatments should be estimated separately in each test and used as separate units of observation in **y**. Family means from the different treatments will have different expected values and different second moments in **C** and **V**.

TOPIC 4: ESTIMATING FIRST MOMENTS OF THE DATA

After the data vector **y** has been specified, the next step is to determine $\boldsymbol{\alpha}$, the expected value of **y**. The $\boldsymbol{\alpha}$ vector is used to express the observations as deviations from their expected values (see Eq. 4.2). In BLP, $\boldsymbol{\alpha}$ is assumed to be known, while in reality it must be estimated. The general approach is to 1) express the observations in terms of the linear model, 2) find $E(\mathbf{y}) = \boldsymbol{\alpha}$, and 3) find functions of the data in **y** (or other data, such as checklots) which yield precise and unbiased estimates of the elements in $\boldsymbol{\alpha}$.

If breeding values, or other genetic values, are being predicted for genetic entries (such as parents) which are all from the same population (i.e. τ is the same for all genotypes), then the expected value of the observed data in **y** will be some function of the general mean μ and the fixed 'nuisance' effects like environments, blocks, etc.

If the genetic entries are not from the same population, then there are different expected values of the genetic values (different τ). Then, the expected value of **y** will be a function of μ, the 'nuisance' effects, and the different fixed genetic effects τ.

Estimating $\boldsymbol{\alpha}$ With Functions of the Observed Data

As an example of $\hat{\boldsymbol{\alpha}}$ that contains precise and unbiased estimates of the fixed effects, consider the grand mean in a half-sib progeny test when that grand mean is based on a large number of randomly selected parents from the same population of parents in **g**. An example would be a test in which half-sib families from say 30 parents are included and these parents are a random sample from the population of say 2,000 parents being tested in all tests (and for which breeding values are being predicted). In Chapter 2, a half-sib family mean from a single test and the grand mean across all half-sib families in the test were shown to have the same expected value,

$$E(\bar{y}_{i\cdot k\cdot}) = E(\bar{y}_{i\cdots}) = \mu + E_i \qquad \textbf{Eq. 7.1}$$

and thus the grand mean is an unbiased estimate of the expected value of each family mean when the observations are means from a single test. To examine the precision of this estimate of the fixed effects, consider the variance of the grand mean:

$$\mathrm{Var}(\bar{y}_{i\cdots}) = \sigma^2_f/s + \sigma^2_{fe}/s + \sigma^2_p/sb + \sigma^2_w/sbn, \qquad \textbf{Eq. 7.2}$$

where s is the number of parents. If a half-sib progeny test includes s = 30 families, b = 6 blocks, and n = 6 trees per plot and the hypothetical (but biologically reasonable) parameters for 10-year-old loblolly pine height presented in Table 5.1 are appropriate, then the variance of the grand mean is 0.0028 m^2, and the standard error is 0.0529 m. With a mean height of 12 meters, this yields a coefficient of variation of 0.0529/12 = 0.0044. In many cases the grand mean in a half-sib progeny test will be sufficiently precise to make reasonable the assumption that $E(\bar{y}_{i\cdot k\cdot})$ is known.

For full-sib tests, the grand mean (i.e. $\bar{y}_{i\cdots\cdot}$) is still an unbiased estimator of every family mean (i.e. $\bar{y}_{i\cdot kl\cdot}$) (see Eq. 6.3), although the precision will probably not be as high as in half-sib tests. Using the linear model described in Eq. 6.1, the variance of the grand mean from full-sib tests is:

$$\begin{aligned}\mathrm{Var}(\bar{y}_{i....}) &= \mathrm{Var}(\mu + E_i + f./s + fe_i./s + h./r + he_i./r + fh../sr + fhe_i../sr \\ &\quad + p_i.../srb + w_i..../srbn) \\ &= \sigma^2_f/s + \sigma^2_{fe}/s + \sigma^2_f/r + \sigma^2_{fe}/r + \sigma^2_d/sr + \sigma^2_{de}/sr \\ &\quad + \sigma^2_p/srb + \sigma^2_w/srbn\end{aligned} \qquad \textbf{Eq. 7.3}$$

and thus the size of this variance depends greatly on the number of females (s) and the number of males (r). A full-sib test might include 30 families, but this could arise from a factorial mating design of 5 males and 6 females. Thus, the breeding values of only 11 parents are represented in this test, compared to the 30 parents in the half-sib example. Use of this approach to estimate fixed effects in full-sib tests should be undertaken with care.

If several sets of full-sib families are randomized in the same blocks of a test (such as might happen if progeny from multiple disconnected diallels or factorials are tested together), then all full-sib family means in a given test might be averaged together to estimate the fixed effects. The objective is to obtain an unbiased and precise estimate of the fixed environmental (site location) effect for each family mean. For this to be nearly free of genetic sampling effects, more males and more females should be included. If family means from one small diallel or factorial set (containing only a few parents) are averaged to estimate the fixed site effect (μ + E_i), this decreases the precision of the estimated site effect (low numbers of males and females in Eq. 7.2). If one small diallel or factorial set happens to include only very good parents, the estimated site effect is too large because of positive genetic sampling effects (good parents). Use of family means from other diallels or factorials in the same test reduces the effects of genetic sampling and hence increases the precision of the estimated site effect. This assumes that all parents are from the same single population i.e. all have the same fixed genetic effects.

Estimating α With Genetic Standards

A second approach to estimating the fixed location (also site, environmental, and treatment) effects is through the use of genetic checks (also called checklots or genetic standards) which are planted in every progeny test. Checks should be genetically broad-based to be stable performers across all environments, and should represent the average of the population of genotypes. The average is an unbiased estimate of the site mean. If checklots do not represent the average of the population, the difference between checks and the average (which can be thought of as τ) must be known, or precisely estimated with other data. This difference cannot be estimated by the observed difference between the checks and select population on a single site because this is effectively the same as estimating the site mean (μ + E_i) as discussed above. This difference between the checklots and the population average could be estimated through a comparison between the two on all progeny tests in the breeding program, possibly leading to the development of some function of a check mean such that the expected value of the function equals the population average.

Even if a broad based check is planted in every test, there will still be a variance associated with the estimate of the check performance in a single test. In other words, there may be a rather broad confidence interval associated with an estimate of the check mean (or function of the check mean) in a particular progeny test. If genetic checks are to be relied upon as estimators of the fixed effects, they should be replicated in a test more often than progeny families, e.g. several plots per block for checks vs. 1 plot per block for progenies. This increases precision of the estimates of the fixed location effects.

How Precise is Precise Enough?

Estimates of $\boldsymbol{\alpha}$ derived using either approach (i.e. grand mean or genetic checks) will have a variance associated with them, i.e. they will not be known exactly. If the estimates are very precise, then the subsequent predictions of breeding or genetic values will very nearly be 'best linear predictions'. The variances of either type of estimator can be calculated, and standard errors compared to the mean to determine if making the assumption that fixed effects are known is reasonable. If neither genetic checks or the overall progeny mean can be used to provide precise estimates of the site effects, then best linear unbiased prediction (BLUP) should be used to develop generalized least squares estimates of the fixed effects (see Chapters 11 and 12). However, since BLUP is computationally more complex, this may require making other assumptions to simplify calculations, and these assumptions might be worse than assuming that fixed effects are known. Although extensive simulation studies would be necessary to demonstrate this, we believe that in many forestry applications fixed effects can be estimated with sufficient precision to allow the assumption that they are known, and therefore utilize BLP.

TOPIC 5: ESTIMATING THE SECOND MOMENTS

The derivation of BLP Equation 4.2 assumed that the second moments, i.e. the variances and covariances, were known; however, these parameters are never known exactly but only estimated. While an extensive discussion of estimating variance and covariance components is beyond the scope of this book (see Hallauer and Miranda 1981; Henderson 1984; Milliken and Johnson 1984; Baker 1986), some brief comments follow. There are two approaches to estimating the appropriate second moments: 1) calculate them directly for the specific set of data which will be used, and 2) use a subset of the data or other external data to estimate the second moments and then apply those estimates in **C** and **V**. Generally, the direct calculation approach will be adequate if the observed data represents a large, random sample of the reference population.

Direct Calculation of the Second Moments

Half-sib tests: The most obvious approach to estimating the second moments is to calculate them directly from the observed data. For example, suppose that you have two progeny tests (i and i') with family means calculated for the same 30 half-sib families. The variance of family means ($\text{Var}(\bar{y}_{i\cdot k\cdot})$) are the diagonal elements of **V** and can be directly estimated for any one progeny test in one of two ways:

1) <u>ANOVA approach</u>
   a) Use analysis of variance (ANOVA) or some other approach (e.g. restricted maximum likelihood) to estimate variance components,
   b) express $\bar{y}_{i\cdot k\cdot}$ in terms of the linear model (e.g. Eq. 2.25),
   c) express $\text{Var}(\bar{y}_{i\cdot k\cdot})$ in model effects (e.g. Eq. 2.26), and
   d) solve for it using variance component estimates from step 1a.
2) <u>Arithmetic approach</u>
   Take the 30 half-sib family means and calculate their variance.

These approaches will yield identical values for $\text{Var}(\bar{y}_{i\cdot k\cdot})$ if the parents are unrelated and data are balanced, but the latter is substantially easier. If the data are unbalanced, there is a different variance associated with family means based on different numbers of observations. That is, the number of trees and/or blocks (i.e. n and/or b in the denominator of Eq. 2.26) can be different for every family and then every family mean can have a different variance. The variance of family means for a given family can be thought of as that variance that would be obtained (upon repeated sampling) if all families were represented by its specific number of blocks and trees. The arithmetic approach can not account for this difference in precision because it results in a single estimate for the variance of family means for each test.

The covariance of family means in test i and test i' (i.e. $Cov(\bar{y}_{i \cdot k \cdot}, \bar{y}_{i' \cdot k \cdot})$) is an element on the off-diagonal of **V**. As for estimating variances above, covariances for any pair of progeny tests can be directly estimated either by 1) analysis of covariance (ANCOVA), or 2) by pairing the observed means in test i and test i' for each of the 30 families, and determining the covariance between those 30 pairs of family means.

Full-sib tests: As with half-sib progeny tests, directly calculating the second moments of full-sib tests could be either done with the ANOVA - ANCOVA approach, or with the arithmetic approach of simply taking the variance (or covariance) of the observed full-sib means (or pairs of means). Depending on the composition of a full-sib test (or tests), estimates of the second moments calculated directly using either approach may be quite imprecise, and for the arithmetic approach, could be substantially biased. For example, consider two replicated progeny tests both containing two half diallels with six different parents in each diallel. Each diallel contains 15 full-sib families, for a total of 30 full-sib families represented on each site. While a random sample of 30 full-sib families might be sufficient to estimate variances and covariances on the two sites, these 30 full-sib families are related (only 12 parents are represented on each site). Using the ANOVA - ANCOVA approach, second moment estimates are unbiased, but may be imprecise due to the small number of parents represented. Using the arithmetic approach, the estimates may also be imprecise, and are biased since the 30 full-sib families are related and not a random sample of all possible full-sib families. Heuristically, there is a covariance among these full-sib families, i.e. they are not as variable as they should be if the parents of each different full-sib family came at random from the population. Simply calculating the variance of the 30 observed full-sib family means will under-estimate the true variance of full-sib family means. Generally speaking, as the number of parents that are represented in the full-sib crosses increase, the degree of under-estimate or bias, decreases.

Precise estimates of second moments developed from many large full-sib tests which are then applied to other tests, may be better than estimates directly calculated from the smaller data sets, although simulation work in this area is needed.

Estimating Second Moments From External Data

The need to estimate second moments from other data that are external to the given test or tests arises when the actual data being used in **y** are too sparse, or too unbalanced, for good estimates to be calculated directly. As an example, to calculate $Cov(\bar{y}_{i\cdot k\cdot},\bar{y}_{i'\cdot k\cdot})$ directly as discussed for half-sib data above, we assumed that the same 30 half-sib families were present in both test i and test i'. Suppose, however, that there are only three common families in the two progeny tests. The covariance between observed family means in those two tests is needed in **V**, but an estimate of $Cov(\bar{y}_{i\cdot k\cdot},\bar{y}_{i'\cdot k\cdot})$ calculated directly using only three observations will be quite imprecise. A better approach in this case would be to estimate $Cov(\bar{y}_{i\cdot k\cdot},\bar{y}_{i'\cdot k\cdot})$ from some other pairs (or groups) of tests which have large numbers of common families, and then assume that those estimates are appropriate for tests where direct calculation of second moments is not satisfactory.

By the same token, when few families are present in a single progeny test, estimates of $Var(\bar{y}_{i\cdot k\cdot})$ calculated directly from the data may be rather imprecise. Using other data to estimate $Var(\bar{y}_{i\cdot k\cdot})$ may be useful in these types of situations. One approach is to develop variance component estimates from external data (with large numbers of families, etc.). Estimates could be developed that apply to certain types (e.g. environments, ages) of tests, and then these estimates could be applied to other tests by calculating the variance of family means from Eq. 2.26. This approach assumes that the variance component estimates from external data are appropriate, but incorporates the design structure (number of blocks and trees) from the actual data being used to make predictions.

If it is necessary to estimate second moments from some progeny tests to apply to other progeny tests, it is important to ensure that the tests used for estimation and the tests to which the estimates will be used are 'alike'. For example, we may be able to calculate the covariance between family means observed for a number of pairs of progeny tests on wet sites. While it may be appropriate to apply the average estimated covariances to other pairs of progeny tests on wet sites, it may not be appropriate to use these estimates for pairs of tests on dry sites, or for two tests when one site is wet and the other is dry. It may be necessary to develop different classes of second moment estimators, or use some kind of modeling approach (see Chapter 8), to ensure that estimates of second moments are reasonable for a particular set of progeny tests.

Type B covariances and correlations: If it is not possible to directly calculate the necessary covariances to enter into **V** and **C**, family mean correlations can often be quite useful (assuming that family means are the units of observation used in **y**). The general approach is to estimate appropriate correlations and variances, and then calculate covariances from these estimates (described in detail below). Since correlations are independent of the scale of measurement, they may be more stable across a wider variety of site types, treatments, etc. To calculate covariances for **C** and off-diagonal elements of **V** once correlations have been estimated (say from external data) variance estimates can be obtained either from external data or from the specific tests measured.

For simplicity, discussion of family mean covariances and correlations is confined to balanced half-sib progeny tests; however, the principles are general and would apply equally well to full-sib situations. For the half-sib case in which the same say 30 families are planted in two different tests and trait y (e.g. height) is measured in test i while trait y* (e.g. diameter) is measured in test i', then the covariance of family means from the two progeny tests can be estimated by first calculating the 30 family means for each trait and then calculating the covariance between the 30 pairs of family means from Eq. 2.14). Assuming the linear model for family means in Eq. 2.25, the expected value of the family mean covariance is (type B covariance, Burdon 1977):

$$
\begin{aligned}
Cov(\bar{y}_{i\cdot k\cdot},\bar{y}^{*}_{i'\cdot k\cdot}) &= Cov(f_k + fe_{ik} + p_{i\cdot k}/b + w_{i\cdot k\cdot}/bn, \\
&\qquad f^{*}_{k} + fe^{*}_{i'k} + p^{*}_{i'\cdot k}/b + w^{*}_{i'\cdot k\cdot}/bn) \\
&= Cov(f_k + fe_{ik}, f^{*}_{k} + fe^{*}_{i'k}) \\
&= Cov(f_k,f^{*}_{k}) \\
&= Cov(\tfrac{1}{2}g_k,\tfrac{1}{2}g^{*}_{k}) \qquad \text{(from Eq. 5.9)} \\
&= \tfrac{1}{4}Cov(g,g^{*}).
\end{aligned}
\qquad \textbf{Eq. 7.4}
$$

This means that for two traits measured in different tests, the expected family mean covariance contains only genetic effects. Note further that the same trait (say height) expressed in two different tests can be treated as two different traits. Thus, this expression (Eq. 7.4) can equally well be used for the same or different traits as long as they are measured in different tests (or treatments, environments, etc.). When the expression is used for the same trait expressed in two different tests, then Eq. 7.4 is a measure of genotype x environment interaction. If there is no g x e, then $\frac{1}{4}Cov(g,g^*) = \frac{1}{4}Cov(g,g) = \frac{1}{4}\sigma^2_A$ for the trait.

The corresponding type B family mean correlation between two tests is:

$$r_{Bf} = Cov(\bar{y}_{i\cdot k\cdot}, \bar{y}^*_{i'\cdot k\cdot}) \;/\; [\; Var(\bar{y}_{i\cdot k\cdot})\; Var(\bar{y}^*_{i'\cdot k\cdot})\;]^{\frac{1}{2}} \qquad \textbf{Eq. 7.5}$$

and could be estimated by calculating the correlation (instead of the covariance) for say 30 pairs of half-sib family means using Eq. 2.15. In an analogous manner, the type B family mean correlation can be calculated for two different traits expressed in two different environments or for the same trait expressed in two different environments. In the latter case, a larger correlation indicates less g x e between the two environments. In either case, the type B family mean correlation is less in absolute value (i.e. closer to zero) than the corresponding genetic correlation (which has one quarter of the corresponding additive genetic variances in the denominator rather than family mean variances). Because family mean variances contain sources of variation due to experimental and environmental factors, in addition to genetic variance, the family mean correlation is always lower than the genetic correlation.

Off-diagonal elements of **V** that are type B covariances can be calculated with estimates of the type B correlation and variances of family means as:

$$Cov(\bar{y}_{i\cdot k\cdot}, \bar{y}^*_{i'\cdot k\cdot}) = r_{Bf}\; [\; Var(\bar{y}_{i\cdot k\cdot})\; Var(\bar{y}^*_{i'\cdot k\cdot})\;]^{\frac{1}{2}}. \qquad \textbf{Eq. 7.6}$$

Note here that the type B family mean correlation might be applied to a large number of pairs of tests, while the respective variances of family means might be estimated individually for each test in each pair.

The elements of **C** are genetic covariances between the observed half-sib family means and the breeding value being predicted and therefore (for the general case where the traits are different) are $\frac{1}{2}Cov(g,g^*)$ (from Eq. 5.5). These can calculated using type B half-sib family mean correlations as:

$$\begin{aligned} 2\, r_{Bf}\; [\; Var(\bar{y}_{i\cdot k\cdot})\; Var(\bar{y}^*_{i'\cdot k\cdot})\;]^{\frac{1}{2}} &= 2\; Cov(\bar{y}_{i\cdot k\cdot}, \bar{y}^*_{i'\cdot k\cdot}) \\ &= 2\; Cov(f_k, f^*_k) \\ &= 2\; \tfrac{1}{4}Cov(g_k, g^*_k) \\ &= \tfrac{1}{2}Cov(g_k, g^*_k) \qquad \textbf{Eq. 7.7} \end{aligned}$$

Thus, twice the type B family mean covariance (from Eq. 7.4) can be used to estimate elements of the **C** matrix.

Type A covariances and correlations: Often two or more traits (such as height and diameter) are measured on the same individuals in the same progeny test. The covariances and correlations calculated between these kinds of measurements are designated type A covariances and correlations (Burdon 1977). When two or more traits are measured in the same progeny tests, type A family mean covariances will be needed on off-diagonals in **V**, and for the balanced, half-sib case (assuming linear model Eq. 2.25) these are:

$$\begin{aligned} Cov(\bar{y}_{i \cdot k \cdot}, \bar{y}^*_{i \cdot k \cdot}) &= Cov(f_k + fe_{ik} + p_{i \cdot k}/b + w_{i \cdot k \cdot}/bn, \\ &\qquad f^*_k + fe^*_{ik} + p^*_{i \cdot k}/b + w^*_{i \cdot k \cdot}/bn) \qquad \textbf{Eq. 7.8} \\ &= Cov(f_k, f^*_k) + Cov(fe_{ik}, fe^*_{ik}) + Cov(p_{i \cdot k}, p^*_{i \cdot k})/b^2 \\ &\qquad + Cov(w_{i \cdot k \cdot}, w^*_{i \cdot k \cdot})/b^2n^2 \\ &= Cov(f,f^*) + Cov(fe,fe^*) + Cov(p,p^*)/b + Cov(w,w^*)/bn \\ &= \tfrac{1}{4}Cov(g,g^*) + \tfrac{1}{4}Cov(ge,ge^*) + Cov(p,p^*)/b + Cov(w,w^*)/bn \end{aligned}$$

Note that a type A family mean covariance has plot and error covariance components, and not just genetic components as do type B covariances. A type A half-sib family mean correlation is defined as:

$$r_{Af} = Cov(\bar{y}_{i \cdot k \cdot}, \bar{y}^*_{i \cdot k \cdot}) \ / \ [\ Var(\bar{y}_{i \cdot k \cdot})\ Var(\bar{y}^*_{i \cdot k \cdot})\ ]^{\frac{1}{2}} \qquad \textbf{Eq. 7.9}$$

and therefore the necessary covariances can be calculated as:

$$Cov(\bar{y}_{i \cdot k \cdot}, \bar{y}^*_{i \cdot k \cdot}) = r_{Af}\ [\ Var(\bar{y}_{i \cdot k \cdot})\ Var(\bar{y}^*_{i \cdot k \cdot})\ ]^{\frac{1}{2}}. \qquad \textbf{Eq. 7.10}$$

So, off-diagonal elements of the **V** matrix that are covariances between different traits measured in the same test on the same trees can be estimated 1) directly by calculating the covariance between the appropriate pairs of family means from the test via Eq. 2.14 or 2) indirectly by obtaining a general type A family mean correlation for the two traits, estimating family mean variances for each trait and applying Eq. 7.12.

Inspection of the Second Moment Estimates

In a sense, there are two issues involved with estimation of second moments. First, precise estimates of each element in **V** and **C** are needed. Secondly, because many variances and covariances are estimated simultaneously, the errors of estimation can "accumulate". The latter can mean that while the individual estimates of the elements of say **V** are of moderate precision, **V** as a whole may be poorly estimated. If **V** as a whole is poorly estimated, unrealistic predictions of genetic values can result. This is somewhat analogous to the comparison of error rates of individual comparisons and experiment-wide error rates in fixed effects comparisons. As for all variance-covariance matrices, **V** must be positive-definite and hence have all positive eigenvalues. After estimation of the elements of **V**, it is useful to calculate and inspect the eigenvalues (Hayes and Hill 1980; Baker 1986).

Use of Correlations and Standardized Data

To help ensure that the overall estimates of **C** and **V** are "good", it is important that estimates of individual elements are consistent among one another. For example, an estimated covariance of an off-diagonal element can not be larger than the square root of the product of the two corresponding estimated variances on the diagonal; this would mean a correlation greater than one between the corresponding two elements of **y**. One way to ensure that this does not happen is to estimate the covariances indirectly as the product of estimated correlations and the corresponding standard deviations (Eqs. 7.6, 7.7 and 7.10). That is, rather than estimating type A and type B covariances directly, estimate type A and type B correlations and then estimate the covariances from Equations 7.6, 7.7 and 7.10 using the estimated variances from the two corresponding diagonals. This approach can be used for the elements of the **C** matrix as well.

Indirect prediction of covariances via correlations and variances does not necessarily ensure consistency among groups of off-diagonal elements. For example, consider three elements of **y**: y_1, y_2 and y_3. If the estimated correlation between y_1 and y_2 is 0.9 and that between y_1 and y_3 is 0.8, then an estimated correlation between y_2 and y_3 of -0.5 is inconsistent.

Another approach which can help facilitate inspection of **V** and ensure consistency is to standardize the data in **y** so that all elements of **y** have a variance of 1 (Hill 1984). Then **V** is a correlation matrix (1's on the diagonals and correlations on off-diagonals) which is easier to inspect visually.

TOPIC 6: CONSEQUENCES OF ERROR IN ESTIMATES OF SECOND MOMENTS

All theory of BLP assumes that the second moments, **C** and **V**, are known when to apply BLP in practice they must be estimated. Since the elements in these matrices are second order statistics (involving quadratic functions of the data), they are subject to large sampling errors and, in general, large samples are needed for precise estimates (Baker 1986). Most of the work (Williams 1962a,b; Harris 1963, 1964; Sales and Hill 1976; Bulmer 1985; Baker 1986) investigating the sensitivity of predictions to errors in estimating **C** and **V** has been for selection indices involving a single index across multiple traits (as discussed in Chapters 9 and 10). Often, simple models assuming homogeneous, balanced data in one way layouts have been employed. Thus, this is an area that needs more work before conclusions can be drawn about the types of applications discussed in previous chapters (e.g. combining data for either multiple or single traits across tests of different ages, designs, precisions, etc). However, some inferences are possible.

In general, rankings of parents based on breeding value predictions for a single trait are relatively less sensitive to errors in estimates of the elements in **V** and **C** than are the predictions themselves (Henderson 1978). In other words, the absolute value of the breeding value predictions are more affected by errors in estimates, than are relative values for any two genotypes. For the single trait case, if selection is done on the basis on rankings calculated with estimates instead of known values, there is less disparity in realized gain than in predicted gain (Harris 1963, 1964; Sales and Hill 1976, Bulmer 1985). That is, predictions of genetic gain based on predicted genetic values are likely to be more affected since the predicted genetic values are affected. However, since the relative rankings of the genotypes are less affected, nearly the same genotypes will actually be selected and thus realized gains will be similar to those achieved using known parameters in **C** and **V**.

Similar conclusions can be drawn for predictions of multiple traits if the traits are positively correlated: rankings of the genotypes are less sensitive to errors than are the actual predictions. If the traits are negatively correlated, rankings will be much more sensitive to errors of estimation.

TOPIC 7: HETEROGENEOUS SECOND MOMENTS

An assumption often made by animal breeders in applications of best linear unbiased prediction (BLUP) is that second moments are homogeneous across environmental groups, which correspond in the forestry case to progeny tests. Second moments will often be heterogeneous in forestry progeny tests. In Case Study 4 in Chapter 5, we examined the effect of different levels of precision among progeny tests. This is one type of heterogenous second moment, i.e. some observations (family means in Case 4) had different variances caused by different relative amounts of plot and error variance in the variance of family means. Another cause for heterogeneous variances could be simply a scale effect, i.e. for many biological traits the variance is proportional to the mean; so, covariances from test locations with larger means (e.g. older or faster growing) may well be larger.

Covariances may also be heterogeneous across progeny tests if the genetic correlation between observations in different environments depends on site factors such as soil, elevation (discussed further in Topic 8). For example, data from fertilized tests may have a different correlation with the genetic value being predicted than data from unfertilized tests. If the covariance structure is known, then BLP will automatically account for any type of heterogeneous second moments (Gianola 1986), regardless of how that heterogeneity arises.

The principles discussed in this section, and for many of the remaining topics, are completely general to any application of BLP using any type of data. To illustrate these principles, however, we present a number of examples assuming that the units of observation (the elements of **y**) are half-sib family means from a single progeny test. We begin with an illustration of the topic of heterogeneous variances, and also demonstrate the use of type B family correlations to estimate covariance parameters.

To simplify notation throughout this section, we write $\bar{y}_{i \cdot k \cdot}$ as y_{ik}. That is, the family mean for the kth family from the ith test is written y_{ik}. Given half-sib progeny tests and the linear model assumed in Eq. 2.23, family means are expressed as deviations from the overall and environmental means as:

$$\bar{y}_{i \cdot k \cdot} = y_{ik} = f_k + fe_{ik} + (p_{i \cdot k}/b) + (w_{i \cdot k \cdot}/bn) \qquad \textbf{Eq. 7.11}$$

Since the family means are from a single test, we define a family effect, F_{ik}, that includes both the additive family effect and family x environment interaction, such that:

$F_{ik} = f_k + fe_{ik},$

$$Var(F_{ik}) = \sigma^2_F = \sigma^2_f + \sigma^2_{fe}. \qquad \textbf{Eq. 7.12}$$

Assume that measurements of height are made in three 12-year-old half-sib progeny tests of loblolly pine and that all families are present in all three tests. The three tests are assumed to have the parameters in Table 7.1. These parameters are hypothetical and have been developed to illustrate how different variance structures can influence best linear predictions. Note in particular that observations from the three progeny tests are equally precise in the sense that the tests have equal heritabilities. Also, since the family mean correlation is constant for all combinations of progeny tests, observations from the three tests will be equally correlated with true breeding values for height in some future environment. The observed family means from the three tests, however, have markedly different variances due to scale effects. The variance of family means, $Var(y_{ik})$, varies with the mean of the test, as do all other variances. Thus, a deviation of a family mean from the grand test mean of 2 feet is less impressive in test 3 (which has a grand mean of 40 feet and a variance of family means of 8.00 ft^2) than in test 1 (which is much slower growing and exhibits less variability among family means).

Because this problem assumes half-sib data from unrelated parents, breeding values can be predicted separately for each parent (see Chapter 8). To predict a breeding value for height for the kth parent using data from the three tests the $\mathbf{V}_k$ matrix is a 3 x 3 matrix of the form:

$$V_k = \begin{bmatrix} Var(y_{1k}) & Cov(y_{1k},y_{2k}) & Cov(y_{1k},y_{3k}) \\ Cov(y_{1k},y_{2k}) & Var(y_{2k}) & Cov(y_{2k},y_{3k}) \\ Cov(y_{1k},y_{3k}) & Cov(y_{2k},y_{3k}) & Var(y_{3k}) \end{bmatrix}$$

Table 7.1 - Hypothetical values for three progeny tests of loblolly pine for Topic 7. The trait is height measured in feet at age 12.[a, b]

Test	$\bar{y}_{i\cdots}$	σ^2_T	σ^2_F	σ^2_p	σ^2_w	$Var(y_{ik})$	h^2_i	h^2_{iF}
1	30	36	2.70	6.30	27.0	4.50	0.3	0.6
2	25	25	1.88	4.39	18.8	3.13	0.3	0.6
3	40	64	4.80	11.2	48.0	8.00	0.3	0.6

[a]$\bar{y}_{i\cdots}$ is the grand test mean measured in feet,
σ^2_T = total phenotypic variance ($\sigma^2_T = \sigma^2_F + \sigma^2_p + \sigma^2_w$),
the calculation of $Var(y_{ik})$ from Eq. 2.26 assumes b=6 and n=6,
h^2_i = individual tree heritability in i[th] test ($h^2_i = 4\sigma^2_F/\sigma^2_T$),
h^2_{iF} = family heritability in i[th] test ($h^2_{iF} = \sigma^2_F/Var(y_{ik})$), and
all other variables defined in text.

[b]The type B family mean correlation (from Eq. 7.5) between height measured in any two environments is r_{Bf} = 0.400. This value applies for any combination of two progeny tests.

The diagonal elements of $\mathbf{V}_k$ are variances of family means from Table 7.1. The element in the first row, second column can be calculated using family mean correlations, and the appropriate variances of family means as in Eq. 7.5 (remember, a family mean ($\bar{y}_{i\cdot k\cdot}$) is being written as y_{ik} to simplify notation):

$$\begin{aligned} Cov(y_{1k},y_{2k}) &= r_{Bf}\,[\,Var(y_{1k})\,Var(y_{2k})\,]^{½} \\ &= 0.400\,[\,(4.50)\,(3.13)\,]^{½} \\ &= 1.49. \\ &= ¼Cov(g^*,g) \end{aligned}$$

where height in test 1 and height in test 2 are considered different traits. All other covariances in $\mathbf{V}_k$ are similarly calculated, and the final $\mathbf{V}_k$ matrix is:

$$\mathbf{V}_k = \begin{bmatrix} 4.50 & 1.49 & 2.40 \\ 1.49 & 3.13 & 2.00 \\ 2.40 & 2.00 & 8.00 \end{bmatrix}$$

To predict breeding value for height growth at age 12, elements of the $\mathbf{C}_k$ matrix are covariances between the observed family means and the true breeding value of the parent (g_k). $\mathbf{C}_k$ is a 3 x 1 vector of the form:

$$\mathbf{C}_k = \begin{bmatrix} Cov(y_{1k},g_k) \\ Cov(y_{2k},g_k) \\ Cov(y_{3k},g_k) \end{bmatrix}$$

$$\mathbf{C}_k = \begin{bmatrix} \frac{1}{2}Cov(g_{1k},g_k) \\ \frac{1}{2}Cov(g_{2k},g_k) \\ \frac{1}{2}Cov(g_{3k},g_k) \end{bmatrix}$$

where g_{ik} is the breeding value of the kth parent as expressed in the ith test. To correctly specify $Cov(y_{ik}, g_k)$ we must specify whether the breeding value g_k is going to be observed in an environment like that of test 1, test 2, or test 3. This is the concept of target environment discussed in detail in the next section. In this case, specifying a target environment amounts to specifying a genetic variance or variance of family means, as we are assuming the type B correlation between measurements of 12 year height in any two environments is constant. For this example, assume we wish to predict a breeding value for the kind of environment found in test 1. Then we define the breeding value g_{Tk} in a target environment which has (from Table 7.1) $\bar{y}_{T\cdots} = 30$ ft, $Var(y_{Tk}) = 4.50$, and $\sigma^2_F = 2.70$. The second element in $\mathbf{C}_k$ is calculated using Eq. 7.9:

$$\begin{aligned} 2\,Cov(y_{2k},g_k) &= 2\, r_{Bf}\,[\,Var(y_{2k})\,Var(y_{Tk})\,]^{\frac{1}{2}} \\ &= 2\,(0.400)\,[\,(3.13)\,(4.50)\,]^{\frac{1}{2}} \\ &= 3.00 \\ &= \tfrac{1}{2}Cov(g^{*}_{k},g_k) \end{aligned}$$

and $\frac{1}{2}Cov(g^{*},g)$ is the familiar element to enter into $\mathbf{C}_k$. All other elements in $\mathbf{C}_k$ are calculated similarly. The final $\mathbf{C}_k$ matrix is:

$$C_k = \begin{bmatrix} 3.60 \\ 3.00 \\ 4.80 \end{bmatrix}$$

Note that while the type B family mean correlation is constant among all pairs of the three tests (i.e. 0.4), the covariances between the observed family means and the breeding values being predicted depend on the variance of family means both in the observed and in the target environment.

Given appropriate variances and covariances in **V** and **C**, BLP adjusts for different variances due to scale effects. In the above example, the predicted breeding values have a variability appropriate to expression in an environment like that of test 1. Had the target planting environment been chosen to be like that of test 3, the variability among predicted breeding values would have been larger to reflect the fact that when all trees have grown more, there is more variability and genetically larger trees are further above the mean.

We do not have to have data and actually make predictions to examine the effects of these different variance structures on best linear predictions. The effect of different variances is seen by examining $\mathbf{b}_k$, the vector of weights which would be applied to $\mathbf{y}_k - \hat{\boldsymbol{\alpha}}_k$. That is, if we had real data, each parent would have three family means (one from each test) and thus $\mathbf{y}_k - \hat{\boldsymbol{\alpha}}_k$ would be a 3 x 1 vector of family means expressed as deviations from the respective test means. The predicted breeding value for parent k is then calculated as $\hat{g}_k = \mathbf{b}_k(\mathbf{y} - \hat{\boldsymbol{\alpha}})$ where $\mathbf{b}_k = \mathbf{C'}_k\mathbf{V}_k^{-1}$ (from Eq. 4.2). For the above example,

$$\mathbf{b}_k = [\ b_1 \quad b_2 \quad b_3\ ]$$
$$\mathbf{b}_k = [\ 0.446 \quad 0.534 \quad 0.333\ ]$$

and therefore the predicted breeding value $\hat{g}_k$ is calculated as

$$\begin{aligned} \hat{g}_k &= b_1\,(y_{1k}) + b_2\,(y_{2k}) + b_3\,(y_{3k}) \\ &= 0.446\,(y_{1k}) + 0.534\,(y_{2k}) + 0.333\,(y_{3k}). \end{aligned}$$

Note that observed family means from test 2 are multiplied by the largest weight, b_2. This is not because observations from test 2 are more precise or more correlated with the true breeding value (all three tests are equal in these respects), but simply because variances are smallest in these tests, and therefore observed family means must be 'scaled up' to receive a weight equal to the other tests in the calculation of the final breeding value.

Thus BLP adjusts for heterogeneous variance structures. To illustrate the influence of scale effects, we assumed for the above example that family means had equal precision and correlation with breeding values being predicted. However, this does not have to be the case. BLP adjusts for all types of heterogeneous second moments simultaneously if **C** and **V** are properly specified.

TOPIC 8: TARGET ENVIRONMENTS

BLP predicts breeding values for a particular trait in the units of measurement and any predicted breeding value applies specifically to a particular target environment, (or class of environments that have the same characteristics). The set of characteristics defining a target environment could be broadly defined (e.g age 10 loblolly pine on Piedmont sites), or quite specific (age 25 Douglas-fir at elevations 1000-1500 feet on moist soils in areas of low disease incidence). In the example in Topic 7 above, the characteristics defining the target environment were an assumed future set of first and second moments, i.e. an environment which has $\bar{y}_{T\cdots} = 30$ ft at 12 years of age, $Var(y_{Tk}) = 4.50$, and $\sigma^2_F = 2.70$ (Throughout this section, $\bar{y}_{i\cdot k\cdot}$ is again written y_{ik} so y_{Tk} represents a family mean for the kth family observed in the target type of test environment).. That target environment may have been chosen because it has the characteristics of an average planting site.

In an example in Chapter 5, breeding values were predicted for loblolly pine for the trait of height (in meters) at age 20, and also at age 5. Homogenous second moments were assumed within an age class (i.e. all tests of a particular age have equal genetic and environmental variances, and equal genetic correlations with one another) and with the true breeding value. The correlation between family means and the true breeding value is needed to determine the covariance elements in **C**. The assumption of equal correlation between family means observed in any test of a given age and the true breeding value implicitly defines the target environment as any progeny test at age 20 in one case, and age 5 in another case.

All progeny tests within an age is a rather broad definition of a target environment and thus the concept of target environments may seem trivial. Suppose, however, that the correlation (or covariance) between any two progeny tests of a given age depends on how alike the two tests are for such things as site quality, soil types, elevation, latitude, geographic region, climatic factors, disease incidence, silvicultural treatments, etc. To predict breeding values, we must first specify one or more planting environments in which those breeding values will be expressed; this is where planted offspring in the future will be observed, i.e. the environment where offspring of selected parents are planted in future commercial plantations. If three different target environments are specified, then it is necessary to predict three breeding values for each parent.

This affects both the dimension of **C** and the elements of **C**. For example, if three breeding values (one for each of three different target environments) are predicted for each of s parents, q = 3s and **g** is a 3s x 1 vector of breeding values to be predicted. Then **C** is n x 3s where n is the total number of observations in **y**. Once the target environments are specified, the covariances between observed data (i.e. **y**) and true breeding values expressed in those target environments can be determined and entered into **C**.

Predictions of the breeding values for the same trait expressed in different target environments will not necessarily rank the s parents in the same order. That is, parents predicted to be best in one target environment are not necessarily those predicted to be best in another target environment. This is because data from different tests receive different weights in the calculation of breeding values for different target environments. This is discussed further in the example below.

As an example, suppose height is measured in four half-sib progeny tests of Douglas-fir at age 20. Tests 1 and 2 are at elevation 1000 ft (1000 ft is designated site type A), and tests 3 and 4 are at elevation 3000 ft (site type C). Good estimates of the second moments are available (Table 7.2) and each family is present in all four tests.

If the four family means in $\mathbf{y}_k$ for the kth parent are ordered such that

$$\mathbf{y}_k' = [\ y_{1k}\ \ y_{2k}\ \ y_{3k}\ \ y_{4k}\ ]$$

then $\mathbf{V}_k$ will be of the form:

$$\mathbf{V}_k = \begin{bmatrix} Var(y_{1k}) & Cov(y_{1k},y_{2k}) & Cov(y_{1k},y_{3k}) & Cov(y_{1k},y_{4k}) \\ Cov(y_{1k},y_{2k}) & Var(y_{2k}) & Cov(y_{2k},y_{3k}) & Cov(y_{2k},y_{4k}) \\ Cov(y_{1k},y_{3k}) & Cov(y_{2k},y_{3k}) & Var(y_{3k}) & Cov(y_{3k},y_{4k}) \\ Cov(y_{1k},y_{4k}) & Cov(y_{2k},y_{4k}) & Cov(y_{3k},y_{4k}) & Var(y_{4k}) \end{bmatrix}$$

$$\mathbf{V}_k = \begin{bmatrix} Var(y_{1k}) & \frac{1}{4}Cov(g_A,g_{A'}) & \frac{1}{4}Cov(g_A,g_C) & \frac{1}{4}Cov(g_A,g_C) \\ \frac{1}{4}Cov(g_A,g_{A'}) & Var(y_{2k}) & \frac{1}{4}Cov(g_A,g_C) & \frac{1}{4}Cov(g_A,g_C) \\ \frac{1}{4}Cov(g_A,g_C) & \frac{1}{4}Cov(g_A,g_C) & Var(y_{3k}) & \frac{1}{4}Cov(g_C,g_{C'}) \\ \frac{1}{4}Cov(g_A,g_C) & \frac{1}{4}Cov(g_A,g_C) & \frac{1}{4}Cov(g_C,g_{C'}) & Var(y_{4k}) \end{bmatrix}$$

$$\mathbf{V}_k = \begin{bmatrix} 18.01 & 7.56 & 2.16 & 2.16 \\ 7.56 & 18.01 & 2.16 & 2.16 \\ 2.16 & 2.16 & 8.00 & 3.36 \\ 2.16 & 2.16 & 3.36 & 8.00 \end{bmatrix}.$$

Table 7.2 - Hypothetical values for Douglas-fir half-sib progeny tests at three different elevations. Trait is height in feet at age 20.[a]

Site Type	$\bar{y}_{i\cdot k\cdot}$	σ^2_T	σ^2_F	σ^2_f	σ^2_{fe}	σ^2_p	σ^2_w	$Var(y_{ik})$
A: Elev 1000 ft	60	144	10.8	7.56	3.24	25.3	107.9	18.01
B: Elev 2000 ft	50	100	7.5	5.25	2.25	17.6	74.9	12.51
C: Elev 3000 ft	40	64	4.8	3.36	1.44	11.2	48.0	8.00

[a]See Table 7.1 for description of all symbols.

The genetic covariance ($Cov(g_i,g_{i'})$) between the observations in two different tests of various types (e.g. calculated as type B family mean covariances from Eq. 7.4):

$Cov(g_A,g_{A'})$ = 30.24 (between two different tests at 1000 ft),
$Cov(g_B,g_{B'})$ = 21.00 (between two different tests at 2000 ft),
$Cov(g_C,g_{C'})$ = 13.44 (between two different tests at 3000 ft),
$Cov(g_A,g_B)$ = 18.00 (between one test at 1000 ft and one at 2000 ft),
$Cov(g_A,g_C)$ = 8.64 (between one test at 1000 ft and one at 3000 ft) and
$Cov(g_B,g_C)$ = 9.60 (between one test at 2000 ft and one at 3000 ft).

To fill in $\mathbf{C}_k$, we need to know the covariance between the observation and the true breeding value, and thus need to specify the site type (elevation in this case), or target environment, in which offspring from selected parents will be planted. To predict breeding values for two target environments, both 1000 ft (g_A) and 3000 ft elevations (g_C), $\mathbf{C}_k$ is of the form:

$$\begin{array}{cc} & \text{BV @ 1000 ft} \quad \text{BV @ 3000 ft} \\ \mathbf{C}_k = & \begin{bmatrix} \text{Cov}(y_{1k},g_A) & \text{Cov}(y_{1k},g_C) \\ \text{Cov}(y_{2k},g_A) & \text{Cov}(y_{2k},g_C) \\ \text{Cov}(y_{3k},g_A) & \text{Cov}(y_{3k},g_C) \\ \text{Cov}(y_{4k},g_A) & \text{Cov}(y_{4k},g_C) \end{bmatrix} \end{array}$$

$$\mathbf{C}_k = \begin{bmatrix} \tfrac{1}{2}\text{Cov}(g_A,g_{A'}) & \tfrac{1}{2}\text{Cov}(g_A,g_C) \\ \tfrac{1}{2}\text{Cov}(g_A,g_{A'}) & \tfrac{1}{2}\text{Cov}(g_A,g_C) \\ \tfrac{1}{2}\text{Cov}(g_C,g_A) & \tfrac{1}{2}\text{Cov}(g_C,g_{C'}) \\ \tfrac{1}{2}\text{Cov}(g_C,g_A) & \tfrac{1}{2}\text{Cov}(g_C,g_{C'}) \end{bmatrix}$$

$$\mathbf{C}_k = \begin{bmatrix} 15.12 & 4.32 \\ 15.12 & 4.32 \\ 4.32 & 6.72 \\ 4.32 & 6.72 \end{bmatrix}$$

With these $\mathbf{V}_k$ and $\mathbf{C}_k$ matrices, predicted breeding values for the kth parent are calculated as $\hat{\mathbf{g}}_k = \mathbf{C}_k'\mathbf{V}_k^{-1}(\mathbf{y}_k - \boldsymbol{\alpha}_k)$ and the resulting vector $\hat{\mathbf{g}}_k$ is 2 x 1 containing 2 predicted breeding values: the first for planting environments at 1000 ft elevation and the other specific for 3000 ft elevation. Note that a target environment of 2000 feet elevation (site type B) could also have been specified and a breeding value predicted for the kth parent at 2000 ft, despite the fact that there are no progeny tests or observations on parent k on those type sites. This is indirect selection, and is possible because the covariances between the actual observations and the true breeding value at 2000 ft are estimated (see Table 7.2).

Breeding value predictions for the two target environments (1000 ft and 3000 ft) will not necessarily rank parents in the same order because test data from 1000 ft will be given more weight in predicting the breeding values for the 1000 ft target environment. Conversely, test data from progeny tests will be given more weight in the prediction of breeding values for the 3000 ft target environment. Parents that perform relatively better at 1000 ft compared to 3000 ft will tend to be more highly ranked for the 1000 ft target environment.

TOPIC 9: PREDICTION OF MULTIPLE TRAITS

In the previous topic, height at 1000 ft was considered a different trait than height at 3000 ft and breeding values were predicted for both traits. Often tree breeders are interested in a number of different traits and consequently wish to predict a number of breeding values for a single parent. To predict not only a breeding value for height at age 12, but for diameter at breast height as well, consider the following example. Again in this section to simplify notation a family mean for the kth family in the ith test (i.e. $\bar{y}_{i \cdot k \cdot}$) will be written y_{ik}. Further, because there are two traits, h_{ik} is used when the family means are for height and d_{ik} is used when the family means are for diameter. Assume that the parameters in Table 7.3 apply to all 12-year-old progeny tests (i.e. homogeneous second moments), and that the kth family is included in two such tests.

Then $\mathbf{y}_k$ is a 4 x 1 vector containing family means for both height and diameter from each of the two tests containing offspring from parent k. If $\mathbf{y}_k$ is ordered so that the family means for height and diameter from test 1 precede those of test 2, the $\mathbf{V}_k$ matrix for the kth parent will be 4 x 4 and of the form:

$$\mathbf{V}_k = \begin{bmatrix} \mathrm{Var}(h_{1k}) & \mathrm{Cov}(h_{1k},d_{1k}) & \mathrm{Cov}(h_{1k},h_{2k}) & \mathrm{Cov}(h_{1k},d_{2k}) \\ \mathrm{Cov}(h_{1k},d_{1k}) & \mathrm{Var}(d_{1k}) & \mathrm{Cov}(d_{1k},h_{2k}) & \mathrm{Cov}(d_{1k},d_{2k}) \\ \mathrm{Cov}(h_{1k},h_{2k}) & \mathrm{Cov}(d_{1k},h_{2k}) & \mathrm{Var}(h_{2k}) & \mathrm{Cov}(h_{2k},d_{2k}) \\ \mathrm{Cov}(h_{1k},d_{2k}) & \mathrm{Cov}(d_{1k},d_{2k}) & \mathrm{Cov}(h_{2k},d_{2k}) & \mathrm{Var}(d_{2k}) \end{bmatrix}$$

$$\mathbf{V}_k = \begin{bmatrix} \mathrm{Var}(h_{ik}) & \mathrm{Cov}(h_{ik},d_{ik}) & \tfrac{1}{4}\mathrm{Cov}(g_{h1},g_{h2}) & \tfrac{1}{4}\mathrm{Cov}(g_{h1},g_{d2}) \\ \mathrm{Cov}(h_{ik},d_{ik}) & \mathrm{Var}(d_{ik}) & \tfrac{1}{4}\mathrm{Cov}(g_{d1},g_{h2}) & \tfrac{1}{4}\mathrm{Cov}(g_{d1},g_{d2}) \\ \tfrac{1}{4}\mathrm{Cov}(g_{hi},g_{h2}) & \tfrac{1}{4}\mathrm{Cov}(g_{d1},g_{h2}) & \mathrm{Var}(h_{ik}) & \mathrm{Cov}(h_{ik},d_{ik}) \\ \tfrac{1}{4}\mathrm{Cov}(g_{hi},g_{d2}) & \tfrac{1}{4}\mathrm{Cov}(g_{d1},g_{d2}) & \mathrm{Cov}(h_{ik},d_{ik}) & \mathrm{Var}(d_{ik}) \end{bmatrix}$$

The only conceptual difference between this $\mathbf{V}_k$ and previous ones is that the elements of the form $\mathrm{Cov}(h_{ik},d_{ik})$ are type A family mean covariances. These are covariances between two traits observed on the same individuals (e.g. height and diameter in the same test). In addition to genetic sources of covariance, the type A

Table 7.3 - Hypothetical values for 12-year-old half-sib progeny tests of loblolly pine for Topic 9.[a]

Trait	$\bar{y}_{i\cdots}$	σ^2_T	σ^2_F	σ^2_f	σ^2_{fe}	σ^2_p	σ^2_w	$Var(y_{ik})$
height (ft)	30	43	2.70	1.80	0.90	6.00	34.3	5.63
DBH (inches)	8	12	1.12	0.56	0.56	1.63	9.25	1.91

[a]$Var(y_{ik})$ assumes b=4 and n=6. See Table 7.1 for description of all symbols.

Assume that the following genetic covariances are also known:

$Cov(g_{hi},g_{hi'})$ = 7.20 (Type B genetic covariance, height in two tests),

$Cov(g_{di},g_{di'})$ = 2.24 (Type B genetic covariance, DBH in two tests),

$Cov(g_{hi},g_{di'})$ = 3.76 (Type B genetic covariance, height and DBH in two different tests), and

$Cov(h_{ik},d_{ik})$ = 1.50 (Type A covariance of family means, height and DBH in one test).

---

includes many environmental and design sources of covariance (and thus are larger than type B family mean covariances for the same two traits). Topic 5 contains a more detailed discussion of type A and B family mean covariances, but note here 1) the type A family mean covariance is used for off-diagonal elements corresponding to family means measured on two different traits in the same test and 2) the type B family mean covariance (which is 1/4 of the genetic covariance as shown in Eq. 7.4) is used for off-diagonal elements corresponding to the same or different traits measured in different tests. Note that the type A family mean covariance for height and diameter (i.e. 1.5) is larger than the corresponding type B family mean covariance (i.e. 0.94).

The final $\mathbf{V}_k$ matrix is:

$$\mathbf{V_k} = \begin{bmatrix} 5.63 & 1.50 & 1.80 & 0.94 \\ 1.50 & 1.91 & 0.94 & 0.56 \\ 1.80 & 0.94 & 5.63 & 1.50 \\ 0.94 & 0.56 & 1.50 & 1.91 \end{bmatrix}$$

To predict two breeding values (height and DBH) in some target environment T for the kth parent from four progeny test observations (two traits measured in each of two tests), $\mathbf{C_k}$ is 4 x 2 and of the general form:

$$\mathbf{C_k} = \begin{bmatrix} Cov(h_{1k}, g_{hTk}) & Cov(h_{1k}, g_{dTk}) \\ Cov(d_{1k}, g_{hTk}) & Cov(d_{1k}, g_{dTk}) \\ Cov(h_{2k}, g_{hTk}) & Cov(h_{2k}, g_{dTk}) \\ Cov(d_{2k}, g_{hTk}) & Cov(d_{2k}, g_{dTk}) \end{bmatrix}$$

where g_{hTk} = the true breeding value for height for the kth parent in target environment T, and

g_{dTk} = the true breeding value for DBH for the kth parent in target environment T.

Note that since we have again assumed a homogeneous variance structure and are predicting breeding values at the same age as the progeny test measurements, the target environment is implicitly defined. Thus the covariance between the observed family means and the breeding value is ½ the genetic covariance.

$$\mathbf{C_k} = \begin{bmatrix} \tfrac{1}{2}Cov(g_{hi}, g_{hi'}) & \tfrac{1}{2}Cov(g_{hi}, g_{di'}) \\ \tfrac{1}{2}Cov(g_{di}, g_{hi'}) & \tfrac{1}{2}Cov(g_{di}, g_{di'}) \\ \tfrac{1}{2}Cov(g_{hi}, g_{hi'}) & \tfrac{1}{2}Cov(g_{hi}, g_{di'}) \\ \tfrac{1}{2}Cov(g_{di}, g_{hi'}) & \tfrac{1}{2}Cov(g_{di}, g_{di'}) \end{bmatrix}$$

$$\mathbf{C_k} = \begin{bmatrix} 3.60 & 1.88 \\ 1.88 & 1.12 \\ 3.60 & 1.88 \\ 1.88 & 1.12 \end{bmatrix}$$

Once the second moments are defined, the predictions are made using Equation 4.2 and the resultant 2 x 1 vector contains predicted breeding values for both height and diameter. In this simple example, several assumptions were made to make the size of the problem small enough for illustration. However, the principles are completely general. For any type of mating design or prediction problem, measurements on multiple traits (or a single trait) can be used to predict multiple traits (or a single trait). To do so, 1) specify the exact genetic values, **g**, to be predicted (e.g. breeding values for height and diameter at 12-years-old and volume at 25 years for 1000 parents represented in both half-sib and full-sib progeny tests)), 2) specify the observations in **y** (e.g. family means for height and diameter at 12 years for every family-test combination), 3) estimate the fixed effects, 4) estimate the elements for **C** and **V** and 5) use Eq. 4.2 to calculate predicted breeding values.

Effect of Multiple Trait Measurements on Precision of Prediction

In the above example, measurements on both height and DBH were used to predict two breeding values for each parent. The $\mathbf{V}_k$ and $\mathbf{C}_k$ matrices are set up to predict the two breeding values simultaneously. Suppose that the product objective is veneer logs and therefore primary interest is in DBH breeding value. The matrices could have been set up to predict a DBH breeding value using only DBH measurements, and the height measurements omitted. This would, however, decrease the precision of prediction on the breeding value for DBH. So, if data are used on multiple traits that are correlated with the trait being predicted, the error variances of predicted breeding values are lower than if only observations on a single trait are used.

To calculate the error variance of prediction (EVP) for the kth parent, we must specify $\mathbf{G}_k$, the matrix of genetic variances and covariances of the breeding values for the kth parent. When predicting breeding values for height and DBH for each parent, $\mathbf{G}_k$ is a 2 x 2 matrix of the general form:

$$\mathbf{G}_k = \begin{bmatrix} \sigma^2_{A,h} & \sigma_{A,hd} \\ \sigma_{A,hd} & \sigma^2_{A,d} \end{bmatrix}$$

where $\sigma^2_{A,h}$ is the additive genetic variance for height,

$\sigma^2_{A,d}$ is the additive genetic variance for DBH,

$\sigma_{A,hd}$ is the additive genetic covariance for height and DBH.

The additive genetic variance for height (Table 7.3) is $\sigma^2_{A,h} = 4\ \sigma^2_{f,h} = 4\ (1.80) = 7.20$. The additive genetic variance for DBH (Table 7.3) is $\sigma^2_{A,d} = 4\ \sigma^2_{f,d} = 4\ (0.56) = 2.24$. The additive genetic covariance for height and DBH ($\sigma_{A,hd}$) is 3.76 (from Table 7.3), and the final $\mathbf{G}_k$ matrix is:

$$\mathbf{G}_k = \begin{bmatrix} 7.20 & 3.76 \\ 3.76 & 2.24 \end{bmatrix}$$

The variance of the predictions (Var($\hat{g}_k$)) is calculated as $\mathbf{C}_k'\mathbf{V}_k^{-1}\mathbf{C}_k$ (from Eq. 4.5), and the error variance of predictions (EVP) is $\mathbf{G}_k$ - Var($\hat{g}_k$) (from Eq. 4.6). The correlation of the predicted ($\hat{g}$) and true breeding value (g) can be calculated as $\{\mathrm{Var}(\hat{g})/\mathrm{Var}(g)\}^{½}$ (from Eq. 4.7). If all data are used (both height and DBH data from two tests as in the previous example), the EVP for the predicted DBH breeding value is 0.99 in^2 and the correlation of g and $\hat{g}$ is 0.75. If only DBH observations (from two tests) are used, the **y** vector reduces from a 4 x 1 to a 2 x 1 for the kth parent, $\mathbf{V}_k$ reduces from a 4 x 4 to a 2 x 2 matrix, $\mathbf{C}_k$ reduces from a 4 x 2 to a 2 x 1, and $\mathbf{G}_k$ reduces to a 1 x 1 (delete all columns and rows pertaining to height observations from all vectors and matrices). Then the EVP for the predicted DBH breeding value is 1.22 in^2, and the correlation of g and $\hat{g}$ is 0.67. Thus, the use of measurements on both height and DBH results in more precise breeding value prediction for DBH than do measurements on DBH alone.

TOPIC 10: USE OF ECONOMIC WEIGHTS TO PREDICT AGGREGATE GENOTYPIC VALUES

Breeders are often interested in predicting an aggregate genotype where a number of different traits are weighted by their economic importance, and then summed to yield a single value. For example, for a product objective of pulpwood, the traits of volume growth and specific gravity of the wood could be combined into one breeding value representing pulp production. Suppose that three traits are important and that the aggregate genotypic value for the kth genotype, h_k is defined as:

$$h_k = m_1\ g_{1k} + m_2\ g_{2k} + m_3\ g_{3k}$$

where m_i = economic weight applied to the ith trait (i = 1, 2, 3 in this example) and
g_{ik} = true breeding value for the ith trait for genotype k.

Now for s genotypes, we wish to predict **h**, an s x 1 vector of aggregate genotypic values (one for each of the k genotypes). The best linear prediction of **h** is $\hat{\mathbf{h}} = \mathbf{C'V^{-1}}(\mathbf{y} - \boldsymbol{\alpha})$ from the BLP Equation 4.2 (see also Eq. 4.3). If desired this could be calculated directly by entering into **C** the covariances of the observations in **y** with the true aggregate genotypic values in **h**. However, it is computationally simpler to make use of Property 5 discussed in Chapter 4: BLP of a linear transformation of genetic values is that same linear transformation of the individual predicted genetic values. Thus:

$$\hat{h}_k = m_1 \hat{g}_{1k} + m_2 \hat{g}_{2k} + m_3 \hat{g}_{3k}$$

where $\hat{g}_{ik}$ = BLP predicted breeding value for the ith trait for genotype k.

Thus, a breeding value for each trait can be predicted with BLP and economic weights applied afterward, and the resulting prediction of aggregate breeding value is BLP of the true aggregate breeding value.

The economic weight for a trait is essentially the economic importance of a unit increase in one trait relative to a unit increase in other traits. Economic weights are often difficult to estimate, and thus sensitivity analyses may be necessary to examine the effects of different economic weight estimates on the final rankings of parents based on the predicted aggregate genotype (Talbert 1986). The approach of predicting breeding values of the individual traits, and subsequently applying economic weights simplifies the process of sensitivity analysis. For examples of determining economic weights in forest trees see Bridgewater and Stonecypher (1979), and Talbert (1984).

TOPIC 11: EVALUATION OF PROGENY TEST STRATEGIES

Often breeders must make specific decisions about progeny test strategies. For example, suppose a tree improvement program includes 2000 parents being progeny tested to predict breeding values for important traits. Questions such as in how many tests should a parent be represented, what are optimal mating and experimental designs, at what age should the tests be measured, and what traits should be measured all must be answered. Indirect or early selection (such as from greenhouse tests) may be possible and useful for certain traits, but should these replace field tests, or complement them? If both field tests and early selection are used together, what proportion of the types is optimum? If estimates of the second moments are available, best linear prediction offers a convenient framework for making these kinds of decisions. Note that neither data (i.e. $\mathbf{y}$) nor first moments (i.e. $\hat{\boldsymbol{\alpha}}$) are needed because the formulas in Chapter 4 that address the precision of the predictions (Eqs. 4.4, 4.6 and 4.7) require knowledge only of the second moments.

The basic approach in the evaluation of testing strategies with BLP is to calculate and compare any one of three variables for each of the alternative testing strategies: the variance of the predictions ($\mathrm{Var}(\hat{\mathbf{g}})$), the error variance of prediction ($\mathrm{Var}(\mathbf{g}-\hat{\mathbf{g}})$ or EVP), or the correlation between the predicted and true breeding value ($\mathrm{Corr}(g,\hat{g})$). In a sense, all three variables measure the 'precision' of a testing strategy, but the ease of interpreting $\mathrm{Corr}(g,\hat{g})$ may make it the most useful of the three. The proposed test strategies with the highest $\mathrm{Var}(\hat{\mathbf{g}})$ and $\mathrm{Corr}(g,\hat{g})$, or the lowest EVP, yield the most precise rankings of parents, and selection based on these rankings result in maximum genetic gain. Increasing the number of progeny tests, of whatever type, will always increase the correlation between the predicted and true breeding value. The marginal increase from an additional test, however, decreases as the number of tests increase. For example, testing each family in 6 field tests instead of 4 increases $\mathrm{Corr}(g,\hat{g})$, but the increase may be relatively insignificant depending on the parameter estimates in $\mathbf{C}$ and $\mathbf{V}$.

For any proposed testing strategy, $\mathbf{V}$, $\mathbf{C}$, and $\mathbf{G}$ can be constructed (if estimates of the appropriate second moments are available), and $\mathrm{Var}(\hat{\mathbf{g}})$ calculated as $\mathbf{C'V^{-1}C}$ (Eq. 4.5), and then $\mathrm{Corr}(g,\hat{g})$ or EVP determined (Eqs. 4.6 and 4.7). Despite the fact that there is no 'real' data from that particular set of progeny tests, and therefore no observed data to enter into $\mathbf{y}$, and no 'real' predicted breeding values, we can still logically calculate the correlation between the predicted and true breeding value. Assuming that there are families in progeny tests with the characteristics represented by $\mathbf{V}$ and $\mathbf{C}$, the calculated correlation (or $\mathrm{Var}(\hat{g})$ or EVP) apply to the population of predictions made from a set of data with the same characteristics. For example, suppose a very good and a very poor family have been tested in exactly the same progeny tests. The $\mathbf{y}_k$ vectors for the two families will be quite different, but both families will have identical $\mathbf{V}_k$ and $\mathbf{C}_k$ matrices. Therefore the two families will have identical $\mathrm{Var}(\hat{\mathbf{g}})$, EVP, and $\mathrm{Corr}(g,\hat{g})$.

So, one approach is to first develop several scenarios that represent combinations of different mating designs, experimental designs, traits to measure, ages to measure, etc. Each scenario will have its own different set of second moments (**C**, **V** and **G**) that will need to be estimated from data on hand (or gathered specifically for this purpose). Once the second moments are estimated for each scenario, the relative precisions of the predictions made from each scenario can be calculated, and the scenarios can be objectively evaluated.

SUMMARY

All of the issues in this chapter have been discussed singly, while in practice many may occur together. For example, it is always necessary to estimate all first moments and all second moments, the second moments are often heterogeneous and it may be desirable to predict breeding values of multiple traits and an aggregate genetic worth simultaneously. This poses no problem to the application of BLP as long as the problem is properly specified and estimates of the first and second moments are accurate and precise. The issue of how precise the estimates of the moments have to be is still somewhat unresolved and probably depends on the details of the specific situation (experimental designs, quality and quantity of data, heritabilities and correlations of both the traits measured and those being predicted, etc). For many forestry applications, data sets for estimating second moments can often be fairly large (a hundred or so genotypes in several test environments). While many of these topics need further investigation, the objective of this chapter has been to highlight some of the power, flexibility and problems encountered in applications of BLP.

PROBLEMS

Problem 7.1: Use the parameters for half-sib tests in Table 7.3 to predict a breeding value for height (g_h) for a single parent assuming you have only DBH measurements in four tests.

a) Set up $\mathbf{V}_k$, $\mathbf{C}_k$, and $\mathbf{G}_k$.

b) Calculate $\mathrm{Corr}(g,\hat{g})$.

Now assume you have height and DBH measurements in four tests, and still wish to predict height breeding value.

c) Set up $\mathbf{V}_k$, $\mathbf{C}_k$, and $\mathbf{G}_k$.

d) Calculate Corr(g,$\hat{g}$).

From a practical standpoint, is it necessary to measure height and DBH, or are measurements on DBH sufficient (given the assumed second moments in Table 7.3)?

Problem 7.2: Given the parameters in Table 7.2, and four progeny tests as described in Topic 8 (Tests 1 and 2 are type A at 1000 feet elevation, and tests 3 and 4 are type C at 3000 feet elevation), you wish to predict breeding values for all three site types.

a) set up $\mathbf{C}_k$.

b) Calculate Corr($\hat{g}_A$,g_A), Corr($\hat{g}_B$,g_B), and Corr($\hat{g}_C$,g_C).

You have three families of interest (D, E, and F), and the family means in the four tests are:

	Test			
Family	1	2	3	4
D	72	72	40	40
E	60	60	48	48
F	66	66	44	44

c) Set up $\mathbf{y}_k$ - α_k for the three parents.

d) Calculate $\hat{g}_A$, $\hat{g}_B$, and $\hat{g}_C$ for each parent.

e) What is the percent gain for each parent on each site type?

You can only afford to establish one production seed orchard, and your lands are equally divided among the three site types. You wish to rank parents for their aggregate performance across all three site types.

f) Calculate an aggregate breeding value for each parent.

g) What is the overall percent gain for each parent?

Problem 7.3: You have estimates of the following parameters for height of loblolly pine in three types of open-pollinated progeny tests. The parameters apply to all tests of a given type:

Progeny Test Type [height (feet)]	$\bar{y}_{i\cdot\cdot\cdot}$	σ^2_T	σ^2_F	σ^2_f	σ^2_{fe}	σ^2_p	σ^2_w
A: 12-year field	30	36	2.70	1.800	0.900	6.300	27.0
B: 5-year field	10	4	0.20	0.130	0.070	0.760	3.04
C: 2-year greenhouse	4	1.25	0.115	0.092	0.023	0.035	1.10

The type B genetic covariances and correlations between observations in tests of various types are:

$Cov(g_A,g_{A'})$ = 7.20 (cov. between two different 12-year field tests)
$r_{Bg}(g_A,g_{A'})$ = 0.667 (corr. between two different 12-year field tests)

$Cov(g_B,g_{B'})$ = 0.520 (cov. between two different 5-year field tests)
$r_{Bg}(g_B,g_{B'})$ = 0.650 (corr. between two different 5-year field tests)

$Cov(g_B,g_{A'})$ = 0.890 (cov. between a 5-year and a 12-year field test)
$r_{Bg}(g_B,g_{A'})$ = 0.300 (corr. between a 5-year and a 12-year field test)

$Cov(g_C,g_{C'})$ = 0.368 (cov. between two different 2-year greenhouse tests)
$r_{Bg}(g_C,g_{C'})$ = 0.800 (corr. between two different 2-year greenhouse tests)

$Cov(g_C,g_{A'})$ = 0.669 (cov. between a 2-year greenhouse test and a 12-year field test)
$r_{Bg}(g_C,g_{A'})$ = 0.300 (corr. between a 2-year greenhouse test and a 12-year field test)

You wish to predict the breeding value for height at age 12 ($g_{A'}$), and you want to determine how many tests are needed to reach a certain correlation between the predicted and true breeding value. For the following test designs, plot Corr($\hat{g}$,g) for 2, 4, 6, 8, 10, and 20 tests (you will need to set up $\mathbf{V}_k$, $\mathbf{C}_k$, and $\mathbf{G}_k$ for each number of tests):

a) 12 year field tests, b=10 (number of blocks),
   n=10 (number of trees/family/block).
b) 12-year field tests, b=3, n=5.
c) 5-year field tests, b=10, n=10.
d) 5-year field tests, b=6, n=8.
e) 5-year field tests, b=3, n=5.
f) 2-year greenhouse tests, b=10, n=10.
g) 2-year greenhouse tests, b=3, n=5.

Compare the curves for the different types of tests. Discuss how the heritability of a family mean in one test (h^2_{iF}), the genetic correlation between repeated tests ($r_{gB}(g_i,g_{i'})$), and between observations from the test and the breeding value ($r_{Bg}(g_X,g_{A'})$) affect the shape of the curve.

Chapter 8
Best Linear Prediction: An Operational Example

INTRODUCTION

Best linear prediction seems well suited to the problem of predicting breeding values and ranking parents in forest tree breeding. In theory, BLP is relatively simple, however, the application of BLP theory to real life situations will generate a number of problems in need of specific solutions. Some of the problems may be somewhat theoretical (for example, obtaining good estimates of the necessary second moments), while others may be more practical (writing a computer program to build the necessary matrices). This chapter will outline our application of BLP for predicting parental breeding values from half-sib progeny test data in a large slash pine breeding program. We discuss a number of specific problems that were encountered, and our attempts to solve these problems. Much of this chapter is adapted from White and Hodge (1988).

BACKGROUND

The Cooperative Forest Genetics Research Program (CFGRP) at the University of Florida has been involved in breeding slash pine for over 30 years. The program includes 15 private corporations and government agencies improving slash pine for approximately 5 million hectares of land throughout the range of slash pine. Mass selection of superior trees from natural stands began in the 1950's. Each cooperator made 50 to 200 selections, and grafted these genotypes into seed orchards. Open-pollinated seed from these seed orchards was used to establish 364 progeny tests from 1960 to 1980. Each test is a randomized complete block, with each family represented in a row-plot in each block. The number of families in a progeny test ranges from 5 to 160 (average = 33), the number of blocks from 3 to 10 (average = 6), and the number of trees per row-plot from 5 to 10 (average = 8). Each test also includes a commercial check which was planted in nearly every test (UC, University check), and up to three additional commercial checks unique to the company planting the test. The total number of trees planted in a test ranged from 500 to 2400 (average = 1600), and survival averaged 78% at age 15.

Progeny tests were measured at ages 5, 10, and/or 15 years. Each tree was measured for height and DBH (diameter at breast height), and scored for the presence (score=1) or absence (score=0) of fusiform rust. The height and DBH measurements were used to calculate volume for each tree (Goddard and Strickland 1968). Family means (i.e. means of the kth family in the ith test) for volume and for rust score (percent infected) were calculated by taking the mean of the plot means (i.e. means across the j blocks for the kth family in the ith test) in each test. In tests where total rust infection was less than 10% only growth traits were measured, while other tests were designed specifically to measure rust resistance and therefore only rust infection was scored.

In total, 1672 parents are represented in these open-pollinated progeny tests and the goal is to predict breeding values for both rust infection and volume at 15 years for each parent. However, the progeny test data set was extremely messy. Some families were tested for only one trait (e.g. rust) in one progeny test, others had family means for both rust and volume growth from more than 20 tests. Also, the quality of the measurements (i.e. the value of the measurements in predicting the trait of interest) varied widely from test to test as well. For example, the ages of the tests ranged from 5 to 15 years, and it is well known that for volume both heritability and juvenile-mature correlations increase with age, and thus family means from older tests are better indicators of the important trait of volume superiority at harvest. Genetic theory also indicates that for a Bernoulli trait (like the rust scores), heritability peaks at 50% incidence. Overall rust infection in the 364 progeny tests ranged from 0 to nearly 100%, and thus some tests provided good information on relative rust resistance, while others provided nearly none. These facts were enough to suggest that BLP or BLUP (Best linear unbiased prediction, see Chapter 11), which could theoretically handle these problems, would be the appropriate methodology to use in analyzing the data.

USE OF BLP VS BLUP

Both BLP and BLUP assume that the second moments (the elements of the **C** and **V** matrices) are known. The primary difference between BLP and BLUP is that BLP also assumes that the first moments are known, while BLUP estimates the first moments (fixed effects) and the breeding values simultaneously (see Chapters 11 and 12). BLUP is the technique used in most animal breeding applications in part because it is difficult to make unbiased and precise estimates of the fixed effects associated with herd, year and season. The CFGRP progeny tests had a large random sample of genotypes in each test location, and from these data we could obtain a very precise, unbiased estimate of the fixed effects associated with any test. It was therefore possible to treat the fixed effects as 'known' which would allow the use of BLP.

A significant practical advantage arose from the decision to use BLP. Assuming that the parents selected in natural stands are all unrelated, open-pollinated progeny test information on any given parent contributes no information about the breeding value of another parent. Thus, the computations of breeding value can be done on a parent-by-parent basis (see Chapter 7) with great reductions in the sizes of the matrices involved, and amount of computer time required.

DEFINING THE $\mathbf{g}_k$ AND $\mathbf{y}_k$ VECTORS

In this instance of open-pollinated data and unrelated parents, breeding values can be predicted for each parent separately. Thus, the data and breeding value vectors are defined separately for each parent. The $\mathbf{g}_k$ vector includes q_k breeding values for the kth parent, where q_k = number of traits being predicted for parent k. We chose to predict breeding values for two traits of economic importance in slash pine: rust resistance at 5 years and individual tree volume at 15 years. Thus $\mathbf{g}_k$ is a 2 x 1 vector containing a rust and a volume breeding value parent k. While rust infection incurred after the age of 5 has relatively little economic impact, infection incurred on the stems of a slash pine tree before the age of 5 years can result in economic loss due to three factors: mortality, product degrade, and growth reduction (Sluder 1977). For volume, we are generally interested in increasing harvest yield at economic maturity (20 to 25 years); however, a large amount of good data (needed to make estimates of the second moments) was available only up to age 15.

We chose to use the family means for percent rust infection and volume from each test in which parent k was included as the elements of the $\mathbf{y}_k$ vector. Remember, since the open-pollinated families are unrelated, only the data from parent k are included in the $\mathbf{y}_k$ vector. The elements of the $\mathbf{y}_k$ vector could have been individual tree measurements, or block means, instead of family means across all blocks in the test. However, neither of these were available in machine readable form, and therefore would require a re-analysis of the 364 tests. Also, compared to family means, using individual tree or block means would increase the size of the matrices substantially. Assuming the open-pollinated families are half-sib families, and adopting the linear model in Eq. 2.23, a family mean for the kth family in the ith test is (see also Eq. 2.25):

$$\bar{y}_{i \cdot k \cdot} = \mu + E_i + f_k + fe_{ik} + p_{i \cdot k}/b_i + w_{i \cdot k \cdot}/b_i n_{ik} \qquad \textbf{Eq. 8.1}$$

where $\bar{y}_{i \cdot k \cdot}$ = the family mean for the kth family in the ith test,
b_i = the number of blocks in the ith test,

n_{ik} = the number of trees/family/block in the ith test,

all other terms are defined in Eq. 2.23, and block effects sum to zero.

For convenience, the element $\bar{y}_{i\cdot k\cdot}$ is written $\bar{v}_{i\cdot k\cdot}$ if the trait is volume, and $\bar{r}_{i\cdot k\cdot}$ if the trait is rust. Thus, if a parent was represented in two tests where both volume and rust were measured, and one where only rust was measured, then n_k = 5 = number of observed family means, and $\mathbf{y}_k$ would be a 5 x 1 vector containing the 5 family means observed on parent k:

$$\mathbf{y}_k = \begin{bmatrix} \bar{v}_{1\cdot k\cdot} \\ \bar{r}_{1\cdot k\cdot} \\ \bar{v}_{2\cdot k\cdot} \\ \bar{r}_{2\cdot k\cdot} \\ \bar{r}_{3\cdot k\cdot} \end{bmatrix}$$

ESTIMATING THE FIXED EFFECTS, α_k

The fixed effects (i.e. μ + E_i) in CFGRP progeny tests are associated with test environments, and are used to express the family means, $\mathbf{y}_k$, as deviations from the test means, α_k, in Eq. 4.2. The derivation showing that μ + E_i are the appropriate fixed effects is in Eq. 5.2 and the topic of estimating fixed effects is discussed at length in Chapter 7.

For an CFGRP average progeny test, the 33 parents included were a random sample from the inference population, and were each represented by from 30 to 100 offspring. There were also a number of checks present in each test. For both rust and volume, a function of the grand family mean (i.e. mean of all family means in the ith test) and the checks was used to estimate the fixed effect associated with each test. Using both the grand mean and the check mean to estimate α_i gives a more precise estimate than using either one by itself.

Rust

Compared across all 364 progeny tests, the mean percent rust infection of the checks was equal to the mean percent infection of the progeny of the selected parents across all levels of rust infection. Thus for any given test, say test i, the mean infection of the offspring of the select parents and the check material both estimate (with varying levels of precision) the fixed effect associated with the test site. So, the fixed effect (overall test mean) for percent rust infection in the ith test was estimated as a weighted average of the two independent estimates of the fixed effect:

$$\alpha_i = (s_i\ \bar{r}_{i\cdots} + k_i\ \bar{c}_i)\ /\ (s_i + k_i) \qquad \textbf{Eq. 8.2}$$

where α_i = fixed effect for percent rust infection for the ith test,
s_i = number of parents (or half-sib families) represented in the test,
k_i = number of checks in the test,
$\bar{r}_{i\cdots}$ = mean percent rust of the offspring of select parents, and
$\bar{c}_i$ = mean percent rust of the checks.

Each family mean ($\bar{r}_{i\cdot k\cdot}$) was calculated as the mean of the plot means, and the grand family mean ($\bar{r}_{i\cdots}$) was calculated as the mean of the family means. The mean of the checks was calculated similarly as the average of all check means (the number of genetic checks in a test ranged from 1 to 6).

Volume

As for rust (described above) there are two estimates of the fixed effects for volume in each test: the mean of the offspring of the select parents, and the mean volume of the check material. Depending on the number of checks and the number of families, the two estimates have differing precisions. Also, across all tests, the progeny of select parents averaged 12% more volume than the checks at age 5 (number of 5 year tests = N_5 = 262), 10% more volume at age 10 (N_{10} = 264), and 12% more volume at age 15 (N_{15} = 36). Since the two estimates do not have the same expected value, the fixed effect for volume was estimated as:

$$\alpha_i = (s_i\ \bar{v}_{i\cdots} + k_i\ 1.11\ \bar{c}_i)\ /\ (s_i + k_i) \qquad \textbf{Eq. 8.3}$$

where α_i = fixed effect for volume for the ith test,
s_i = number of parents (or half-sib families) represented in the test,
k_i = number of checks in the test,
$\bar{v}_{i\cdots}$ = mean volume of the offspring of parents, and
$\bar{c}_i$ = mean percent rust of the checks.

The 1.11 was included in the formula to bring the checklot volume up to that expected of the progeny of select parents. The overall progeny mean and the check mean are calculated as indicated above for rust.

DEFINING THE C_k MATRIX

The elements of the C_k matrix are all covariances between an observed family mean from the y_k vector and an unobservable breeding value. To predict two breeding values from n_k observed family means for the kth parent, C_k is an n_k x 2 matrix. From Eq. 5.5, each element of the C_k matrix equals ½Cov(g*,g) for the general case where the trait observed is not the trait for which breeding values are being predicted. When predicting breeding values, we are measuring a trait (volume or rust infection) on the offspring from a parent on a specific site and at a specific age, and trying to use that information to predict how future offspring from that parent will perform on a different site, and possibly at a different age. Thus, the appropriate covariances to use in estimating elements of C_k are type B covariances (since the two traits in type B covariances are measured on different individuals, see Chapter 7), and each element will equal ½Cov(g*,g) where g* is the trait being measured and g is the trait being predicted (see Eq. 5.5).

DEFINING THE V_k MATRIX

Each V_k matrix is an n_k x n_k symmetric matrix containing the variances and covariances among the n_k observed family means in the y_k vector. The diagonal elements of the V_k matrix are always variances of the observations, which in this case are family means. The off-diagonal elements are all covariances between the observed family means, and are of two types:

1) Type A covariances between rust and volume family means from the same test, of the form (from Eq. 7.8):

$$\tfrac{1}{4}\mathrm{Cov}(g^*,g) + \tfrac{1}{4}\mathrm{Cov}(g^*e,ge) + \mathrm{Cov}(p^*,p)/b + \mathrm{Cov}(w^*,w)/bn.$$

2) Type B covariances between family means for volume (or rust) in one test, and volume (or rust) in another test. Thus, there will be elements estimating the volume-volume, rust-rust, and volume-rust family mean covariances of the form (from Eq. 7.4):

$$\tfrac{1}{4}\mathrm{Cov}(g^*,g)$$

ESTIMATING ELEMENTS OF THE C_k AND V_k MATRICES

Direct vs Indirect Estimation of Second Moments

Any element in the C_k or V_k matrices can be either estimated directly from the data, or predicted indirectly using relationships developed from other data. Neither approach is likely to be optimal in all circumstances.

> **Example 8.1**: Suppose we wish to estimate the covariance between family means for percent rust infection and family means for volume in the same test. One approach would be to use the data on the sample of families in that test to calculate the covariance directly for that one test. This is only an estimate of the true covariance, however, and is likely to be quite imprecise. A second approach might be to use another (external) data set with a much larger sample of families to estimate the covariance much more precisely, and then use that estimated covariance from the large data set as a prediction of the covariance from the smaller test. However, although the second estimate (from the external data set) is more precise, it may not be an unbiased estimate of the covariance in the small data set since the families are growing in a different location. If rust infection is higher or growth is quite different, then the true covariance between the family means may also be quite different on the external site than on the original site of interest.

There are four types of second moments to be estimated: variances of family means, type A covariances, type B covariances, and genetic variances of the two traits being predicted for the G_k matrix of Eq. 4.6. Either direct or indirect estimation could have been utilized for variances of family means, type A covariances and genetic variances; however, in our case only the indirect approach was possible for the type B covariances. Direct estimation of the type B covariances for two progeny tests requires that the same families be represented in the two tests. For example, if family X is represented in tests 1 and 2, the type B covariance between family means in tests 1 and 2 is needed as an element in the **V** matrix. Often, however, family X may be the only common family in the two tests, making it impossible to directly estimate the necessary covariance. Even when there are a few common families in the two tests, direct estimates of type B covariances are highly imprecise. Since type B covariances make up all the elements of the C_k matrix, and most of the elements of the V_k matrix, it was necessary to develop models or estimates of type B covariances from a large, balanced data set to be used in estimating all type B elements. We also developed models for variances of family means and type A covariances.

Development of Models for Estimation of Second Moments

A set of 28 OP progeny tests was identified to serve as a base for developing models to predict the necessary second moments. All 28 tests had been measured at ages 10 and 15 years, and 18 of the tests had been measured at age 5 years. There were three steps involved in the process:

1) Estimate variance and covariance components for each of the 28 tests,
2) Calculate family mean variances and covariances assuming a constant test design, and
3) Develop linear regression models to predict family mean variances and covariances for each test (or pair of tests) using easily measured independent parameters.

Estimation of variance and covariance components: Each of the 28 tests was subjected to an unweighted analysis of variance of plot means for both rust and volume at each age (Snedecor and Cochran 1967, p 475). Estimates of within-plot variances were calculated from the individual tree measurements and divided by the harmonic mean to put them on a plot-mean basis. SAS Type I observed mean squares (Freund and ittell 1981) were then equated to their expected values and solved for the estimated variance components, which is analagous to Henderson's Method III (Milliken and Johnson 1984, p 231). No transformations were made on percent rust incidence as analyses indicated that the relative values of the variance components were not affected when test mean incidence was between 10 and 90% (see also Rockwood and Goddard 1973, Sohn and Goddard 1979). Tests outside the 10 to 90% range were not used. The linear model used for the analysis of variance was that in Eq. 2.23 with the exception that all factors (except μ) were considered random for the purposes of variance component estimation.

Assumption of a constant test design: The variance and covariance components from the analysis of variance were then recombined into an expression of variance (or covariance) of family means assuming a constant test design of 6 blocks (b=6) and 8 trees per family in each block (n=8). Thus, from Eq. 2.26,

$$\mathrm{Var}(\bar{y}_{i\cdot k\cdot}) = \sigma^2_f + \sigma^2_{fe} + \sigma^2_p/6 + \sigma^2_w/48$$
$$= \sigma^2_F + \sigma^2_p/6 + \sigma^2_w/48 \qquad \textbf{Eq. 8.4}$$

where $\sigma^2_F = \sigma^2_f + \sigma^2_{fe}$.

Note that the variance component associated with the family effect (σ^2_F) is defined to include σ^2_f and σ^2_{fe}. Type A covariances of family means for rust and volume were estimated similarly with covariance components in place of variance components in Eq. 8.4. Finally, using the assumed constant test design and the estimated variance components, a family heritability (h^2_F) was calculated for each test as:

$$h^2_F = \frac{\sigma^2_F}{\sigma^2_F + \sigma^2_p/6 + \sigma^2_w/48.} \qquad \textbf{Eq. 8.5}$$

The assumption of a constant design was made to allow us to make predictions for the variances or covariances of family means for an average test, instead of having to include design components in the independent parameters. An alternative approach would have been to predict each variance component individually, and then combine them with the design parameters appropriate for the particular test. However, this would have required three separate predictions (e.g. σ^2_F, σ^2_p, and σ^2_w for a variance of family means) which would then be combined, instead of a single prediction of the variance (or covariance) of family means. Most of the 364 progeny tests to which these estimates are applied do not deviate greatly from the average values of b=6 and n=8.

To estimate type B covariances of family means, tests that contained common families were paired and covariances estimated from Eq. 2.14 (using family means as the observation) for all possible pairs of traits and ages. For example, if two tests contained 15 common families, there would be 15 pairs of family means for rust at any given age. These 15 observations were used to calculate a type B covariance of family means for rust in two different tests. One pair of tests with common families provided 9 rust-rust and 9 volume-volume covariances (as each test was measured at three ages), and 18 rust-volume covariances.

Linear regression models of second moments: Variances and covariances varied dramatically from age to age, and from test to test within an age. The assumption of homogeneous second moments across all tests generally made in animal breeding

(Henderson 1984) was not appropriate for this slash pine data set. We therefore developed linear regression models for the 28 tests which would predict the necessary second moments as functions of easily measured variables that would be available for all 364 tests. The elements needed for any $\mathbf{V}_k$ or $\mathbf{C}_k$ matrix could then be predicted from the measured variables.

Many different transformations of the measured variables were tried in an attempt to linearize relationships and homogenize variances (Mosteller and Tukey 1977). To be adopted as a final predictive equation, a model had to be biologically interpretable, be significant at $P=0.01$, and have each regressor in the model be significant at $P=0.05$. Where needed, limits on the regressors and/or predictions were sometimes specified.

Both type A and type B covariances were transformed into family mean correlations before modeling by dividing by the standard deviations of family means, and models were then fit to the correlations. We felt better models could be fit to correlations than covariances because 1) correlations are bounded by 1 and -1, and even imprecise estimates would be forced within the theoretical range, and 2) eliminating the heterogeneous scale effect could reduce the number of variables needed in the model. The predicted type A and type B correlations could then be back-transformed into covariances by multiplying by the appropriate predicted variances of family means (see Chapter 7). This approach had the secondary advantage of keeping the covariances and the variances of the **V** matrices 'on the same scale'. Predicting covariances apart from variances could lead to a **V** matrix that is not positive definite (see Graybill 1976). That is, if the relationships between the diagonals and off-diagonals are not as they should be, the inverse matrix may end up with negative values on the diagonals and this may result in nonsense predictions of g_k and negative estimates of $\mathrm{Var}(\hat{g})$.

Final Models for Prediction of Second Moments

Second moments of nearly all types vary substantially for different tests or pairs of tests. We needed to predict four basic types of second moments for the operational application of BLP to the CFGRP slash pine data (Tables 8.1 and 8.2). They were:

a) Variance of Family Means in One Test - Volume, Rust used to 1) estimate diagonal elements of **V**, 2) in conjunction with correlation models to estimate off-diagonal elements of **V** and **C**, and 3) in conjunction with family heritability models to estimate elements of **G**,

Table 8.1 - Definitions and abbreviations of independent variables used in building regression models.

Abbreviation	Definition
R	Mean rust infection level of a given test, $0 \le R \le 1$.
RDIF	Absolute value of the difference in rust between two tests, = abs(R_i - $R_{i'}$).
RIMS	Block by family interaction mean square from an analysis of variance of plot means for rust in a given test.
RSUM	Sum of mean rust infection levels for a pair of tests, = (R_i + $R_{i'}$).
V	Mean volume of a test, dm^3.
VCV	Coefficient of variation of a test for volume, = (sqrt(VIMS))/ V.
VCVSUM	Sum of volume coefficients of variation for a pair of tests, = VCV_i + $VCV_{i'}$.
VDIF	Absolute difference of standardized volumes for two tests, = abs($VSTAN_i$ - $VSTAN_{i'}$).
VIMS	Block by family interaction mean square from an analysis of variance of plot means for volume of a test.
VSTAN	Mean test volume expressed as a standardized deviate from the overall mean for that age. So for each age, the mean volumes from the 28 tests were standardized to have mean = 0 and variance = 1 ($\bar{y}_5$ = 0.2191, s_5 = 0.1387, $\bar{y}_{10}$ = 1.8869, s_{10} = 1.0290, $\bar{y}_{15}$ = 4.9835, s_{15} = 1.9274).

b) Type A Correlation of Family Means in One Test - Rust-Volume used in conjunction with variance of family means models to estimate off-diagonal elements in **V**,

c) Type B Correlation of Family Means in Two Tests - Rust-Rust, Volume-Volume, Rust-Volume used in conjunction with variance of family means models to estimate off-diagonal elements in **V** and all elements of **C**, and

d) Family Heritability in One Test - Rust, Volume at 15 Years used in conjunction with variance of family means models to estimate elements in **G**.

Variance of family means for volume: For volume growth, the variance of family means was primarily influenced by test age, although the average volume of a test within an age class, and its coefficient of variation (CV) were also important (model 1, Table 8.2). Variance for volume increases dramatically as age (and therefore mean volume) increases.

Mean coefficients of variation were similar for the three age classes (0.26, 0.23, 0.19 for 5, 10 and 15 years, respectively), but varied within age classes, particularly for the 5 year class where CV ranged from 0.13 to 0.50. The models predict that older tests, tests with larger mean volumes within an age, and tests with higher CV's have larger variances of family means.

Variance of family means for rust: The variance of family means for rust was primarily affected by the mean level of rust infection in the test, and to a lesser extent by RIMS (block x family interaction mean square for rust infection from an analysis of variance of plot means) (model 2, Table 8.2). The variance is a quadratic function of the mean rust level, and is highest between 40 and 70% infection. Similar results were found by Sohn and Goddard (1979). Tests with higher RIMS also tend to have higher variances of family means. In contrast to volume, age of the test had no effect on variances for rust.

Type A correlation of family means, rust-volume in one test: No adequate regression models were found, however, the mean correlations were significantly different than zero (0.06, 0.13, and 0.14 for ages 5, 10 and 15, respectively) and were therefore used (model 3, Table 8.2).

Table 8.2 - Back-transformed regression models and their fit statistics for estimating the second moments needed for the C_k and V_k matrices. All models are significant at P = 0.001 and the reported R^2 values are for the final back-transformed model accounting for the imposed limits. Abbreviations for regressor variables are defined in Table 8.1.

Model Number and Dependent Variable	Age	Mean	Regression Model	Imposed Limits	R^2	n
1. Variance of Family Means for Volume (dm^3)	All				0.78	76
	5	1	3.17(10exp(80.53+2.89(VCV)-83.04(Vexp -0.01)))	$0.02 \le \hat{y} \le 4.00$		
	10	85	3.17(10exp(80.90+1.40(VCV)-83.04(Vexp -0.01)))	$2.41 \le \hat{y} \le 280$		
	15	600	3.17(10exp(81.14+0.93(VCV)-83.04(Vexp -0.01)))	$32.1 \le \hat{y} \le 1350$		
2. Variance of Family Means for Rust ((%)/100)^2	All	0.015	0.00355+0.06517(R)-0.06296(R^2)-0.4750(RIMS) +1.9715(R)(RIMS)-1.9048(R^2)(RIMS)	$0.1 \le R \le 0.9$ $0.001 \le \hat{y} \le 0.025$	0.56	78
3. Type A Correlation, Volume and Rust	5	0.06	No Significant Model Found	NA	NA	18
	10	0.13				28
	15	0.14				28
4. Type B Correlation, Rust in Two tests	All	0.50	0.2471+0.6414(RSUM)-0.2680(RSUM^2) -0.5006(RDIF)	$0.1 \le R \le 0.9$	0.25	316
5. Type B Correlation, Volume in Two Tests	5:All	0.15	0.7288-1.029(VCVSUM)-0.0931(VDIF)	$0.00 \le \hat{y}$	0.20	115
6. Type B Correlation, Volume in Two Tests	10:10	0.34	0.8213-0.6924(VCVSUM)-0.2110(VDIF)	$0.00 \le \hat{y}$	0.24	171
	10:15	0.40	0.8547-0.6924(VCVSUM)-0.2110(VDIF)			
	15:15	0.44	0.8700-0.6924(VCVSUM)-0.2110(VDIF)			
7. Family Heritability, Rust in One Test	All	0.66	0.3278+1.3579(R)-1.1778(R^2)	$0.1 \le R \le 0.9$	0.16	78
8. Family Heritability, Volume in One Test	15	0.62	-0.0843+9.3437(VCV)-27.3401(VCV^2)	$0.15 \le \hat{y} \le 0.72$	0.71	28

Type B correlation of family means, rust in two tests: The type B correlation for rust is a function of the mean rust infection in both tests (model 4, Table 8.2). Predicted correlations are highest when both tests are in the 40 to 70% infection range, moderately high when both tests have low infection or both tests have high infection, and lowest when one test has high infection and the other has low infection (Figure 8.1). The ages of the two tests had no effect on the type B correlation.

Type B correlation of family means, volume in two tests: The predictive model for type B correlations for volume is a function of the age of the two tests, their CV's, and their relative mean volume for their age (see VSTAN, Table 8.1) (models 5 and 6, Table 8.2). For pairs of tests that are older, more precise (lower CV's), and of the same relative mean volume, the model predicts higher type B correlations. The age of the two tests is the primary factor: if either one of the two tests is 5-years-old, the mean type B correlation for volume between the two tests is only 0.15, while the mean correlation for two 10-year-old tests is 0.34, and for two 15-year-old tests is 0.44.

Type B correlation of family means, rust-volume in two tests: No adequate regression models were found, and the mean correlations did not significantly differ from zero. All second moments of this type were estimated at zero.

Family heritability, rust in one test: The family heritability for rust is a quadratic function of the mean level of rust infection in the test (model 7 Table 8.2), and peaks at a mean infection level of 58% with a family heritability of 0.72. As was the case for variances of rust, age of the test was unimportant. By rearranging Eq. 8.5, this model was used to calculate the additive genetic variance for rust as:

$$\sigma^2_A = 4\ \sigma^2_F = 4\ \mathrm{Var}(\bar{r}_{i\cdot k\cdot})\ h^2_F$$

where σ^2_A = additive genetic variance,

σ^2_F = family variance component in one test (Eq. 8.4),

$\mathrm{Var}(\bar{r}_{i\cdot k\cdot})$ = variance of family means for rust estimated by model 2, Table 8.1, and

h^2_F = family heritability estimated by model 7, Table 8.1.

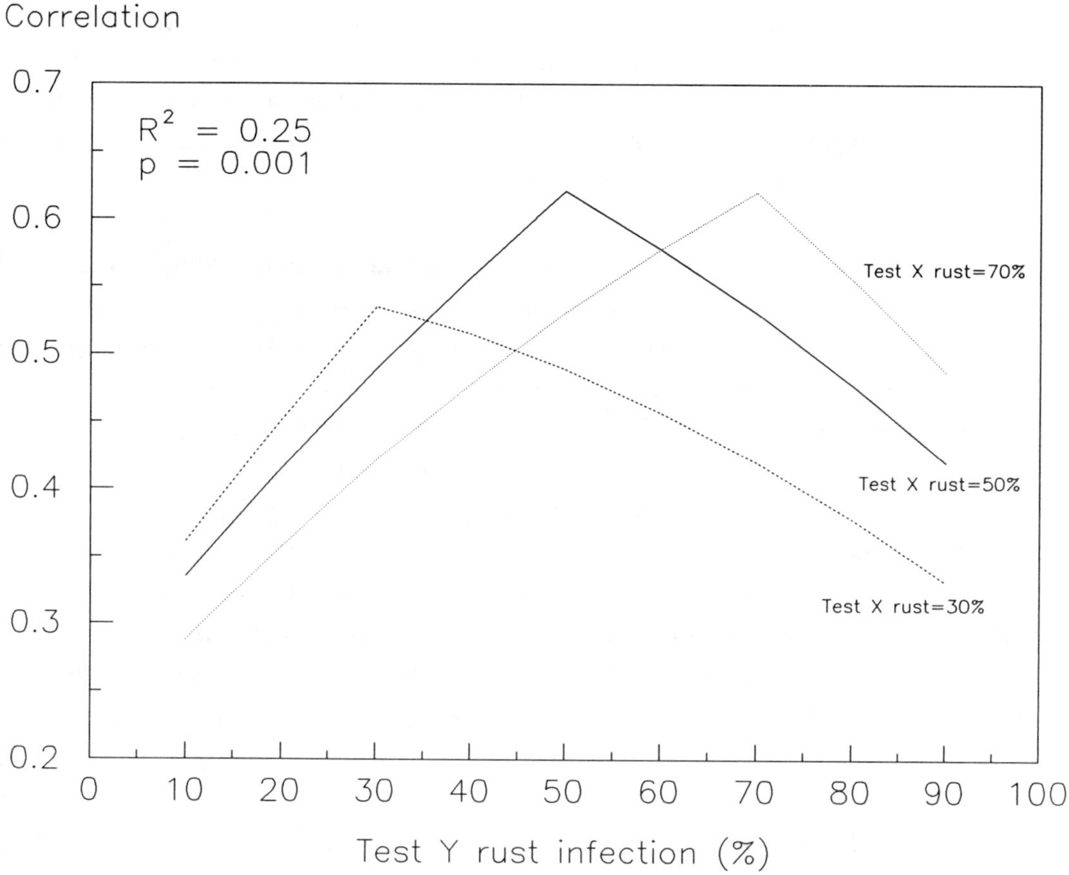

Figure 8.1. Type B family mean correlation for percent rust infection in two tests.

Family heritability, volume at 15 years in one test: The family heritability for volume at 15 years was a quadratic function of the CV of the test. As CV increased, predicted heritabilities decreased (model 8, Table 8.2). The predictions were bounded at 0.15 and 0.72. models 1 and 8, Table 8.2 were used to calculate the additive genetic variance for volume as for rust above.

Weighted Average of Predicted and Observed Variances of Family Means

In contrast to type B correlations, variances of family means could be directly estimated for each of the 364 tests. If a test had a large number of families and/or blocks, a direct estimate of the variance of family means might be better than a prediction from the models. If a test had a small number of families and/or blocks, the prediction from models 1 and 2 in Table 8.2 (based on the relatively well-estimated test mean, and for volume, the test age) might be better than an imprecise direct estimate from the test data. In an attempt to estimate the variance in each test with maximum precision, we developed a function that used a weighted average of the predicted variance of family means ($Var_p(\bar{y}_{i \cdot k \cdot})$) and the observed variance of family means ($Var_o(\bar{y}_{i \cdot k \cdot})$) (i.e. a variance estimate directly calculated from the data) as the final estimate of the variance of family means ($Var(\bar{y}_{i \cdot k \cdot})$) to be used in constructing matrices:

$$Var(\bar{y}_{i \cdot k \cdot}) = (Var_o(\bar{y}_{i \cdot k \cdot}) \cdot s)/50 + (Var_p(\bar{y}_{i \cdot k \cdot}) \cdot (50 - s))/50 \qquad \textbf{Eq. 8.6}$$

where s = number of families in the test, and s is set to 50 if there are more than 50 families.

The observed estimate of the variance of family means, $Var_o(\bar{y}_{i \cdot k \cdot})$, in the ith test has a variance which is approximately inversely proportional to the number of families s in that test (Figure 8.2). For tests with few families, the observed variance ($Var_o(\bar{y}_{i \cdot k \cdot})$) is an imprecise estimate, and thus for small values of s, more weight is placed on the predicted variance of family means. As the number of families increases, the observed variance becomes more precise, thus as s approaches 50, more weight is placed on the observed estimate of family means. Increasing s above 50 does not significantly decrease the variance of $Var_o(\bar{y}_{i \cdot k \cdot})$. In calculating a final estimate of the variance of family means to be used in constructing matrices, more weight was placed on the observed variance in approximately half of the 364 tests, and half had more weight placed on the predicted variance.

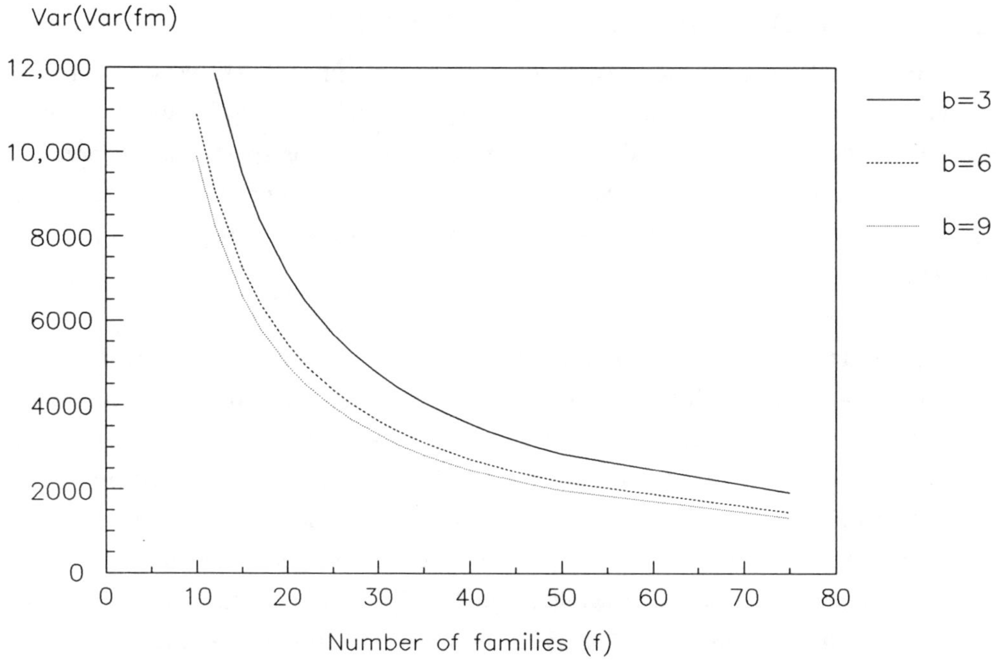

Figure 8.2. Variance of the observed estimate of the variance of family means (Var(Var(fm)).

Derivation:

Var(fm)=variance of family mean in one test,
Var(Var(fm)=variance of estimate of Var(fm),
Var(f)=family variance=200,
Var(p)=plot variance=100,
Var(w)=within plot variance=700,
b=number of blocks,
f=number of families,
n=number of trees per plot=7,
MSF=family mean square from ANOVA
=b Var(f) + Var(p) + Var(w)/n.

$$V(fm) = Var(f) + Var(p)/b + Var(w)/bn = MSF/b.$$

$$Var(Var(fm)) = Var(MSF/b) = 2\ MSF^2/b^2 f = 2\ (200b + 100 + 700/7)^2 / b^2 f = 80{,}000(b+1)^2 / b^2 f$$

TARGET ENVIRONMENTS FOR PREDICTION

Because the genetic expressions of both volume and growth vary with the test environment (Table 8.2), it becomes necessary to specify an environment for which breeding values will be predicted. If a target environment is specified, the predictive models can be used to calculate the covariances between the observed measurements ($\bar{y}_{i \cdot k \cdot}$ from Eq. 8.1) and the predictand (the trait for which breeding values are being predicted).

> **Example 8.2:** Family X is measured for rust in test 1 where the mean rust infection is 25%. The type B rust-rust correlation (used in determining the type B covariance which is entered into the $\mathbf{C_k}$ matrix) between measurements in two tests (i.e. two environments) is a function of the mean rust infection in the two tests (or environments). Specifically, the model uses the sum of the mean rust infection levels for the two tests, and the absolute value of the difference of the infection levels (model 4, Table 8.2). Although we know that the mean infection in test 1 is 25%, the mean infection in the other test, i.e. the target environment, is not known and therefore must be specified in order to predict a breeding value for rust resistance in the target environment.

We chose to predict rust breeding values (BLPR50) at a target environment of 50% mean rust infection, i.e. an environment where unimproved material would incur 50% infection. For 15-year volume breeding values (BLPVOL), the target environment was specified to be the average 15-year test volume (4.21 ft^3/tree) and CV (0.1924).

SPECIAL ISSUES

In applying BLP, the above procedures were utilized for the majority of the progeny test data, however there were two unusual situations which required slight modifications in the estimation procedures for the first and/or second moments: 1) a subset of the population had been mass selected only for rust resistance, and 2) based on initial progeny test results, a selected subset of the population was included in special 'definitive rust' tests designed to intensively assess rust resistance.

Mass Selection For Rust Resistance

There were 1672 parents included in the open-pollinated progeny tests. Most of these parents (1396) had been mass selected in natural stands with very little rust infection, with a primary emphasis on volume growth and quality traits, and consequently little gain in rust resistance was made. However, in the early 1960's it became apparent that rust was a serious problem in slash pine and one cooperator, Brunswick Pulp Land Company, developed a seed production area from a natural stand of slash pine which had well over 90% infection. Removal of all infected trees left over 300 rust-free trees in the SPA, and the offspring from 276 of these trees (designated as C-lots, for Cronartium resistant lots) were included in a number of OP progeny tests including only offspring from C-lot trees. In these progeny tests, across all levels of infection, the offspring from the C-lot trees were significantly more resistant than the unimproved check. The percent rust infection for C-lot offspring in any test could be estimated as:

$$\bar{r}^{c}_{i\ldots} = -4.98 + 0.8124\,\bar{c} \qquad (n=30 \text{ and } R^2=0.81) \qquad \textbf{Eq. 8.7}$$

where $\bar{r}^{c}_{i\ldots}$ = mean percent infection for C-lot offspring, and
$\bar{c}$ = mean percent infection for unimproved check.

Volume measurements were also made, and the mean volume of offspring from C-lot trees was not significantly different from the unimproved check. Thus, we had two distinct populations: 1) 1396 trees with 11% more volume growth than unimproved checks, but equal rust resistance, and 2) 276 C-lot trees with volume growth equal to the check, but more rust resistant. Thus $\boldsymbol{\tau}$, the expected value of **g**, is different for the two populations, i.e. the two populations had different fixed genetic effects. Since the tests of the two populations were mutually exclusive (standard parents and C-lot parents were never in the same test), we predicted breeding values for the two populations separately. When predicting C-lot breeding values, we assumed that E(**g**) = 0, however, when we compared those to standard parent breeding values, we adjusted those by the difference between the two values of $\boldsymbol{\tau}$, i.e. the two population means. If we let $\boldsymbol{\tau}$ for the 1396 'standard' parents be zero for rust and volume, we calculated appropriate values of $\boldsymbol{\tau}$ for the 276 C-lot parents as described below.

Rust: Recall that the target environment for rust was a site where unimproved checks would incur a mean of 50% infection. This would be equal to the infection incurred by the offspring of the 1396 standard parents. Offspring of the C-lot parents would incur a mean 36% infection in the target environment (when the check is 50% infected) (from Eq. 8.7). Therefore, we predicted the expected deviations of the C-lot breeding values around a mean of 36% infection, and subsequently adjusted those breeding values for the 14% difference in mean breeding value (τ) between the two population.

> **Example 8.3.** Suppose for C-lot parent k we predicted a breeding value of +5% deviation from the C-lot mean in the target environment (mean C-lot infection is 36%, and mean unimproved check infection rate is 50%). Standard parent j is predicted to have a breeding value of -5% deviation from the standard population mean in the target environment (mean standard population infection is 50%, and mean unimproved check infection is 50%). To compare rust breeding values for C-lot parents and standard parents, add the appropriate τ, 36% for C-lot, and 50% for standard breeding values. Thus, C-lot parent k has a breeding value of 41% (5% + 36%) and standard parent j has a breeding value of 45% (-5% + 50%) in the 50% target environment. Thus, τ for rust for the C-lot parents is 36%, while it is 50% for the standard parents.

Volume: The volume target environment was a site where the offspring of the 1396 'standard' parents would produce a mean of 4.21 ft^3 per tree at 15 years. This volume would be 11% more than the checks, would produce, thus checks (and C-lots) would have a mean of 3.79 ft^3, a difference of 0.42 ft^3 per tree. Therefore if τ for volume for standard parents is zero, τ for volume for C-lot parents is -0.42 ft^3 per tree, and breeding value predictions must be adjusted accordingly before comparing parents from the two populations.

Definitive Rust Tests of a Selected Population

Based on initial results of progeny tests, a subset of the 1396 standard parents was identified as being potentially rust resistant. Offspring from these parents were then included in a series of 'definitive' rust tests which were specially designed to measure rust resistance, and were planted in high rust hazard sites. Normal progeny tests included a random sample of the 1396 parents, and were planted across a wide range of infection levels. When calculating breeding values for the standard parents, we wished to use both data from the normal and the definitive tests. However, because the parents represented in the definitive tests were a selected subset, as opposed to a random sample, a problem arose in calculating the approriate first and second moments to use for definitive tests because the two types of tests had different first and second moments applicable to them.

> **Example 8.4:** Suppose parent k was judged moderately rust resistant on the basis of results from two normal progeny tests (including a random sample of parents) where in both tests its offspring had 22% infection (compared to the mean infection of all offspring in the test of 30%, and the infection of the unimproved check of 30%). On this basis, it was subsequently included in four definitive rust tests. In each of these four tests, offspring of parent k had 45% infection, the offspring of the selected sample of resistant parents (the test mean) was 36%, and the unimproved check infection was 58%. If we calculate α_i for each of these four tests as a simple weighted average of the check and progeny means (Eq. 8.2), parent k will appear below average for rust resistance because the other families in the definitive tests are on average substantially more resistant than a random sample of the population.

Because they included a selected population, both the fixed effects ($\boldsymbol{\alpha}$) and the second moments (**V**, **C**) of the definitive rust tests were different than they would have been if the tests had included a random population. In the situation with the C-lots, we simply specified the appropriate values of τ for the two distinct, and separately tested, populations. However, τ for standard parent k is the same whether it is in normal or definitive rust tests, or both. To apply BLP for a parent in both types of tests and simultaneously use all the data available on that parent, both the fixed effects and second moments of the definitive rust tests were expressed as if there were a random sample of families in that test.

Estimating the fixed effect in definitive rust tests: The mean infection expected if a random sample of families had been included in a definitive rust test was estimated as:

$$\bar{c} = \bar{r}_{i\cdots} = -8.31 + 1.3575\,\bar{r}^{d}_{i\cdots} \qquad (n=65 \text{ and } R^2=0.77) \qquad \textbf{Eq. 8.8}$$

where $\bar{c}$ = percent infection of the checks, which is assumed equal to $\bar{r}_{i...}$,

$\bar{r}_{i...}$ = estimated mean percent infection for a random sample of families, and

$\bar{r}^{d}_{i...}$ = observed mean percent infection of offspring of the selected families in the definitive test.

To estimate the fixed effect for rust, α_i, associated with the ith test, a weighted average of the check and estimated progeny means was calculated using Eq. 8.2 and the estimated $\bar{r}_{i...}$ from Eq. 8.8.

Estimating the variance of family means in definitive rust tests: We calculated an estimate of the variance of family means for rust (assuming a random sample of families in the test) by weighting an observed variance and a predicted variance as in Eq. 8.6. This was done in five steps:

1) Assuming a random sample of families in the test, a predicted variance of family means ($Var_p(\bar{y}_{i \cdot k \cdot})$) was calculated using model 2, Table 8.2, with the estimated fixed effect, α_i (site mean for rust) .
2) We then calculated a predicted variance of family means for the selected population ($Var^{*}_{p}(\bar{y}_{i \cdot k \cdot})$) using the actual observed progeny mean infection of the selected families, P_2, in model 2, Table 8.2, and also calculated the actual observed variance of family means for the selected population in the test ($Var^{*}_{o}(\bar{y}_{i \cdot k \cdot})$). The * superscript indicates that the estimates apply to the selected family population in the definitive test, not an assumed random sample.
3) We calculated a factor k = $(Var^{*}_{o}(\bar{y}_{i \cdot k \cdot})) / (Var^{*}_{p}(\bar{y}_{i \cdot k \cdot}))$ which measures if the test was more or less variable than would be expected for the actual infection level of the selected families.
4) An estimated observed variance of family means (assuming a random sample of families in the test) ($Var_o(\bar{y}_{i \cdot k \cdot})$) was calculated by adjusting the predicted variance of family means (assuming a random sample of families) as:

   $$Var_o(\bar{y}_{i \cdot k \cdot}) = k\ (Var_p(\bar{y}_{i \cdot k \cdot}))$$
5) A weighted average of $Var_o(\bar{y}_{i \cdot k \cdot})$ and $Var_p(\bar{y}_{i \cdot k \cdot})$ was calculated using Eq. 8.6.

MATRIX CONSTRUCTION EXAMPLE

The following small example illustrates the use of the models and estimation procedures to construct the necessary matrices and vectors and calculate breeding values.

Data and Matrix Definition

Offspring from parent 10356 were included in two 10-year-old progeny tests in which rust and volume were measured. Fixed effects estimating the environmental means, family means, coefficient of variation for volume (CVVOL), and the weighted average of the predicted and observed variance of family means ($\mathrm{Var}(y_{i\cdot k\cdot})$) have all been calculated (Table 8.3).

Table 8.3 - Open-pollinated progeny test data for family 10356 which was planted in two tests measured at 10-years-old.

Test	Trait	Environmental Mean (α_i)[a]	Family 10356 Mean[b]	CVVOL	$\mathrm{Var}(\bar{y}_{i\cdot k\cdot})$[c]
9-32	10-year-vol.	2.330 ft^3/tree	2.524 ft^3/tree	0.272	0.08110 ft^6/tree
9-32	10-year-rust	0.695	0.761	--	0.00591
9-38	10-year-vol.	0.850 ft^3/tree	0.927 ft^3/tree	0.307	0.02786 ft^6/tree
9-38	10-year-rust	0.663	0.665	--	0.00204

[a]Environmental means ($\alpha_i = \mu + E_i$) were estimated from Eq. 8.2 (rust) and Eq. 8.3 (volume).

[b]The family mean is $\bar{y}_{i\cdot k\cdot}$ from Eq. 2.25. The units of family means for rust are the proportion of trees infected.

[c]The variance of family means was estimated from Eq. 8.6.

Thus for parent 10356, $\mathbf{y}_k - \boldsymbol{\alpha}_k$ is 4 x 1, $\mathbf{V}_k$ is 4 x 4, $\mathbf{C}_k$ is 4 x 2, and $\mathbf{G}_k$ is 2 x 2 as below:

$$\mathbf{y}_k - \boldsymbol{\alpha}_k = \begin{bmatrix} y_1 \\ y_2 \\ y_3 \\ y_4 \end{bmatrix} \quad \mathbf{V}_k = \begin{bmatrix} v_{11} & v_{12} & v_{13} & v_{14} \\ v_{21} & v_{22} & v_{23} & v_{24} \\ v_{31} & v_{32} & v_{33} & v_{34} \\ v_{41} & v_{42} & v_{43} & v_{44} \end{bmatrix} \quad \mathbf{C}_k = \begin{bmatrix} c_{11} & c_{12} \\ c_{21} & c_{22} \\ c_{31} & c_{32} \\ c_{41} & c_{42} \end{bmatrix}$$

$$\mathbf{G}_k = \begin{bmatrix} G_{11} & G_{12} \\ G_{21} & G_{22} \end{bmatrix}$$

Filling in the $\mathbf{y}_k$ Vector

The elements of $\mathbf{y}_k - \boldsymbol{\alpha}_k$ are family deviations, so $\mathbf{y}_k - \boldsymbol{\alpha}_k$ is:

$$\mathbf{y}_k - \boldsymbol{\alpha}_k = \begin{bmatrix} 2.524 \\ 0.761 \\ 0.927 \\ 0.665 \end{bmatrix} - \begin{bmatrix} 2.330 \\ 0.695 \\ 0.850 \\ 0.663 \end{bmatrix} = \begin{bmatrix} 0.194 \\ 0.066 \\ 0.077 \\ 0.002 \end{bmatrix}$$

Filling in the $\mathbf{V}_k$ Matrix

Diagonal elements: The diagonal elements of $\mathbf{V}_k$ are the appropriate variances of family means ($\mathrm{Var}(\bar{y}_{i \cdot k \cdot})$). For example v_{11} is the variance of the first observation in the $\mathbf{y}_k$ vector, y_1, so $v_{11} = 0.08110$, etc.

$$\mathbf{V}_k = \begin{bmatrix} .08110 & v_{12} & v_{13} & v_{14} \\ v_{21} & .00591 & v_{23} & v_{24} \\ v_{31} & v_{32} & .02786 & v_{34} \\ v_{41} & v_{42} & v_{43} & .00204 \end{bmatrix}$$

Off-diagonal elements: To estimate the off-diagonal elements, first calculate the appropriate correlation, and then multiply that by the square roots of the variances to calculate the covariance.

Elements v_{12} and v_{21} are the covariance between the first and second observations in $\mathbf{y}_k$, which are family means for rust and volume in the same ten-year-old test (test 9-32). From model 3, Table 8.2, the type A family mean correlation between rust and volume at 10 years is 0.13. Thus $v_{12} = v_{21} = 0.13 \times (0.08110 \times 0.00591)^{½} = 0.00285$.

Elements v_{34} and element v_{43} are also the covariance between rust and volume in a given test (test 9-38). From model 3, Table 8.2, the type A correlation between rust and volume at 10 years is 0.13. Thus $v_{34} = v_{43} = 0.13 \times (0.02786 \times 0.00204)^{½} = 0.00098$. Note that even though the correlation is the same as for v_{12} and v_{21}, the covariance is much smaller. Covariances are scale dependent and the variances of family means are smaller for both growth and rust in test 9-38 than in test 9-32.

Elements v_{13} and v_{31} are the covariance between family means for volume in test 9-32 and volume in test 9-38. To determine the type B family mean correlation we use model 6, Table 8.2, which requires VCVSUM, the sum of the volume CV's in the two tests, and VDIF, the absolute value of the difference between the site volume means standardized for the appropriate age (see VSTAN, Table 8.1). The standardized volume for test 9-32 is (2.33 - 1.8869)/1.0290 = 0.4306, and the standardized volume for test 9-38 is (0.85 - 1.8869)/1.0290 = -1.0077. Thus, VDIF = abs(-1.0077 - 0.4306) = 1.4383. For tests 9-32 and 9-38, VCVSUM = 0.272 + 0.307 = 0.579. From model 6, Table 8.2, the type B family mean correlation between two 10-year-old tests is 0.8213 - 0.6924(0.579) - 0.2110(1.4383) = 0.1169. Finally, element $v_{13} = v_{31} = 0.1169 \times (0.08110 \times 0.02786)^{½} = 0.00556$.

Elements v_{24} and v_{42} are the covariance between family means for rust in test 9-32 and 9-38. Using the two rust means (0.695, 0.663) in model 4, Table 8.2, the type B family mean correlation is estimated at 0.608. Thus $v_{24} = v_{42} = 0.608 \times (0.00591 \times 0.00204)^{½} = 0.00211$.

Elements v_{14} and v_{41} are the covariance between rust in test 9-32 and volume in test 9-38. Similarly, elements v_{23} and v_{32} are the covariance between rust in test 9-38 and volume in test 9-32. The type B correlation between rust and volume in two tests is zero therefore all of these elements are zero.

The final $\mathbf{V}_k$ matrix is:

$$\mathbf{V}_k = \begin{bmatrix} .08110 & .00285 & .00556 & 0 \\ .00285 & .00591 & 0 & .00211 \\ .00556 & 0 & .02786 & .00098 \\ 0 & .00211 & .00098 & .00204 \end{bmatrix}$$

Filling in the $\mathbf{C}_k$ Matrix

All elements in $\mathbf{C}_k$ are estimated in the same manner as the off-diagonal elements of $\mathbf{V}_k$: first calculate the correlation, and then convert it into a covariance. Element c_{11} is the covariance between the family mean for volume in test 9-32 and the breeding value for volume at 15 years in the defined target environment (VOL). The target environment is a site where the average 15-year-old tree produces 4.21 ft^3 which when standardized is (4.21 - 4.9835)/1.9724 = -0.3921. For test 9-32 the standardized volume mean for 10 years is (2.33 - 1.8869)/1.0290 = 0.4306. Therefore, VDIF = abs(-.3921 - 0.4306) = 0.8227. The volume CV in the target environment is 0.1924 (average for all 15-year tests in the program), thus VCVSUM in this case is 0.4644 (0.1924 + 0.272). Substituting these values into the age10:age15 equation, model 6, Table 8.2, the type B family mean correlation can be estimated as 0.8547 - 0.6924(0.4644) - 0.2110(0.8227) = 0.3596. The covariance can now be calculated as the product of the correlation and the square roots of the variances multiplied by two because the open-pollinated family means represent half the breeding value (see Eq. 7.7). Thus, c_{11} = 2 x 0.3596 x $(0.08110 \times 0.32221)^{½}$ = 0.11626.

Element c_{31} is also the covariance between the family mean for volume in a test (9-38) and the breeding value for volume (VOL). The test 9-38 volume mean standardized for 10 years is (0.85 - 1.8869)/1.0290 = -1.0077. Therefore, VDIF = abs(-.3921 - -1.0077) = 0.6156, and VCVSUM in this case is 0.4994. Substituting these values into the age 10:age 15 equation, model 6, Table 8.2, the type B family mean correlation can be estimated as 0.8547 - 0.6924(0.4994) - 0.2110(0.6156) = 0.3790. The covariance can now be calculated as c_{31} = 2 x 0.3790 x $(0.02786 \times 0.32221)^{½}$ = 0.07182.

Element c_{22} is the covariance between the family mean for rust measured in test 9-32 and the breeding value for rust in the defined target environment (R50). The target environment is a 0.50 (50%) infection rate, and the mean infection in test 9-32 is 0.695 (69.5%). Using the two rust means (0.50, 0.695) in model 4, Table 8.2, the type B correlation is calculated as 0.533. From model 2, Table 8.2, the variance of family means in an environment of 50% rust (with average RIMS=0.04175) is estimated at 0.02184. Thus, $c_{22} = 2 \times 0.533 \times (0.00591 \times 0.02184)^{½} = 0.01211$.

Element c_{42} is the covariance between the family mean for rust measured in test 9-38 and R50. Using 0.50 and 0.663 (the rust mean infection in test 9-38), model 4 from Table 8.2 estimates the type B correlation as 0.549. Thus, the covariance estimate for $c_{42} = 2 \times 0.549 \times (0.00204 \times 0.02184)^{½} = 0.00733$.

Element c_{21} is the covariance between the family mean for rust measured in test 9-32 and the breeding value for volume in the defined target environment. Remember that the type B correlation between rust in one environment and volume in another is zero, thus element c_{21} is zero, and similarly, element c_{41} is zero. Element c_{12} is the covariance between volume measured in test 9-32 and the breeding value for rust in the target environment. Using the same logic as above, element c_{12} is zero, and element c_{32} is also zero.

The final $\mathbf{C}_k$ matrix is:

$$\mathbf{C}_k = \begin{bmatrix} 0.11626 & 0 \\ 0 & 0.01211 \\ 0.07182 & 0 \\ 0 & 0.00733 \end{bmatrix}$$

Filling in the $\mathbf{G}_k$ Matrix

For all parents, $\mathbf{G}_k$ is the matrix of genetic variances and covariances for the traits being predicted. Element G_{11} is the variance for VOL, which is the breeding value for individual tree volume on a site where the average tree produces 4.21 ft^3. By definition, the variance of breeding values is the additive genetic variance. With a CV=0.1924, from age 15 equation, model 1, Table 8.2, the variance of family means for volume ($Var_p(\bar{v}_{i\cdot k\cdot})$) is 0.32221. From model 8, Table 8.2, and with CV=0.1924, the family heritablility for volume at age 15 (h^2_F) is 0.7014. The additive genetic variance is then estimated as $4\ h^2_F\ Var_p(\bar{v}_{i\cdot k\cdot}) = 4\ (0.7014)\ (0.32221) = 0.904$.

Element G_{22} is the variance of R50, which is the breeding value for rust infection on a site where an unimproved population averages 50% infection. Using model 2, Table 8.2, with a value for rust mean of 50% (R=0.50), and RIMS=0.04175, the variance of family means for rust ($Var_p(\bar{r}_{i\cdot k\cdot})$) is estimated as 0.02184. From model 7, Table 8.2, with R=0.50, the family heritability is estimated at 0.7123. The additive genetic variance is then estimated as 4 h^2_F $Var_p(\overline{r}_{i\cdot k\cdot})$ = 4 (0.7123) (0.02184) = 0.06222.

Elements G_{12} and G_{21} are the additive genetic covariances between VOL and R50 which are equivalent to the type B covariances between rust and volume that are zero. Thus elements G_{12} and G_{21} are zero.

The final $\mathbf{G}_k$ matrix is:

$$\mathbf{G}_k = \begin{bmatrix} 0.904 & 0 \\ 0 & 0.06222 \end{bmatrix}$$

Calculating Breeding Values

Using Eq. 4.2 (and assuming $\boldsymbol{\tau}_k=\mathbf{0}$) for parent 10356, $\hat{\mathbf{g}}_k = \mathbf{C}_k\mathbf{V}_k^{-1}(\mathbf{y}_k-\boldsymbol{\alpha}_k)$, and

$$\hat{\mathbf{g}}_k = \begin{bmatrix} 0.41 \text{ ft}^3 \\ 0.072 \end{bmatrix}$$

In the above vector, the predicted breeding values are centered around zero (the assumed $\boldsymbol{\tau}$), and are specific to the target environments specified. The element g_1 predicts that the breeding value of parent 10356 is 0.41 ft^3 of wood per tree above the average parent in a target environment that averages 4.21 ft^3 at age 15. The element g_2 is the predicted breeding value for rust, and indicates the breeding value of parent 10356 is 7.2% higher than the average genotype in a target environment of 50% rust infection. Remember that $E(g)=\boldsymbol{\tau}$, and represents the breeding value of the 'average' genotype. Note that we assumed $\boldsymbol{\tau}_k=\mathbf{0}$ and thus the breeding values were centered around zero. The predicted breeding values may be easier to interpret if a meaningful $\boldsymbol{\tau}$ (representative of the target environment) is added back to the predictions. For example, in the target environment the average genotype will produce 4.21 ft^3 of wood per tree at age 15, thus $\boldsymbol{\tau}$ for volume is 4.21 ft^3. In the target environment for rust, the average genotype will incur 50% infection, thus $\boldsymbol{\tau}$ is 50% (parent 10356 is not a C-lot parent, which would have a $\boldsymbol{\tau}$ of 34%). With this change in assumptions, $\hat{\mathbf{g}}_k$ from Eq. 4.2 is $\hat{\mathbf{g}}_k = \boldsymbol{\tau}_k + \mathbf{C}_k\mathbf{V}_k^{-1}(\mathbf{y}_k-\boldsymbol{\alpha}_k)$, and

$$\mathbf{g}_k = \begin{bmatrix} 0.41 + 4.21\ \text{ft}^3 \\ 0.072 + 0.500 \end{bmatrix} = \begin{bmatrix} 4.62\ \text{ft}^3 \\ 0.572 \end{bmatrix}$$

In the specified target environments, parent 10356 is predicted to produce 4.61 ft^3 of wood per tree at age 15, a gain of 9.7% = ((4.62 - 4.21)/4.21), and incur 57.2% rust infection, 7.2% more infection (57.2% - 50.0%) than the average genotype. Various measures of the precision of the predicted breeding values can be calculated using Eqs. 4.4, 4.6, and 4.7. These are interpreted as discussed in Chapters 4 - 7, and for parent 10356 are

EVPVOL = 0.586 ft^6	EVPR50 = 301 $\%^2$ = (0.0301) 100^2
Var(BLPVOL) = 0.318 ft^6	Var(BLPR50) = 321 $\%^2$ = (0.0321) 100^2
CORR(BLPVOL,VOL) = 0.593	CORR(BLPR50,R50) = 0.719

COMPUTER APPLICATION OF BLP

With the models and estimation procedures outlined above, it was possible to apply BLP to the CFGRP open-pollinated slash pine progeny test data. There were 1672 parents represented in 364 different tests. Each of the 1672 parents were in different subsets of the 364 tests, and thus a particular parent was likely to have unique $\mathbf{C}_k$ and $\mathbf{V}_k$ matrices. Filling in the $\mathbf{C}_k$ and $\mathbf{V}_k$ matrices can be quite complex.

> **Example 8.5**: Suppose parent k is represented in 4 tests in which volume and rust are measured, to make a total of 8 observations which will be entered into the $\mathbf{y}_k$ vector. We wish to predict a breeding value for two traits, volume at 15 years (VOL), and rust resistance (R50). The $\mathbf{V}_k$ matrix is a symmetric 8 x 8 matrix with 8 diagonal elements (variances of family means), and 28 different off-diagonal elements (covariances between two family means). The $\mathbf{C}_k$ matrix is an 8 x 2 matrix for a total of 16 elements. To construct the $\mathbf{V}_k$ and $\mathbf{C}_k$ matrices, a total of 52 second moments must be estimated from the data and models, and entered into the correct cells of the matrices. This process is repeated for every parent, some of which may have 40 or more observations.

Constructing $\mathbf{V}_k$ and $\mathbf{C}_k$ by hand for each of the 1672 parents would have been an extremely time consuming task, and the resulting matrices would most likely have been filled with errors. A short computer program using SAS/IML on the PC (Appendix 2) was written to build the matrices and calculate the breeding values for 1672 slash pine parents. The program uses a SAS data set which contains one line for each observation on each family: the deviations, $\mathbf{y}_k-\boldsymbol{\alpha}_k$, are already calculated, as are the variances of the family means, which will be be the diagonal elements of the $\mathbf{V}_k$ matrices. The program then calculates and enters the remaining elements of the $\mathbf{V}_k$ matrices and the $\mathbf{C}_k$ matrices, and calculates the breeding values.

PREDICTED BREEDING VALUES FOR SLASH PINE PARENTS

Range and Distribution

The predicted breeding values for volume (BLPVOL) range from -2.37 to 2.60 ft^3, with a mean very near the expected value of zero (mean = -.03 ft^3). The estimated genetic variance for VOL is 0.904 ft^6, and therefore approximately 96% of the true breeding values should lie between -1.90 and 1.90 ft^3. The predicted breeding values appear to be normally distributed with over 98% between -1.80 and 1.80 (Figure 8.3).

Best linear predictions of breeding value are estimates of the sum the additive effects of all pairs of alleles across all loci. Thus, one can interpret breeding values as the mean value of the progeny of a selfed cross of a parent (assuming no inbreeding depression). Recall that VOL was predicted for a target environment where the average tree would produce 4.21 ft^3 at 15 years. If its alleles were the gametic array used to generate a new population, a parent with BLPVOL equal to 1.0 ft^3 would yield offspring which would be expected to produce an average of 5.21 ft^3 per tree instead of 4.21 ft^3, a gain of 24%. Note that this prediction applies specifically to individual tree volumes in 15-year-old row-plot progeny tests.

Rust breeding values (R50) are predicted for a target environment of 50% rust hazard, i.e. a site where unimproved material incurs 50% infection. To aid in interpreting the values, after prediction the target environment mean (α) of 50% was added to each BLPR50. If a population was derived from the gametic array from a parent with an BLPR50 of 25%, that population would be expected to average 25% rust infection where an unimproved population would incur 50%. The BLPR50's of the 1672 parents range from -50 to 141% (mean = 50%). The estimated genetic

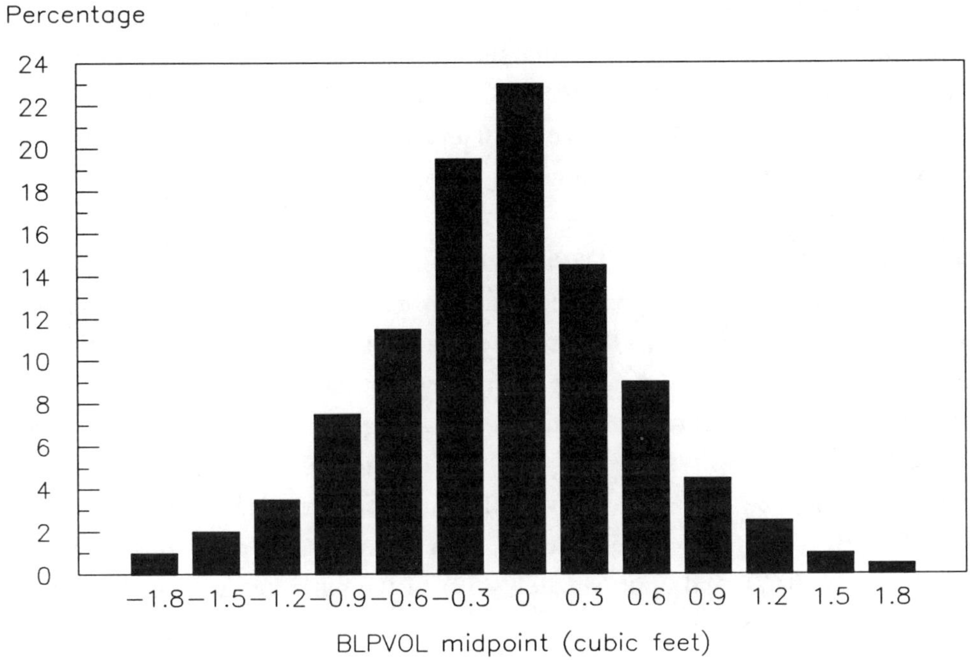

Figure 8.3. Frequency distribution of BLPVOL for 1672 parents.

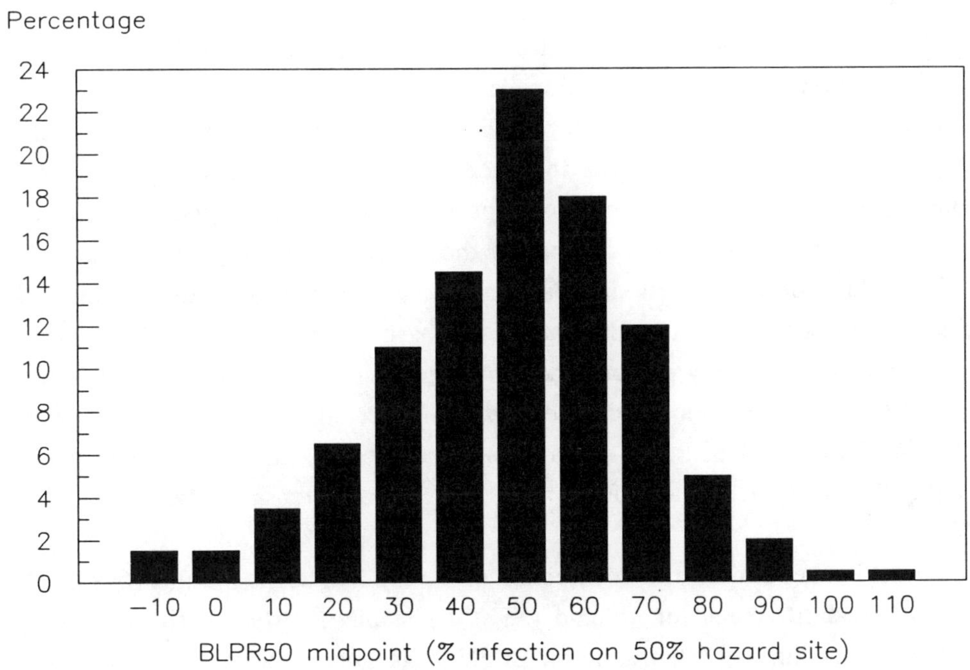

Figure 8.4. Frequency distribution of BLPR50 for 1672 parents.

variance for true R50 breeding values is 622, and thus approximately 96% of the true breeding values should lie between 0 and 100%. Over 98% of the predicted breeding values lie between 0 and 100% (Figure 8.4). Although the bounds to phenotypic expression of the trait of rust resistance are 0 and 100%, the underlying genetic values, which BLPR50 predicts, may quite logically exceed these limits.

Correlation of Predicted and True Breeding Values

One elegant way to express the relative precision of the predicted breeding values is the correlation between the predicted and the true breeding value ($Corr(g,\hat{g})$) (Eq. 4.7). As the quality and quantity of progeny test data increases, the variance of predictors ($Var(\hat{g})$) and $Corr(g,\hat{g})$ increase. For parents measured for rust in at least one test, $Corr(g,\hat{g})$ ranges from 0.44 to 0.88 (mean = 0.75). Thus, for the 1672 parents, approximately 56% of the variation in true breeding value R50 can be explained by BLPR50 ($R^2 = 0.75^2 = 0.56$). For volume, $Corr(g,\hat{g})$ ranges from 0.11 to 0.89 (mean = 0.67), and approximately 45% of the variation in VOL is explained by BLPVOL.

BENEFITS OF USING BLP

BLP breeding values are estimates of genetic value, and therefore can be used to calculate expected genetic gain. For example, if selected parents are grafted into a seed orchard, the average expected breeding value of the trees grown from orchard seed (the expected genetic gain) is the average of the BLP breeding values of the parents in the orchard (for a monoecious species, and assuming equal fecundity of all parents). This is true regardless of whether the selection is done on the basis of BLP breeding values or a classical approach such as averaged standard scores. One traditional approach to ranking parents on the basis of progeny test information has been to standardize family performance in each test, and average a family's scores across all tests. To examine differences between selecting parents from a population of 1396 on the basis of BLP and averaged standard scores, expected genetic gains for volume and rust were calculated for a range of selection intensities (White and Hodge 1988).

For 15-year volume growth, selection of the top half of the parents (699 of 1396) based on standard scores for volume (SSVOL) results in 91% of the genetic gain expected if selection had been done on BLPVOL (Figure 8.5). As selection intensity

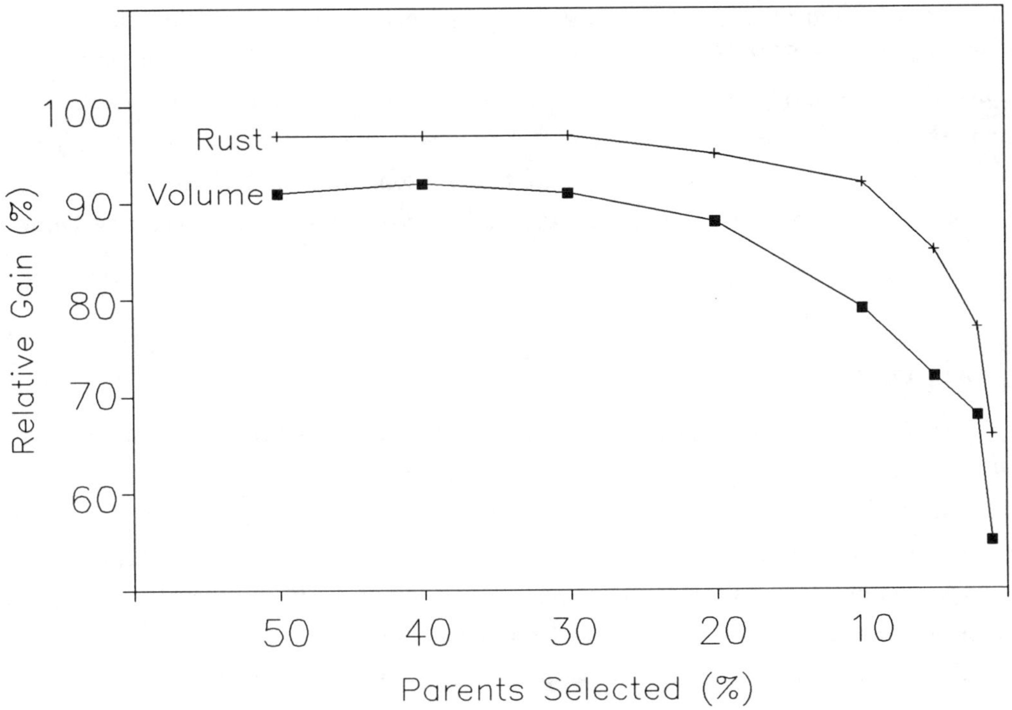

Figure 8.5. Relative genetic gain efficiency of parental selection based on standard score versus BLP breeding value for different levels of selection intensity.

increases, selection on SSVOL becomes progressively less efficient. For an intensity of 1%, selection on SSVOL is only 55% as efficient as on BLPVOL. The same pattern occurs when comparing BLPR50 and standard scores for rust (SSRUST). This pattern occurs primarily because of the tendency for variances of standard scores to be highest when parents are in few tests, and the contrasting tendency for BLP breeding values to have the highest variance when parents are in a large number of tests (see Chapter 5).

SUMMARY

BLP can be successfully applied to a wide variety of tree breeding programs, and will be of most advantage when dealing with diverse and messy data, i.e. different information available on different parents, and the quality of information varying widely from test to test. In theory, BLP is rather simple, but an operational application may be quite complex. To apply BLP operationally in a breeding program will require numerous decisions about the data to be utilized, how to estimate first and second moments, what genetic values will be predicted, and how to actually calculate the predictions. Often certain assumptions can or will be made to either simplify the process, or because no other approach is possible. For example we assumed data from a single progeny test was balanced and homogeneous. That is, differences between blocks were confined to first moments, while second moments were assumed identical across blocks. We also assumed all families were equally represented in the test. For other breeding programs, this assumption, and others that we made with our data, may not be appropriate. Rather different approaches might be required if the bulk of the data is from cross-pollinated progeny tests as opposed to open-pollinated tests as described in this chapter. Breeders should be aware of the implicit and explicit assumptions made when using BLP in order to evaluate whether they are appropriate and understand the consequences of violations of those assumptions.

Chapter 9
Selection Index Theory

INTRODUCTION

Previous chapters have emphasized the prediction of breeding values of parents based on the performance of their offspring. In practice these predicted breeding values are used to select the very best parents to carry forward in a breeding program and to select parents to retain in the production population (e.g. by roguing parents with low predicted breeding values from seed orchards). It is also often necessary to select among the individual offspring resulting from the matings among many parents. Here the objective is to select among individual plants or animals. These forward selections are then retained in the breeding program and possibly put into a production population (e.g. grafted into a seed orchard). The task is to rank all individuals available for selection according to some selection criterion so that those with the highest values of the criterion can be selected.

For each individual we may have information on several traits of varying economic importance and from several different sources (e.g. from the individual's siblings growing in different plots of the same progeny test or in different tests). We wish to develop a system of ranking all of the candidates that makes "optimal" use of all the data. It turns out that all of the theory developed in Chapter 4 applies directly to this problem. If all of the data are completely balanced (i.e. equal family sizes in tests of equal age and precision), then a single selection index can be developed that results in the best linear prediction of genetic worth values of all individuals. This is the Smith-Hazel index (Smith 1936, Hazel 1943) which has been commonly used in forestry (e.g. Baradat 1976, Namkoong 1976, Burdon 1979, 1982, Cotterill and Jackson 1981, Bridgwater et al. 1983, Harvey and Townsend 1985, Land et al. 1986, Talbert 1986). The coefficients of the index are the same for all individuals; so, it is common to first estimate the vector of coefficients and then multiply the observed data for each individual by the estimated coefficients to obtain a predicted genetic worth for that individual. These predicted genetic worths for all candidates serve as the selection criteria upon which to base selection. If the data are not balanced (i.e. some individuals have more, older, better, and/or divergent sources of observed data), then a single index with a single set of coefficients cannot possibly be developed that would be appropriate for use on all candidates. In these cases, best linear predictions of the candidates' genetic worths can be developed, but will require as many "selection indices" as there are classes of data. For example, suppose that there are two open-pollinated progeny tests with each test containing a different group of families. If each test is completely balanced, but the two tests differ in precision, age, and/or design, then two selection indices would be needed to predict the breeding values of candidates in test 1 and test 2, respectively. Once predicted, all candidates from both tests can be ranked together. These would be best linear predictions.

In theory then, selection index is equivalent to best linear prediction as long as the added assumptions of equal quality and quantity of data hold for all candidates. If type, quality or quantity of data vary among candidates, then use of a single selection index for all candidates (as is common) is not best linear prediction. The theory of classical selection index has been well described in the literature (for general discussions see: Henderson 1963, Lin 1978, Nordskog 1978, Falconer 1982, Bulmer 1985; for general forestry discussions see: Stonecypher and Arbez 1976, Burdon 1979, 1982, Harvey and Townsend 1985, Bridgwater and Squillace 1986, Land et al. 1986, Cotterill and Dean 1989). The classical selection index has been widely applied in breeding programs of many species (for animals see: Van Vleck 1970, 1976, Nordskog 1978; for plants see: Robinson et al. 1951, Elgin et al. 1970, Eagles and Frey 1974, Openshaw and Hadley 1984, Baker 1986; for forest trees see: Bridgwater and Stonecypher 1979, Burdon 1979, 1982, Shelbourne and Low 1980, Cotterill and Jackson 1981, 1985, Christophe and Birot 1983, Dean et al. 1983, 1986, Talbert 1984, 1986, Harvey and Townsend 1985, Land et al. 1986).

The purpose of this chapter is to provide a complete development of the theory and properties of selection indices using the same approach, nomenclature and terminology as in previous chapters. In one sense, this is somewhat redundant because the BLP theory of Chapter 4 completely applies to selection indices. However, this chapter is framed in terms of a single index developed for predicting genetic worths of individuals (as opposed to predicting breeding values of parents as in Chapter 4). This is the common way to present selection index theory and the complete correspondence with BLP theory of previous chapters is highlighted. In addition, we try to describe all assumptions which must hold for a single selection index to be best linear prediction and discuss some problems which arise in the actual use of selection indices. In this chapter we assume that the quality and quantity of information are the same for all candidates. The next chapter provides examples of common selection index applications and also discusses some possible approaches if the data are unbalanced or of unequal precision.

SELECTION INDEX THEORY

Specification of the Problem

In previous applications of BLP where emphasis was on predicting breeding values of parents (Chapters 5-8), the most common unit of observation has been a family mean obtained by averaging the data for a given family across all individuals and plots within a given progeny test (see discussion in Chapter 7 on the unit of observation). When the goal is selection of individuals, several levels of phenotypic data may be available (depending upon the experimental design) that provide information about any given individual. For example for an open-pollinated progeny test at several test locations, levels include the individual phenotype, the average of all individuals of that family in that block (i.e. plot mean), the average of all individuals of that family in that test (i.e. family mean), and the average of all individuals of that family at other test locations. Each level of data provides information about the individual of interest and each of these levels of information may exist for several traits of varying economic importance. The problem is then to predict the aggregate genetic worth of an individual where that genetic worth is aggregated over all traits with each trait weighted by its relative economic importance. In a heuristic sense, the index developed aims to weight each piece of data (i.e. each level of observation for each trait) in accord with its value in predicting the aggregate genetic worth of the individual.

Several vectors and matrices need to be defined to specify the problem. The definitions are identical to those in Chapter 4 with one subtle difference. In Chapter 4 the BLP problem was set up to make all predictions in one calculation and therefore all data from all candidates were included in each vector and matrix. For example, the **y** vector was n x 1 containing all observations on all candidates. This was imperative to the derivation. In the development of classical selection index theory where a single index is assumed to apply to all candidates, each vector and matrix can be specified for each individual candidate. This is because the index equation coefficients can be developed for any single individual candidate and then applied to all candidates (since a single index is assumed suitable). So, the **y** vector of observations is much smaller, m x 1, containing the data from any one individual candidate. With this in mind,

y is an m x 1 vector of phenotypic observations pertaining to a single candidate for selection (this may include several types of observations such as individual measurements and family means on several traits),

$\boldsymbol{\alpha}$ = E(**y**) is an m x 1 vector of expected values of the observed data pertaining to each candidate,

b is an m x 1 vector of index coefficients to be estimated,

g is a q x 1 vector of unobservable genetic values, and

a is a q x 1 vector of economic weights.

Both $\boldsymbol{\alpha}$ and **a** are vectors of constants while **y** and **g** are random vectors. We assume that E(**g**) = **0**, and that the following second moments are known (see Chapter 4 for further explanations of these second moments):

Var(**y**) = **V** is an m x m matrix of variances and covariances among the observations for a single candidate,

Var(**g**) = **G** is a q x q matrix of variances among the q genetic values in **g**, and

Cov(**y**,**g**) = **C** is an m x q matrix of covariances between the observations on a single candidate and the genetic values.

Then we want to predict the genetic worth, w, of each candidate where the genetic worth function is

$$\begin{aligned} w &= a_1g_1 + a_2g_2 + \ldots + a_qg_q \\ &= \Sigma a_ig_i \\ &= \mathbf{a}'\mathbf{g}. \end{aligned} \qquad \textbf{Eq. 9.1}$$

Note that the genetic worth (w) is the sum of genetic value (g_i) for each trait weighted by the economic importance (a_i) of that trait. Thus, w is equivalent to the aggregate genotype (h) discussed in BLP Property 5 of Chapter 4, and also in Topic 10 of Chapter 7. Similarly, the economic weights (a_i) are equivalent to the weights (m_i) discussed in those same sections. In this chapter, we have used w to denote aggregate genetic worth and **a** for the vector of economic weights to be consistent with the bulk of the selection index literature.

By restricting ourselves to linear functions of the observed data pertaining to each candidate, the selection index, I, becomes

$$
\begin{aligned}
I &= b_1(y_1-\alpha_1) + b_2(y_2-\alpha_2) + \ldots + b_m(y_m-\alpha_m) \\
&= \Sigma b_j(y_j-\alpha_j) \\
&= \mathbf{b}'(\mathbf{y} - \boldsymbol{\alpha}).
\end{aligned}
$$

Eq. 9.2

Note that each phenotypic observation is expressed as a deviation from its fixed effects (such as block and test location effects). The different y_j's are different levels of information for different traits measured on a given candidate. The index, I, is a linear function of the observed data and is the prediction of the genetic worth, w, of that individual. Thus, $I = \hat{w}$ and the goal is to develop a set of index coefficients such that the predicted genetic worths are very close to the true genetic worths. The problem then is to derive estimates of the coefficients, **b**, so that the predictions of genetic worth are "optimal" in some sense. This is identical to the best linear prediction derived in Chapter 4 except we have phrased the problem in terms of estimating the coefficients instead of predicting the breeding values. This way of framing the problem gives the appearance of limiting us to a single set of coefficients that will be used for each and every candidate regardless of the quality and quantity of data pertaining to that candidate. In actual practice, this is the way classical selection theory has most often been applied; however, as seen in Chapter 4 we could develop a different set of index coefficients for each candidate if the data are unbalanced. We discuss this further in later sections of this chapter.

Example 9.1: Suppose we have measured ten-year height and stem straightness in an open-pollinated progeny test consisting of 40 families planted in a randomized complete block design with 6 blocks and 10 trees per family in each block (i.e. 60 total trees per family). The goal is to predict the genetic worth of each tree in this test and then to select a subset of trees from this test on the basis of these predictions such that the trees selected have the highest predicted genetic worth. Suppose also that we know that the relative economic weights for ten-year height and stem straightness are a_h and a_s. Then the genetic worth of each tree in the test is:

$$w = a_h g_h + a_s g_s$$

where g_h and g_s are the breeding values of that tree for height and stem straightness at ten years. The index, I, to be calculated for each tree is:

$$I = b_1(h_{ijkl}-\alpha_1) + b_2(s_{ijkl}-\alpha_2) + b_3(\overline{h}_{ijk\cdot}-\alpha_3) + b_4(\overline{s}_{ijk\cdot}-\alpha_4) + b_5(\overline{h}_{i\cdot k\cdot}-\alpha_5) + b_6(\overline{s}_{i\cdot k\cdot}-\alpha_6)$$

where the observation y_{ijkl} is denoted h_{ijkl} if the trait is height, and s_{ijkl} if the trait is straightness, and similar notation is used for plot and family means. The expected values, α_i, are different elements for each individual tree measurement, plot mean or family mean. There are six index coefficients required, one for each of the six types of information available on each candidate (i.e. 3 levels of information for each of two traits).

Employing the linear model of Eq 2.23 which treats test location and block effects as fixed, the expected values can be calculated and we discuss this in detail in the next chapter. However, as an example, $h_{ijkl}-\alpha_1$ expresses each ten-year height measurement as a deviation from the block mean in which the candidate tree occurs.

For this example, **y** is a 6 x 1 vector of ten-year observations on each candidate and its relatives (i.e. the plot and family means are averaged over the candidate's siblings). The objective is to estimate **b**, a 6 x 1 vector of coefficients, that will assign "optimal" weights to the various observations to predict w for each candidate. Once the six coefficients are estimated, they are used to calculate the genetic worth of each of the 2400 trees in the test by calculating the index 2400 times using the appropriate data from each tree. In this example, the 6 x 6 **V** matrix would contain variances of individual measurements, plot means and family means as diagonal elements and covariances among these observations for off-diagonal elements. The 2 x 2 **G** matrix contains additive genetic variances and covariances and the 6 x 2 **C** matrix contains covariances between the six types of observations on each candidate and the two breeding values for that candidate e.g. $\text{cov}(h_{ijkl},g_h)$ is the first element.

Derivation of Selection Index Formula

Remember that the classical selection index formula is best linear prediction (with the added requirement that one index is suitable for all candidates because the data are completely balanced and of equal types, ages and precisions for all candidates). Thus, the theory of BLP in Chapter 4 is completely applicable and we really need no new formula. However, we present a slightly different derivation here that is more suitable to the selection index problem specified above (see Henderson

1963, Lin 1978). Estimates of the index coefficients are derived such that either the correlation between the genetic worth and the index, i.e. Corr(w,I), is maximized or such that the expected squared deviations between the genetic worth and the index, i.e. $E(w - I)^2$, is minimized. Both approaches result in the same estimates for the coefficients; the latter approach was taken in Chapter 4 so here we proceed with the former. By definition of a correlation (see Eq. 2.15):

$$\text{Corr}(w,I) = \text{Cov}(w,I)/[\text{Var}(w)^{1/2}\ \text{Var}(I)^{1/2}]\ .$$

By substituting from Eqs. 9.1 and 9.2, each of the three terms in this correlation are evaluated in turn.

$$\begin{aligned} \text{Cov}(w,I) &= \text{Cov}[\mathbf{a}'\mathbf{g},\mathbf{b}'(\mathbf{y}\text{-}\boldsymbol{\alpha})] & \\ &= \text{Cov}(\mathbf{a}'\mathbf{g},\mathbf{b}'\mathbf{y}) & \text{(since } \boldsymbol{\alpha} \text{ is a vector of constants)} \\ &= \mathbf{a}'\text{Cov}(\mathbf{g},\mathbf{y})\mathbf{b} & \text{(from Eq. 2.31)} \\ &= \mathbf{a}'\mathbf{C}'\mathbf{b} & \text{(by the definition of } \mathbf{C}\text{)} \\ &= \mathbf{b}'\mathbf{C}\mathbf{a}. & \qquad\qquad \textbf{Eq. 9.3} \end{aligned}$$

The last step makes use of the fact that **a'C'b** is a scalar (which it must be since if w and I are both scalars then so must be the covariance) and that the transpose of a scalar is itself.

For the next term in the correlation,

$$\begin{aligned} \text{Var}(w) &= \text{Var}(\mathbf{a}'\mathbf{g}) & \text{(from Eq. 9.1)} \\ &= \mathbf{a}'\text{Var}(\mathbf{g})\mathbf{a} & \text{(since } \mathbf{a} \text{ is a vector of constants)} \\ &= \mathbf{a}'\mathbf{G}\mathbf{a}. & \qquad\qquad \textbf{Eq. 9.4} \end{aligned}$$

For the final term,

$$\begin{aligned} \text{Var}(I) &= \text{Var}[\mathbf{b}'(\mathbf{y}\text{-}\boldsymbol{\alpha})] & \text{(from Eq. 9.2)} \\ &= \text{Var}(\mathbf{b}'\mathbf{y}) & \text{(since } \boldsymbol{\alpha} \text{ is a vector of constants)} \\ &= \mathbf{b}'\text{Var}(\mathbf{y})\mathbf{b} & \text{(since } \mathbf{b} \text{ is a vector of constants)} \\ &= \mathbf{b}'\mathbf{V}\mathbf{b}. & \qquad\qquad \textbf{Eq. 9.5} \end{aligned}$$

Then,

$$\text{Corr(w,I)} = \mathbf{b'Ca}/[(\mathbf{a'Ga})^{½}(\mathbf{b'Vb})^{½}].$$

Maximizing the log of this expression is equivalent to maximizing the correlation itself and the calculus proceeds more smoothly. So,

$$\ln[\text{Corr(w,I)}] = \ln(\mathbf{b'Ca}) - ½\ln(\mathbf{a'Ga}) - ½\ln(\mathbf{b'Vb}).$$

The next step is to take the partial derivatives of this expression with respect to the index coefficients, **b**. For this step, the following results from calculus are used. First, $\partial(\ln x)/\partial x = 1/x$ for any variable x. Second, the chain rule states $\partial(\ln cx^2)/\partial x = 2cx/cx^2$. The third term in the above correlation, ½ln(**b'Vb**), contains a quadratic function in **b** and **b** is analogous to x in the chain rule. Finally, the second term, ½ln(**a'Ga**), contains only constants when the partial derivatives are taken with respect to **b**; its derivative is therefore **0**. Thus,

$$\frac{\partial\ \{\ln[\text{Corr(w,I)}]\}}{\partial\ \mathbf{b}} = \frac{\mathbf{Ca}}{\mathbf{b'Ca}} - \frac{½(2\mathbf{Vb})}{\mathbf{b'Vb}}$$

Remember that division is <u>not</u> defined in matrix algebra (multiplication by the inverse is the analogous operation). However, both **b'Ca** and **b'Vb** are scalars, so this is equivalent to the operations of division and multiplication by scalars defined in Chapter 1.

Equating this expression to zero and rearranging terms,

$$\mathbf{Vb} = \mathbf{Ca} \cdot (\mathbf{b'Vb}/\mathbf{b'Ca}).$$

The last term on the right, **b'Vb/b'Ca**, is a scalar and can be dropped without affecting the proportionality of the b's. That is, suppose for some solution to the above expression, the last term on the right turns out to be some scalar constant, say c. Then the coefficients (i.e. the **b** vector) will all be multiplied by c, so the relative proportions of the coefficients to each other do not depend on c. So we can drop the last term (which is equivalent to setting it to 1) and

$$\mathbf{Vb} = \mathbf{Ca}.$$

The solution is

$$\mathbf{b} = \mathbf{V}^{-1}\mathbf{Ca} . \qquad \textbf{Eq. 9.6}$$

Thus, **b** is an m x 1 vector of index coefficients chosen to maximize the correlation between the actual and predicted genetic worth.

To highlight the equality of Eq. 9.6 with the BLP coefficient estimates of Eq. 4.2, rewrite the index equation (Eq. 9.2) with the coefficient estimates from Eq. 9.6,

$$\begin{aligned}\hat{w} &= \mathbf{b}'(\mathbf{y} - \boldsymbol{\alpha}) \\ &= \mathbf{a}'\mathbf{C}'\mathbf{V}^{-1}(\mathbf{y} - \boldsymbol{\alpha}). \qquad \textbf{Eq. 9.7}\end{aligned}$$

This is identical to the BLP Eq. 4.2 except that 1) the premultiplication by **a** above creates a linear combination of the predicted genetic values into a single aggregated genetic worth (Property 5 in Chapter 4 shows this for BLP) and 2) the vectors and matrices in Equations 9.6 and 9.7 apply to a single candidate and are therefore much smaller than those of BLP Eq. 4.2 which apply to all observations on all candidates. If the data were unbalanced, different **C**, **V** and $\boldsymbol{\alpha}$ could be used for each candidate and so a different set of coefficients developed for each candidate. Thus in theory, BLP and selection index are identical; selection index is best linear prediction. In practice, BLP develops a different set of coefficients for each candidate, while selection index makes the added assumption that a single set of coefficients is appropriate for all candidates. While some authors (Henderson 1973, 1977) use the terms synonomously, we prefer to think of selection index as a special case of BLP.

Properties of Selection Indices

The properties of selection indices are those detailed for BLP in Chapter 4; they are described here in the context of selection indices for predicting genetic worth of individual candidates. These properties depend on the following assumptions holding: a) the first and second moments (i.e. $\boldsymbol{\alpha}$, **C** and **V**) are known constants and therefore do not require estimation from the data, b) the data are completely balanced and of equal type, quality and quantity for each and every candidate (so that a single selection index is suitable), and c) the economic weights are accurate. In practice the first and second moments must be estimated from the data and when the data are not completely balanced, more than one selection index is required. If the economic weights are misspecified, selection index is still best linear prediction (if the first two assumptions hold), but it is best linear prediction of the wrong genetic worth function. We present the properties of selection index here as if these three sets of assumptions hold and examine the assumptions in more detail in the section, Topics in Selection Index Theory.

The predictions of genetic worth using a selection index are unbiased and minimize the error variance of predictions among all possible linear functions of the observed data. If the joint distribution between **g** and **y** is multivariate normal, then a) selection index provides the best (i.e. minimum error variance) predictions among all possible functions and transformations (not just linear functions) of the data, b) selection on the basis of the index maximizes (over repeated sampling) the probability of selecting the better of any two candidates, and c) selection on the basis of the index maximizes (over repeated sampling) the expected genetic progress for a fixed number of selections made. Other useful properties are detailed below.

1. The variance among the predictions can be estimated: Applying the index to all candidates will result in a spread among the predictions of genetic worth, w_i. It turns out that the predictions will tend to spread out more (i.e. have a larger variance) as they become more precise. If there are small amounts of data on each candidate (e.g. few blocks, few individuals per block, etc) or the data are of low quality (e.g. high amounts of experimental error), then the variance among predictions of genetic worth is smaller; that is the predictions will tend to cluster more closely around zero. This variance is a scalar corresponding to each and every prediction. Since the calculated index for each candidate serves as the prediction of genetic worth, $I = \hat{w}$ and therefore $\mathrm{Var}(I) = \mathrm{Var}(\hat{w}) = \mathbf{b'Vb}$. This is calculated in Eq. 9.5 but is repeated here now that we have an estimate of **b**. Note that the units of this variance will be the square of the units of I.

Derivation:

$$\begin{aligned}
\mathrm{Var}(\hat{w}) &= \mathrm{Var}(I) = \mathrm{Var}[\mathbf{b}'(\mathbf{y} - \boldsymbol{\alpha})] \\
&= \mathrm{Var}[\mathbf{a'C'V^{-1}}(\mathbf{y} - \boldsymbol{\alpha})] && \text{(from Eq. 9.7)} \\
&= \mathbf{a'C'V^{-1}}[\mathrm{Var}(\mathbf{y} - \boldsymbol{\alpha})]\mathbf{V^{-1}Ca} && \text{(using Eq. 2.30)} \\
&= \mathbf{a'C'V^{-1}VV^{-1}Ca} && \text{(since } \mathbf{V} = \mathrm{Var}(\mathbf{y}) \text{ \& } \boldsymbol{\alpha} \text{ constant)} \\
&= \mathbf{a'C'V^{-1}Ca} && \textbf{Eq. 9.8} \\
&= \mathbf{b'Vb} && \text{(from Eq. 9.6).}
\end{aligned}$$

2. The covariance between the predicted and true genetic worth can be estimated: This covariance, Cov(w,I), is a scalar indicating the correspondence between the predicted genetic worth (i.e. I = $\hat{w}$) and the true genetic worth, w, that would be expected upon repeated sampling. It is useful for calculating Property 4 below. It has units that are the square of the units of I because I and w have the same units. This is derived above in Eq. 9.3, but is shown here to demonstrate the equality of Cov(w,I) and Var(I) for the estimation formula for **b** shown in Eq. 9.6. Depending upon convenience, any of the last four lines of the derivation below may be used for the calculation of this covariance.

Derivation:

$$
\begin{aligned}
\mathrm{Cov}(w,\hat{w}) &= \mathrm{Cov}(w,I) \\
&= \mathrm{Cov}[\mathbf{a'g},\mathbf{a'C'V^{-1}}(\mathbf{y} - \boldsymbol{\alpha})] && \text{(substituting from Eqs. 9.1,9.7)} \\
&= \mathbf{a'}\{\mathrm{Cov}[\mathbf{g},(\mathbf{y} - \boldsymbol{\alpha})]\}\mathbf{V^{-1}Ca} && \text{(using Eq. 2.31 with } \mathbf{a}, \mathbf{C} \,\&\, \mathbf{V} \text{ constants)} \\
&= \mathbf{a'C'V^{-1}Ca} && \text{(by definition of C, } \boldsymbol{\alpha} \text{ constant)} \\
&= \mathbf{b'Ca} && \text{(from Eq. 9.6)} \\
&= \mathbf{b'Vb} && \text{(Substituting again using Eq. 9.6)} \\
&= \mathrm{Var}(I) = \mathrm{Var}(\hat{w}).
\end{aligned}
$$

3. The error variance of the predictions can be estimated: The estimated error variance of a prediction, Var($\hat{w}$ - w), that would occur in repeated sampling can be calculated. It provides information about the precision of the prediction. Since all candidates' genetic worths are calculated with the same index, all have the same error variance. The error variance responds in the opposite direction to the variance among the predictions (Property 1, above) in that the error variance decreases as the quality and quantity of the data increases. The last three lines of the derivation provide alternative computational formulae.

Derivation:

$$
\begin{aligned}
\mathrm{Var}(\hat{w}-w) &= \mathrm{Var}(I - w) \\
&= \mathrm{Var}(I) + \mathrm{Var}(w) - 2\mathrm{Cov}(I,w) && \text{(from Eq. 2.11)} \\
&= \mathbf{b'Vb} + \mathbf{a'Ga} - 2\mathbf{b'Vb} && \text{(from Property 2 \& Eq. 9.4)} \\
&= \mathbf{a'Ga} - \mathbf{b'Vb} && \textbf{Eq. 9.9} \\
&= \mathbf{a'Ga} - \mathbf{a'C'V^{-1}Ca} && \text{(Substituting for } \mathbf{b}) \\
&= \mathbf{a'[G - C'V^{-1}C]a}.
\end{aligned}
$$

4. Correlations between true and predicted genetic worths can be estimated: This correlation, Corr(w,I) = Corr(w,$\hat{w}$), will be a maximum among all possible linear combinations of the data because this was the basis of the derivation of Equation 9.6. It is often useful to actually estimate this correlation as an easily interpreted measure of precision of the predictions. If large quantities of high quality data are used, then this will tend to be nearer 1.

Derivation:

$$\begin{aligned} \mathrm{Corr}(w,\hat{w}) &= \mathrm{Corr}(w,I) \\ &= \mathrm{Cov}(w,I)/[\mathrm{Var}(w)\mathrm{Var}(I)]^{½} \\ &= \mathrm{Var}(I)/[\mathrm{Var}(w)\mathrm{Var}(I)]^{½} \qquad \text{(from Property 2, above)} \\ &= [\mathrm{Var}(I)/\mathrm{Var}(w)]^{½} \\ &= [\mathbf{a'C'V^{-1}Ca/a'Ga}]^{½}. \qquad \textbf{Eq. 9.10} \end{aligned}$$

5. Expected genetic progress can be estimated for the index: The expected progress from one generation of truncation selection (ΔI, in units of the index) on the basis of the predicted genetic worths can be estimated. If the index value, I_i, is calculated for each candidate and a subset of the candidates with the highest index values are retained, then the gain in the index is shown below to be I_s, where I_s is the average index value for the selected candidates. This expresses gain above the previous population average (since we assumed that E(**g**) = **0**). That is, if this is the third generation of selection, then this expresses gain above the second generation, not above the initial unselected population. Note that $\boldsymbol{\alpha}$, **C** and **V** must apply to the entire population of third generation candidates. For the derivation below, I_t is the average index value across all candidates and $i_s = (I_s - I_t)/[\mathrm{Var}(I)^{½}]$ is defined as the intensity of selection (Falconer 1982). Note that $I_t = 0$ because $E[\mathbf{b'}(\mathbf{y} - \boldsymbol{\alpha})] = 0$ for each and every candidate; so the index values averaged over all candidates will be zero. Also, from Becker (1975), we adapt the general expression for gain, Gain = [Cov(selection units, genetic worth)/Var(selection units)] • selection differential. The last two lines of the derivation provide alternative computational formulae. I_s is the predicted gain in the index obtained by averaging the index values of all selected individuals. This can be calculated only after the data have been collected and the index applied. The last line states that the gain in the index can also be calculated in the absence of real data if i_s and Var(I) are known. With the further assumption of normality, Falconer (1982) describes how i_s is related to the percentage of candidates selected (which can be anticipated in advance of selection).

Derivation:

$$\begin{aligned} \Delta I &= [Cov(w,I)/Var(I)] \cdot (I_s - I_t) && \text{(from Becker, 1975)} \\ &= I_s - I_t && \text{(from Property 3, above)} \\ &= I_s && \textbf{Eq. 9.11} \\ &= i_s \cdot Var(I)^{1/2}. \end{aligned}$$

6. Genetic progress for each individual trait in the index can be estimated:
Selection on the basis of an index results in genetic changes in each of the individual traits and the magnitude and direction of these changes is not always immediately apparent by inspection. Thus, it is always necessary to examine the genetic progress made in each trait to ensure that selection on the basis of the index has not resulted in unacceptable change (e.g. in the wrong direction) in one or more of the traits. For a group of selected individuals whose average selection index value is I_s, the estimated genetic progress in each trait is derived below as a q x 1 vector, $\hat{\mathbf{g}}_s$. There are q elements in $\hat{\mathbf{g}}_s$ because there are q traits in the genetic worth function, w. The ith element of $\hat{\mathbf{g}}_s$ is the predicted genetic value of the ith trait (in units of the ith trait) averaged over the selected group of individuals. This averaged genetic value for each trait is expressed as a deviation from the unselected mean in the units of that trait. The goal of this derivation, then, is to predict the vector of genetic values when the selection units are the index values for all individual candidates. This is the problem in Property 5 except that instead of predicting gain for the aggregate genetic worth, w, we predict gain for the vector of traits in the genetic worth. Again, i is the intensity of selection, $i_s = I_s/Var(I)^{1/2}$ and the selection differential on the index is $I_s - I_t = I_s$ (since $I_t = 0$). The last five lines below represent alternative computational formulae.

Derivation:

$$\begin{aligned} \hat{\mathbf{g}}_s &= [Cov(\mathbf{g},I)/Var(I)] \cdot I_s && \text{(adapted from Becker, 1975)} \\ &= \{Cov[\mathbf{g}, \mathbf{a'C'V^{-1}}(\mathbf{y} - \boldsymbol{\alpha})]/Var(I)\} \cdot I_s && \text{(from Eq. 9.7)} \\ &= \{[Cov(\mathbf{g,y})\mathbf{V^{-1}Ca}]/Var(I)\} \cdot I_s && (\mathbf{a, C, V}, \& \ \boldsymbol{\alpha} \text{ are constants, Eq. 2.31}) \\ &= \{[\mathbf{C'V^{-1}Ca}]/Var(I)\} \cdot I_s && \text{(by the definition of } \mathbf{C}) \\ &= \{[\mathbf{C'V^{-1}Ca}]/Var(I)\} \cdot i_s \cdot Var(I)^{1/2} && \text{(substituting for } I_s) \end{aligned}$$

$$= i_s \cdot \{[\mathbf{C'V^{-1}Ca}]/\mathrm{Var}(I)^{½}\} \qquad \textbf{Eq. 9.12}$$
$$= i_s \cdot [\mathbf{C'b}/\mathrm{Var}(I)^{½}] \qquad \text{(from Eq. 9.6)}$$
$$= i_s \cdot [\mathbf{C'b}/(\mathbf{b'Vb})^{½}] \qquad \text{(from Eq. 9.5).}$$

TOPICS IN SELECTION INDEX THEORY

Determination of the Economic Weights

The q weights, $a_1, a_2, \ldots a_q$, in Equation 9.1 are usually defined relative to each other and a_i is the relative value of a one unit increase in the ith trait. For example, suppose there are two traits in the genetic worth function, height (in meters) at 20 years and rust resistance (in %) at 20 years. Say that a growth and yield model projects that a 1 m increase in height at 20 years is equivalent (in dollars or yield) to a 10% decrease in the incidence of rust. Because of the scale in which the two traits are measured (i.e. m and %), a unit gain in rust is only 1/10 as valuable as a unit gain in height. In this case, we want the ratio of the weights of height to rust, a_h/a_r to be 10. Setting $a_h = 1$, then $a_r = 0.1$; or setting $a_r = 1$, $a_h = 10$. If height were measured in feet instead of meters, the relative weights would change. One foot is equivalent to 0.3048m so now one unit (i.e. 1 foot) increases in height are easier to come by so they must be less valuable than one unit increases in height when measured in meters. Now, the ratio of $a_h/a_r = 3.048$.

The process of determining relative economic weights is beyond the scope of this book, but several authors have discussed various approaches (Bridgwater and Stonecypher 1979, Busby 1983, Cotterill and Jackson 1985, Bridgwater and Squillace 1986). For various reasons, reliable economic weights are often difficult to determine and the inability to make precise estimates of economic weights is sometimes cited as a reason to avoid using selection indices. We agree with the conclusion that the difficulty in estimating economic weights is not an argument against use of selection indices. Regardless of the weights that are chosen, the index results in a best linear prediction of that linear function of the genetic values. That is, suppose two sets of weights are chosen, $\mathbf{a'}_1 = \{a_{11} \ a_{12} \ldots a_{1q}\}$ and $\mathbf{a'}_2 = \{a_{21} \ a_{22} \ldots a_{2q}\}$. Then there will be two definitions of genetic worth, w_1 and w_2, and two sets of index coefficients from Equation 9.6, $\mathbf{b}_1$ and $\mathbf{b}_2$. So, $\hat{w}_1$ (obtained by applying $\mathbf{b}_1$ in Equation 9.7) is the best linear prediction of w_1, and $\hat{w}_2$ is the best linear prediction of w_2 (assuming all the assumptions for selection indices hold).

If precise economic weights are difficult to determine, then several vectors of weights can be specified that span the range of reasonable economic values. This is analogous to conducting a sensitivity analysis in economics. For each alternative vector of weights, Equation 9.7 can be calculated for each candidate for selection. Individuals that are outstanding for all or most of the alternative weighting schemes have a high predicted genetic worth for all or most of the economic scenarios. These individuals should be selected.

While the number of calculations involved in predicting genetic worth for many different vectors of weights may appear formidable, it is actually quite feasible. In Equation 9.7, $\mathbf{C'V^{-1}(y - \alpha)}$ can be calculated once for each candidate for selection; this results in a q x 1 vector, $\hat{\mathbf{g}}$, of predicted genetic values for the q traits for that individual (see Property 5 of Chapter 4). Once calculated, these predicted genetic values can be easily aggregated into numerous genetic worth functions by choosing numerous weight vectors with which to premultiply $\hat{\mathbf{g}}$. For example, $\hat{w}_1 = \mathbf{a'}_1\hat{\mathbf{g}}$ and $\hat{w}_2 = \mathbf{a'}_2\hat{\mathbf{g}}$. This type of sensitivity analysis seems useful when economic weights are difficult or impossible to determine.

Multiple sets of economic weights may also be needed if there are multiple breeding objectives (see Namkoong 1976, Bridgwater and Squillace 1986). For example, if a breeding cooperative composed of several organizations desired to breed for both pulpwood and solid wood products, the cooperative could 1) develop a compromise set of weights, 2) develop two indices and only select candidates which are outstanding for both, or 3) develop two indices and use them to select for two different subpopulations or varieties (i.e. one variety for pulpwood and one for solid wood).

Estimating Fixed Effects

When developing a selection index and estimating the coefficients, some authors use estimated effects as the observations rather than means. For example, in the linear model (Eq. 2.23) for an open-pollinated progeny test, there is a plot effect, p_{ijk}, associated with the plot containing trees from family k in block j and location i. We could either use as observations, the plot mean or an estimate of p_{ijk}. The latter would require subtracting and adding several terms to the plot mean. Burdon (1982) and Harvey and Townsend (1985) describe applications of selection index using estimated effects rather than means.

To be consistent with all applications of BLP and linear models in this book, we always treat $\mathbf{y}$ as unadjusted observations (i.e. not expressed as deviations). Types of observations then include individual measurements (i.e. phenotypic observations not corrected for the mean), plot means, family means for a given site, and family means averaged across all sites. Index Equation 9.2 assumes that the fixed effects, $\boldsymbol{\alpha}$, are known and these are used to express the observations as deviations from their expected values. In practice, these are never known and rather are estimated from the data. To estimate fixed effects, $\boldsymbol{\alpha}$, it is necessary to express each type of observation in **y** in terms of a specific linear model, take the expected value of each type of observation, and then find a mean from the data that is a precise estimate of this expected value. For the plot mean of family k in block j, i.e. $\bar{y}_{ijk\cdot}$, the expected value is $\mu + E_i + B_j$ and an estimate of this is the jth block mean (averaged overall families in block j, i.e. $\bar{y}_{ij\cdot\cdot}$). Many examples of obtaining estimates of the fixed effects are presented in the next chapter.

Finite Sample Size Corrections for Second Moments

Because the fixed effects are estimated from the data rather than known without error, variances and covariances required for the **V** matrix are affected by sampling. That is, the elements appropriate for the **V** matrix are defined as the variances and covariances associated with the phenotypic data. In practice these are always observations expressed as deviations from their estimated fixed effects (i.e. $\mathbf{y} - \hat{\boldsymbol{\alpha}}$ is substituted for $\mathbf{y} - \boldsymbol{\alpha}$ in Eqs. 4.2 and 9.2). For Chapters 4-8 we have considered the fixed effects as known, not estimated, and then $\text{Var}(\mathbf{y} - \boldsymbol{\alpha}) = \text{Var}(\mathbf{y}) = \mathbf{V}$. This is because the fixed effects are considered constants and thus have no variance. However, to be more precise, the fixed effects are estimated, so $\hat{\boldsymbol{\alpha}}$ is a random variable and $\text{Var}(\mathbf{y} - \hat{\boldsymbol{\alpha}}) = \text{Var}(\mathbf{y}) + \text{Var}(\hat{\boldsymbol{\alpha}}) - 2\text{Cov}(\mathbf{y},\hat{\boldsymbol{\alpha}})$. This latter expression requires that the sampling precision associated with the estimates of the fixed effects be factored into the variances used in the **V** matrix. This means that finite sample size corrections are needed to alter the usual variances and covariances we have used that assume infinite sample sizes. Methods for deriving finite sample size correction factors are presented in Chapter 2 and their use in selection indices is detailed elsewhere: a) for general applications Falconer (1981) and Bulmer (1985); for forestry applications Burdon (1982), Harvey and Townsend (1985) and Land et al. (1986).

For example applications of selection indices, we continue to assume that the fixed effects are known constants and that variances and covariances can therefore be written without finite sample size corrections. This assumption allows the needed variances and covariances to be written much more simply and hopefully facilitates presentation of the concepts. As along as sample sizes are fairly large, this assumption is a good one. In particular, if the number of unrelated families in a test is more than 20, most sample size corrections will have minimal influence on the variances and covariances.

Consequences of Error in Estimates of Second Moments

In practice, elements of the **C** and **V** matrices will be estimates of the variances and covariances rather than the known population parameters themselves. As discussed in Chapter 6, rankings based on single trait breeding value predictions seem relatively insensitive to errors in estimating these parameters (Henderson 1978); however, much more work is needed in this area. With multiple trait indices, it appears that the rankings based on predicted genetic worth will be much more sensitive to errors in estimating **C** and **V** if the traits in the index are negatively correlated than if they are positively correlated (Harris 1963, 1964, Sales and Hill 1976, Bulmer 1985).

These simulation studies just cited deal with simple experimental designs and are not directly applicable to most situations in forestry. It would appear, however, that the size of most data sets available for estimating variances and covariances in forestry provide estimates of the parameters that are more reliable than many breeding disciplines (such as breeding dairy cows). And, based on these simulation studies, we believe that predictions of genetic worth can be relatively precise; however, simulation studies appropriate to forest tree experimental designs are needed.

Analogous to the procedure described for assessing sensitivity to errors in estimating economic weights, a range of estimates for the **C** and **V** matrices can be used to generate a several predictions of genetic worth for each candidate. Candidates which rank at the top for the most of the **C** and **V** estimates can be interpreted as winners.

Restrictions on Selection Indices

If two traits in an index are adversely correlated, then gain in one trait may be made while the other actually declines in genetic value. For example, Dean et al. (1986) reported index selection in *P. caribaea* in Australia based on an index incorporating stem diameter, straightness and branch diameter. Straightness and branch diameter are positively correlated in *P. caribaea*, but both of these form traits are negatively correlated with the growth trait, stem diameter. When the economic emphasis was placed on growth, strong positive gains were made in stem diameter, but only at the expense of both form traits. That is, selection on the basis of an index placing high weight on growth resulted in genetic deterioration in the form traits.

Unless the economic weights are very precise and are assurably stable through time and across market conditions, it may be unacceptable to allow genetic deterioration in any trait. In this type of situation, the index can be derived with a restriction (Kempthorne and Nordskog 1959) which assures no change (zero genetic gain) in one trait. Cotterill and Dean (in press) give an excellent accounting of the rationale behind restricted indices and have reported on the their use (Cotterill and Jackson 1981, Dean et al. 1986). In the above example, they developed a restricted index that ensured no genetic change in branch diameter. This resulted in a different set of index coefficients and different candidates selected. Selection on the basis of this restricted index, in fact, resulted in no change in branch diameter. Gain in the growth trait, stem diameter, was reduced to half that from selection based on the unrestricted index. So, use of the restricted index ensured no negative genetic change in the restricted trait (branch diameter), but only at the expense of reduced gain in the growth trait.

As seen above, use of a restricted index does not come free. There is reduced gain for some other traits in the index compared to that from use of an unrestricted index. Independent culling could also be used for a similar purpose. In the case of the example above, the unrestricted (high growth emphasis) index could have been used to predict genetic worth for each candidate, but then a culling level established for branch diameter. Candidates would have to meet or exceed the culling level to be selected regardless of their predicted genetic worth. The value of the restricted index is that it maximizes gain in the other traits subject to the restriction of no genetic change for the restricted trait. So, by using the restricted index above to ensure no genetic change in branch diameter, the loss in stem diameter (the growth trait) was minimized.

Index Traits Not in the Genetic Worth Function

In forestry, it is often useful to make predictions for a linear function of mature genetic values when the traits measured (and therefore in the index) are juvenile. Suppose, for example, that we measure height and straightness in a series of progeny tests that are 10-years old, but wish to predict genetic worth at a harvest age of 30 years. The traits and economic weights in the genetic worth function must then be at harvest age. The traits of interest might be harvest-age volume and straightness. So, the problem then is to predict for each candidate for selection, a genetic worth composed of harvest-age volume and straightness based on an index from measurements on 10-year height and straightness.

This type of indirect selection problem has been discussed in previous chapters for BLP of parental breeding values and in the literature in terms of selection index of individuals (Binet 1965, Lin 1978, Cotterill and Dean in press). No new theory is required. All matrices and vectors remain as specified at the beginning of the chapter: 1) phenotypic observations of the measured traits form the **y** vector (for the example above, individual measurements, plot means and family means of the 10-year measurements are the elements of **y**), 2) $\boldsymbol{\alpha}$ is the vector of expected values of these phenotypic observations, 3) **V** is the variance-covariance matrix among these measured phenotypic observations and 4) **C** is the matrix of covariances between the genetic values of the traits being predicted (the 30-year genetic values) and the phenotypic observations (the 10-year data).

The **y**, $\boldsymbol{\alpha}$ and **V** matrices all deal with the data as measured and any pertinent traits (i.e. traits correlated with those in the genetic worth function being predicted) can conceivably be included. The **C** matrix changes depending upon which traits are in the genetic worth function. In practice it is often necessary to estimate the covariances needed for the **C** matrix using genetic theory and external (to the measurements) sources of data. For example, the 10-year test measurements will not provide estimates of covariances between 10-year measurements and 30-year genetic values. However, there may be data available from other tests to provide these estimates. The process of estimating these covariances is discussed in Chapters 5,6,7, and 8 in the context of BLP of parental breeding values and in the next chapter for selection indices.

Selection indices for this type of indirect selection have several uses. First, use of harvest-age economic weights dictates use of harvest-age traits in the genetic worth function and this is certainly the ultimate breeding objective in many tree improvement programs. Another use of indirect selection is in selection from tests measured at different ages. Suppose there are 10-year data from one progeny test containing individuals from 30 open-pollinated families and 15-year data from another test with trees from 25 different open-pollinated families. Making selections of the best trees from all 55 families is greatly facilitated by developing two selection indices, one for each test, that predict genetic worth in terms of the same harvest-age traits. Then the data from each test gets weighted in accord with the ultimate breeding objective (gain in mature genetic worth). Since the predictions of genetic worth from Equation 9.7 will be in the same units for both indices (harvest-age units of I), the predictions of genetic worth for all individuals can be ranked and selections made on that basis. All other things being equal, there should be more selections out of the older test. The older data are more correlated with the breeding objective and therefore, superior genotypes are more easily identified.

SUMMARY AND CONCLUSIONS

Given certain assumptions, use of a selection index results in best linear predictions of genetic worths of the candidates for selection. These assumptions are: 1) the data are balanced (i.e. all candidates for selection have equal amounts of information), 2) the data are of equal precision (i.e. all candidates have equal quality data), 3) the economic weights are accurate, and 4) the genetic variances and covariances needed for the **C** and **V** matrices are known parameters. When these assumptions hold there is no better (in the sense of maximizing genetic progress) method of selection. In practice, these assumptions are always violated to one degree or another. We do not believe that violation of these assumptions necessarily argues against the use of selection indices. In fact, we believe just the opposite. It is often useful to calculate many alternative indices to test alternative scenarios about the above assumptions. The scenarios can explore the ranges in quality and quantity of information, in different economic weights and in different estimates of **C** and **V**. This allows a realistic appraisal of the sensitivity of the index (and hence of the selections made based on index values) to the assumptions made. In most cases it will not be necessary to explore all of these variables. For example, if the data are balanced and of equal precision and if the **C** and **V** estimates are relatively precise, then it is only necessary to explore sensitivity to various estimates of the economic weights.

The value of the selection index approach is that it requires a specified model (i.e. **a**, **C** and **V** must be specified) and thereby quantifies the assumptions made during selection. All selection processes require assumptions. When using subjective methods of selection the breeder is implicitly making assumptions and integrating information on each candidate in some manner. Use of selection indices does not remove all subjectivity (since the breeder must specify the model), it does however require that the model be written down. Then, since that model incorporates the best estimates and assumptions available at the time, selection index then makes the best single aggregation of the information.

Another advantage of selection index over subjective approaches to selection is that because the model is written down, it can be revised as new data (better estimates of **a**, **C**, **V**) become available. This points the way to which types of research are needed and allows the results of that research to be immediately incorporated into the next cycle of selection. The new estimates (and hence the updated index) can also be used to examine the possible "mistakes" made using the old index. For example, if new data indicates too much weight placed on a given trait in the old index, perhaps this could be compensated for in the new index.

Finally, if candidates for selection have unequal quantity or quality of information, then one selection index may not suffice. To be best linear prediction, there needs to be a different selection index for each class of data. That is, individuals with like quantities and qualities of data can be served by a single index. In the extreme case where a different index is developed for each candidate, then this is best linear prediction as developed in Chapter 4. For even moderate sized selection programs, it will become a computationally intractable problem to invert a **V** matrix that incorporates all information from all candidates. However, there may be many instances when suitable approximations are found. For example, if the data fall into a few discrete classes within which the data are of similar quality and quantity for each candidate, then a different selection index may developed for each class. Approximations may be suitable in other instances and approaches to the use of selection indices with "messy" data are discussed in the next chapter.

PROBLEMS

Problem 9.1: For Equations 9.1 - 9.11, use the dimensions of the vectors and matrices in each equation to calculate the dimension of the result. For example, in 9.1, the result is w which is a scalar with dimension 1 x 1.

Problem 9.2: Consider Example 9.1 and suppose that 10-year height is measured in feet and that straightness is measured in degrees.

a. Specify the units of the various type of various variances and covariances needed for the **C** and **V** matrices.

b. Now suppose that the units of all height measurements were converted to meters instead of feet, what would happen to the magnitudes of the variances and covariances?

Problem 9.3: Use the calculation formula for **b** in Equation 9.6 and do the matrix algebra to show the following are equalities.

a. $\mathbf{a}'\mathbf{C}'\mathbf{V}^{-1}\mathbf{C}\mathbf{a} = \mathbf{b}'\mathbf{V}\mathbf{b}$ (from Eq. 9.8).

b. $\mathbf{a}'\mathbf{C}'\mathbf{V}^{-1}\mathbf{C}\mathbf{a} = \mathbf{b}'\mathbf{C}\mathbf{a}$ (from Property 2).

c. $\mathbf{b}'\mathbf{V}\mathbf{b} = \mathbf{b}'\mathbf{C}\mathbf{a}$.

Chapter 10
Selection Index Applications

INTRODUCTION

In this chapter, selection index principles are illustrated through the use of some numerical examples. Initially we consider selection index as classically applied: with an assumption of balanced data with homogeneous second moments. We first examine a simple case of index selection for a single trait in an open-pollinated progeny test, and then consider selection index for a single trait applied over a number of open-pollinated tests. This is followed by an application in a single cross-pollinated test. Finally, we present an application of selection index for multiple traits, and consider how to examine sensitivity to economic weights. For each case, the objectives are two fold: 1) to illustrate how to set up the vectors and matrices necessary to apply a selection index, and 2) to demonstrate a few interesting principles about how selection indices operate. We then discuss selection index in cases where the assumptions of balanced data and homogeneous second moments may be inappropriate, and conclude by presenting approximate approaches to the selection of individuals which should be nearly optimum (and may be substantially easier to apply operationally). This chapter contains fairly detailed algebraic manipulations with linear models containing several subscripts. The reader is referred to Chapter 2 and the excellent paper by Comstock and Moll (1963) for further discussion and examples of this type of algebra.

CLASSICAL APPLICATIONS OF SELECTION INDEX

Selection index can be used in a variety of ways, e.g. to predict the aggregate genetic worth of an individual in a progeny test. Aggregate genetic worth ($\hat{w}$) may be composed of a single trait or multiple traits. Various types of information can be used in the index, from observations on the individual candidate to diverse types of information on ancestors and relatives. Regardless of the how the aggregate worth is determined or what data are used, $\hat{w}$ is always calculated as

$$
\begin{aligned}
I = \hat{w} &= \mathbf{b}'(\mathbf{y} - \boldsymbol{\alpha}) \\
&= \mathbf{a}'\mathbf{C}'\mathbf{V}^{-1}(\mathbf{y} - \boldsymbol{\alpha})
\end{aligned}
$$

from Eq. 9.7. Thus the application of selection index always follows the same general steps:

1) Define a linear model appropriate for the data.
2) Decide what kinds of observations will make up the **y** vector (for example, individual measurements on several traits from the candidate tree, and related family means).
3) Determine the expected values ($\boldsymbol{\alpha}$) of the observations in **y** and find means of the data that have the same expected values and precisely estimate these expected values.
4) Determine the phenotypic variances and covariances of the observations in **y**, and construct **V**.
5) Consider what traits will compose the aggregate worth w, calculate the genetic covariances between the observations and those traits, and construct **C**.
6) Determine the genetic covariances between all the traits in w, and construct **G**.
7) Estimate **a**, the relative economic weights of the traits in w.
8) Calculate $I = \hat{w}$ from Eq. 9.6 or Eq. 9.7.

These steps are essentially the same ones necessary to apply BLP as discussed in Chapters 4 through 8, although then the interest was in predicting parental breeding values from progeny test data. Now the interest is in predicting breeding values for single individuals based on data on that individual and on its relatives.

Case 1: Single Half-Sib Test, Single Trait

The first example is a simple application of selection index for a single trait in a single open-pollinated progeny test. Assume there is a 7.5-year-old row-plot progeny test of *Pinus caribaea* with 6 blocks and 6 trees per plot, and the goal is to select the top 5% of the individuals in the test on the basis of height. The linear model (Eq. 2.23) defined for half-sib data is appropriate for this use, and the various first and second moments applicable to this test are assumed known (Table 10.1, based on Dean et al. 1987).

Table 10.1 - Means, variances, and covariances for height growth, stem diameter, and branch diameter in 7.5 year-old open-pollinated *P. caribaea* progeny tests.[a,b]

Means and Variances						
Trait	Mean	h^2	σ^2_f	σ^2_{fe}	σ^2_p	σ^2_w
Height (m)	9.3	0.20	0.1330	0.0443	0.3750	2.1250
Stem Diameter (cm)	13.7	0.25	0.4667	0.1556	1.0425	5.9075
Branch Diameter (4 unit scale)[c]	2.8	0.10	0.0053	0.0018	0.0300	0.1700

Covariances and Correlations					
Traits	σ_f	σ_{fe}	σ_p	σ_w	r_A[d]
Height, Stem Diam.	0.1943	0.0648	0.5082	2.8798	0.780
Height, Branch Diam.	-0.0141	-0.0047	-0.0430	-0.2435	-0.531
Stem Diam., Branch Diam.	-0.0358	-0.0121	-0.0942	-0.5341	-0.720

[a]Data based on Dean et al. (1986), Test D.
[b]Values for h^2 are on an individual tree basis: $h^2 = 4\sigma^2_f \,/\, (\sigma^2_f + \sigma^2_{fe} + \sigma^2_p \;\; \sigma^2_w)$.
[c]Branch diameter is measured on a 4 unit scale, where 1 is for thick branches, and 4 is for thin branches.
[d]r_A = genetic correlation = $\sigma_{f,uv} / (\sigma^2_{f,u} \cdot \sigma^2_{f,v})^{½}$ for the traits u and v.

Specifying the y and α vectors: When index selection is applied in a single open-pollinated row-plot progeny test there are generally three types of observations which can be used as elements of **y**: individual tree observations, plot means, and family means. For this example:

$$\mathbf{y} = \begin{bmatrix} \text{ind. obs.} \\ \text{plot mean} \\ \text{family mean} \end{bmatrix} = \begin{bmatrix} y_{ijkl} \\ \bar{y}_{ijk\cdot} \\ \bar{y}_{i\cdot k\cdot} \end{bmatrix}$$

$$\boldsymbol{\alpha} = \begin{bmatrix} \text{E(ind. obs.)} \\ \text{E(plot mean)} \\ \text{E(family mean)} \end{bmatrix} = \begin{bmatrix} E(y_{ijkl}) \\ E(\bar{y}_{ijk\cdot}) \\ E(\bar{y}_{i\cdot k\cdot}) \end{bmatrix}.$$

The expected value of each type of observation is needed in order to determine the elements of $\boldsymbol{\alpha}$. Taking the first type of observation in **y**, an individual tree observation is modeled as

$$y_{ijkl} = \mu + E_i + B_{ij} + f_k + fe_{ik} + p_{ijk} + w_{ijkl}, \qquad \textbf{Eq. 10.1}$$

and the expected value of this observation is

$$\begin{aligned} E(y_{ijkl}) &= E(\mu + E_i + B_{ij} + f_k + fe_{ik} + p_{ijk} + w_{ijkl}) \\ &= \mu + E_i + B_{ij} \end{aligned} \qquad \textbf{Eq. 10.2}$$

since the first three terms are constants (fixed effects) and the expectations of all other effects are zero as defined in the model Eq. 2.23. A plot mean (the second element in **y**) is the average of all progeny of a given family in a particular block, and can be written as

$$\begin{aligned} \bar{y}_{ijk\cdot} &= \sum_{1}^{n} (\mu + E_i + B_{ij} + f_k + fe_{ik} + p_{ijk} + w_{ijkl})/n \qquad \textbf{Eq. 10.3} \\ &= \mu + E_i + B_{ij} + f_k + fe_{ik} + p_{ijk} + w_{ijk\cdot}/n. \end{aligned}$$

The expected value of a plot mean is

$$\begin{aligned} E(\bar{y}_{ijk\cdot}) &= E(\mu + E_i + B_{ij} + f_k + fe_{ik} + p_{ijk} + w_{ijk\cdot}/n) \\ &= \mu + E_i + B_{ij}, \end{aligned} \qquad \textbf{Eq. 10.4}$$

thus, the first and second elements of **y** have the same expected value.

A family mean (the third element of **y**) is the average of all individual trees in the family across the entire test, and can be written as

$$\bar{y}_{i \cdot k \cdot} = \sum_{j}^{b} \sum_{l}^{n} (\mu + E_i + B_{ij} + f_k + fe_{ik} + p_{ijk} + w_{ijkl})/bn \qquad \textbf{Eq. 10.5}$$

$$= \mu + E_i + f_k + fe_{ik} + p_{i \cdot k}/b + w_{i \cdot k \cdot}/bn.$$

The expectation of the family mean is

$$E(\bar{y}_{i \cdot k \cdot}) = E(\mu + E_i + B_{ij} + f_k + fe_{ik} + p_{i \cdot k}/b + w_{i \cdot k \cdot}/bn) \qquad \textbf{Eq. 10.6}$$
$$= \mu + E_i$$

since across the whole test the B_{ij} effects sum to zero (assuming balanced data), and the expected value of all other effects is zero as defined in Eq. 2.23.

In the derivation of selection index theory, μ, E_i, and B_{ij} (which comprise the elements of $\boldsymbol{\alpha}$), are known; however, in practice we utilize $\hat{\boldsymbol{\alpha}}$, which contains estimates. Note that both individual observations and plot means (y_1 and y_2) have an expected value of $\mu + E_i + B_{ij}$. A precise and accurate estimate of $\mu + E_i + B_{ij}$ is needed for the estimates of α_1 and α_2. A block mean is defined as

$$\bar{y}_{ij \cdot \cdot} = \sum_{k}^{s} \sum_{l}^{n} (\mu + E_i + B_{ij} + f_k + fe_{ik} + p_{ijk} + w_{ijkl})/sn, \qquad \textbf{Eq. 10.7}$$

where s = number of parents. The expectation of a block mean is

$$E(\bar{y}_{ij \cdot \cdot}) = E(\mu + E_i + B_{ij} + f./s + fe_{i \cdot}/s + p_{ij \cdot}/s + w_{ij \cdot \cdot}/sn)$$
$$= \mu + E_i + B_{ij}. \qquad \textbf{Eq. 10.8}$$

Thus, y_{ijkl}, $\bar{y}_{ijk \cdot}$ and $\bar{y}_{ij \cdot \cdot}$ all have the same expectation. Since the overall block mean ($\bar{y}_{ij \cdot \cdot}$) has the lowest error variance (because it is averaged over the most trees), it is a good estimate for the first two elements of $\boldsymbol{\alpha}$ (i.e. for individual observations and plot means).

Similarly, the expected value for the overall test mean ($\bar{y}_{i\cdots}$) is equal to the expectation for a family mean ($\bar{y}_{i\cdot k\cdot}$),

$$E(\bar{y}_{i\cdots}) = \mu + E_i \qquad \textbf{Eq. 10.9}$$

and thus $\bar{y}_{i\cdots}$ can be used to estimate the expected value for family means, and is used as α_3.

Example 10.1: Consider y_{i243}, the 3rd tree in the 4th family in the 2nd block in the ith test. If y_{i243} = 10.0m (the tree measurement), $\bar{y}_{i24\cdot}$ = 9.2m (the plot mean), $\bar{y}_{i\cdot 4\cdot}$ = 10.5m (the family mean), $\bar{y}_{i2\cdot\cdot}$ = 9.9m (the block mean), and $\bar{y}_{i\cdots}$ = 9.3m (the test mean), then **y** and **α** will be:

$$\mathbf{y} = \begin{bmatrix} \text{ind. obs.} \\ \text{plot mean} \\ \text{family mean} \end{bmatrix} = \begin{bmatrix} y_{i243} \\ \bar{y}_{i24\cdot} \\ \bar{y}_{i\cdot 4\cdot} \end{bmatrix} = \begin{bmatrix} 10.0 \\ 9.2 \\ 10.5 \end{bmatrix}$$

$$\boldsymbol{\alpha} = \begin{bmatrix} E(\text{ind. obs.}) \\ E(\text{plot mean}) \\ E(\text{family mean}) \end{bmatrix} = \begin{bmatrix} E(y_{i243}) \\ E(\bar{y}_{i24\cdot}) \\ E(\bar{y}_{i\cdot 4\cdot}) \end{bmatrix} = \begin{bmatrix} E(\bar{y}_{i2\cdot\cdot}) \\ E(\bar{y}_{i2\cdot\cdot}) \\ E(\bar{y}_{i\cdots}) \end{bmatrix} \approx \begin{bmatrix} 9.9 \\ 9.9 \\ 9.3 \end{bmatrix}$$

Variances and covariances of observations: To construct the 3 x 3 **V** matrix, variances and covariances among the three types of observations in **y** are needed. Using the linear model and the parameters in Table 10.1, the variance of individual observations (element v_{11} in **V**) is

$$\begin{aligned} Var(y_{ijkl}) &= Var(\mu + E_i + B_{ij} + f_k + fe_{ik} + p_{ijk} + w_{ijkl}) \\ &= \sigma^2_f + \sigma^2_{fe} + \sigma^2_p + \sigma^2_w \end{aligned} \qquad \textbf{Eq. 10.10}$$

Note that μ, E_i, and B_{ij} are constants and therefore do not have variances. The variance of plot means (element v_{22} in **V**) is

$$\begin{aligned} Var(\bar{y}_{ijk\cdot}) &= Var(\mu + E_i + B_{ij} + f_k + fe_{ik} + p_{ijk} + w_{ijk\cdot}/n) \\ &= \sigma^2_f + \sigma^2_{fe} + \sigma^2_p + \sigma^2_w/n \end{aligned} \qquad \textbf{Eq 10.11}$$

The variance of family means (element v_{33} in **V**) is

$$\begin{aligned} \mathrm{Var}(\bar{y}_{i\cdot k\cdot}) &= \mathrm{Var}(\mu + E_i + f_k + fe_{ik} + p_{i\cdot k}/b + w_{i\cdot k\cdot}/bn) \\ &= \sigma^2_f + \sigma^2_{fe} + \sigma^2_p/b + \sigma^2_w/bn \end{aligned} \qquad \textbf{Eq 10.12}$$

The covariance of an individual observation and the mean of the plot containing that observation (elements v_{12} and v_{21} in **V**) is equal to the variance of the plot mean:

$$\begin{aligned} \mathrm{Cov}(y_{ijkl},\bar{y}_{ijk\cdot}) &= \mathrm{Cov}(\mu + E_i + B_{ij} + f_k + fe_{ik} + p_{ijk} + w_{ijkl}, \\ &\qquad \mu + E_i + B_{ij} + f_k + fe_{ik} + p_{ijk} + w_{ijk\cdot}/n) \\ &= \mathrm{Cov}(f_k,f_k) + \mathrm{Cov}(fe_{ik},fe_{ik}) + \mathrm{Cov}(p_{ijk},p_{ijk}) \\ &\quad + \mathrm{Cov}(w_{ijkl},w_{ijkl})/n \\ &= \sigma^2_f + \sigma^2_{fe} + \sigma^2_p + \sigma^2_w/n \qquad \textbf{Eq. 10.13} \\ &= \mathrm{Var}(\bar{y}_{ijk\cdot}) \end{aligned}$$

The covariance of an individual observation and the family mean containing that observation (elements v_{13} and v_{31} in **V**) is equal to the variance of the family mean:

$$\begin{aligned} \mathrm{Cov}(y_{ijkl}, \bar{y}_{i\cdot k\cdot}) &= \mathrm{Cov}(\mu + E_i + B_{ij.} + f_k + fe_{ik} + p_{ijk} + w_{ijkl}, \\ &\qquad \mu + E_i + f_k + fe_{ik} + p_{i\cdot k}/b + w_{i\cdot k\cdot}/bn) \\ &= \mathrm{Cov}(f_k, f_k) + \mathrm{Cov}(fe_{ik}, fe_{ik}) + \mathrm{Cov}(p_{ijk}, p_{ijk})/b \\ &\quad + \mathrm{Cov}(w_{ijkl}, w_{ijkl})/bn \\ &= \sigma^2_f + \sigma^2_{fe} + \sigma^2_p/b + \sigma^2_w/bn \qquad \textbf{Eq. 10.14} \\ &= \mathrm{Var}(\bar{y}_{i\cdot k\cdot}) \end{aligned}$$

Finally, the covariance of a plot mean and the family mean containing that plot mean (elements v_{23} and v_{32} in **V**) is also equal to the variance of the family mean:

$$\begin{aligned}\mathrm{Cov}(\bar{y}_{ijk\cdot}, \bar{y}_{i\cdot k\cdot}) &= \mathrm{Cov}(\mu + E_i + B_{ij} + f_k + fe_{ik} + p_{ijk} + w_{ijk\cdot}/n, \\ &\qquad \mu + E_i + f_k + fe_{ik} + p_{i\cdot k}/b + w_{i\cdot k\cdot}/bn) \\ &= \mathrm{Cov}(f_k, f_k) + \mathrm{Cov}(fe_{ik}, fe_{ik}) + \mathrm{Cov}(p_{ijk}, p_{i\cdot k}/b) \\ &\quad + \mathrm{Cov}(w_{ijk\cdot}/n, w_{i\cdot k\cdot}/bn) \\ &= \sigma^2_f + \sigma^2_{fe} + \sigma^2_p/b + \sigma^2_w/bn \qquad \textbf{Eq. 10.15} \\ &= \mathrm{Var}(\bar{y}_{i\cdot k\cdot})\end{aligned}$$

The third step of the derivation of Eq. 10.15 makes use of the fact that $\mathrm{Cov}(w_{ijk\cdot}/n, \bar{w}_{i\cdot k\cdot}/bn) = (1/bn)\ \sigma^2_{w}$. If $j=2$ then this plot mean is averaged over all trees from the kth family in block 2 and

$$\mathrm{Cov}(w_{i2k\cdot}/n, w_{i\cdot k\cdot}/bn) = (1/bn^2)\ \mathrm{Cov}(\sum_{1}^{n} w_{i2kl}, \sum_{j}^{b}\sum_{1}^{n} w_{ijkl})$$

$$= (1/bn^2)\ \mathrm{Cov}(\sum_{1}^{n} w_{i2kl}, \sum_{1}^{n} w_{i2kl}) \qquad \text{(because when } j \neq 2\text{, all covariances} = 0)$$

$$\begin{aligned}= (1/bn^2)\ [\ &\mathrm{Cov}(w_{i2k1},w_{i2k1}) + \mathrm{Cov}(w_{i2k2},w_{i2k2}) \\ &+ \ldots + \mathrm{Cov}(w_{i2kn},w_{i2kn})\ ] \qquad \text{(expand summation notation)}\end{aligned}$$

$$= (n/bn^2)\ \mathrm{Cov}(w_{i2kl},w_{i2kl}) \qquad \text{(since there are n like terms)}$$

$$= (1/bn)\ \sigma^2_{w}.$$

Using the parameters in Table 10.1 and Eqs. 10.10 - 10.15, **V** is

$$\mathbf{V} = \begin{bmatrix} \mathrm{Var}(y_{ijkl}) & \mathrm{Cov}(y_{ijkl},\bar{y}_{ijk\cdot}) & \mathrm{Cov}(y_{ijkl},\bar{y}_{i\cdot k\cdot}) \\ \mathrm{Cov}(y_{ijkl},\bar{y}_{ijk\cdot}) & \mathrm{Var}(\bar{y}_{ijk\cdot}) & \mathrm{Cov}(\bar{y}_{ijk\cdot},\bar{y}_{i\cdot k\cdot}) \\ \mathrm{Cov}(y_{ijkl},\bar{y}_{i\cdot k\cdot}) & \mathrm{Cov}(\bar{y}_{ijk\cdot},\bar{y}_{i\cdot k\cdot}) & \mathrm{Var}(\bar{y}_{i\cdot k\cdot}) \end{bmatrix}$$

$$\mathbf{V} = \begin{bmatrix} 2.6773 & 0.9065 & 0.2988 \\ 0.9065 & 0.9065 & 0.2988 \\ 0.2988 & 0.2988 & 0.2988 \end{bmatrix}$$

Determining traits in w and constructing C: In this example, selection is for the single trait of height growth, or in other words, aggregate genetic worth (w) is equivalent to breeding value for the one trait of height growth at 7.5 years. To predict a breeding value (g_{ijkl}) for the ijklth tree using as the data in **y** the phenotypic measurement on that individual, the mean of the plot which contains that individual, and the family mean, the only covariances needed are between these observations and the breeding value for height. So, in this example, **C** is a 3 x 1 matrix containing these three covariances. To derive these requires the use of genetic theory to define the breeding value g_{ijkl} in terms of the linear model.

Consider that the breeding value of an individual in the kth half-sib family is partly composed of f_k, the average effect of the kth parent. Consider also that $Var(f_k) = \sigma^2_f = (1/4)\ \sigma^2_A$, which leaves $(3/4)\ \sigma^2_A$ expressed within a half-sib family. We can therefore define g_{ijkl} as follows:

$$g_{ijkl} = f_k + d_{ijkl} \qquad \textbf{Eq. 10.16}$$

where g_{ijkl} = the breeding value of the lth individual in the kth family in the jth block of the ith test,

f_k = random effect of the kth family, $E(f_k)=0$, $Var(f_k) = \sigma^2_f = \frac{1}{4}\sigma^2_A$,

d_{ijkl} = random additive genetic effect of the lth individual in the jth block of the kth family in the ith test, $E(d_{ijkl})=0$, $Var(d_{ijkl}) = 3\sigma^2_f = (3/4)\sigma^2_A$.

The term d_{ijkl} accounts for the part of the breeding value of the ijklth individual which cannot be accounted for by f_k, the average effect of the kth parent. In the linear model in Eq. 2.23, the term w_{ijkl} contains d_{ijkl}, the within-family deviation, as well as other random error effects. We can expand the linear model by defining w_{ijkl} as follows:

$$w_{ijkl} = d_{ijkl} + e_{ijkl} \qquad \textbf{Eq. 10.17}$$

where e_{ijkl} = random non-additive genetic error and experimental tree error of the lth individual in the ijkth plot, $E(e_{ijkl})=0$, $Var(e_{ijkl}) = \sigma^2_w - 3\sigma^2_f = \sigma^2_w - (3/4)\sigma^2_A$.

With Eqs. 10.16 and 10.17, the covariances between the observations in **y** and the breeding value, g_{ijkl}, can now be determined. The covariance between an individual tree observation and g_{ijkl} (the breeding value of that tree) is

$$\begin{aligned}
Cov(y_{ijkl}, g_{ijkl}) &= Cov(\mu + E_i + B_{ij} + f_k + fe_{ik} + p_{ijk} + w_{ijkl}, g_{ijkl}) \\
&= Cov(f_k, g_{ijkl}) + Cov(w_{ijkl}, g_{ijkl}) \\
&= Cov(f_k, f_k + d_{ijkl}) + Cov(e_{ijkl} + d_{ijkl}, f_k + d_{ijkl}) \\
&= Cov(f_k, f_k) + Cov(d_{ijkl}, d_{ijkl}) \\
&= \sigma^2_f + 3\sigma^2_f \qquad \textbf{Eq. 10.18} \\
&= \sigma^2_A.
\end{aligned}$$

Thus, the covariance between an individual tree observation and that tree's breeding value is σ^2_A. This is the first element of **C** because it is the covariance between the first element of **y** (y_{ijkl}) and the breeding value being predicted.

The covariance between a plot mean and g_{ijkl} (the second element of **C**) is

$$\begin{aligned}
Cov(\bar{y}_{ijk\cdot}, g_{ijkl}) &= Cov(\mu + E_i + B_{ij} + f_k + fe_{ik} + p_{ijk} + w_{ijk\cdot}/n, g_{ijkl}) \\
&= Cov(f_k, g_{ijkl}) + Cov(w_{ijk\cdot}/n, g_{ijkl}) \\
&= Cov(f_k, f_k + d_{ijkl}) + Cov(e_{ijk\cdot}/n + d_{ijk\cdot}/n, f_k + d_{ijkl}) \\
&= Cov(f_k, f_k) + (1/n)\, Cov(d_{ijkl}, d_{ijkl}) \\
&= \sigma^2_f + (3/n)\, \sigma^2_f \qquad \textbf{Eq. 10.19} \\
&= (1/4)\, \sigma^2_A + (3/4n)\, \sigma^2_A \\
&= (1/4)\, \sigma^2_A\, (1 + 3/n)
\end{aligned}$$

The covariance between a family mean and g_{ijkl} (the third element of **C**) is

$$\begin{aligned}
Cov(\bar{y}_{i\cdot k\cdot}, g_{ijkl}) &= Cov(\mu + E_i + f_k + fe_{ik} + p_{i\cdot k}/b + w_{i\cdot k\cdot}/bn, g_{ijkl}) \\
&= Cov(f_k, g_{ijkl}) + Cov(w_{i\cdot k\cdot}/bn, g_{ijkl}) \\
&= Cov(f_k, f_k + d_{ijkl}) + Cov(e_{i\cdot k\cdot}/bn + d_{i\cdot k\cdot}/bn, f_k + d_{ijkl}) \\
&= Cov(f_k, f_k) + (1/bn)\, Cov(d_{ijkl}, d_{ijkl}) \\
&= \sigma^2_f + (3/bn)\, \sigma^2_f \qquad \textbf{Eq. 10.20} \\
&= (1/4)\, \sigma^2_A + (3/4bn)\, \sigma^2_A \\
&= (1/4)\, \sigma^2_A\, (1 + 3/bn)
\end{aligned}$$

Using the parameters found in Table 10.1, and Eqs. 10.18 - 10.20, **C** is:

$$\mathbf{C} = \begin{bmatrix} \text{Cov}(\underline{y}_{ijk\cdot}, g_{ijkl}) \\ \text{Cov}(\underline{y}_{ijk\cdot}, g_{ijkl}) \\ \text{Cov}(y_{i\cdot k\cdot}, g_{ijkl}) \end{bmatrix} = \begin{bmatrix} 0.5320 \\ 0.1995 \\ 0.1441 \end{bmatrix}$$

Defining G: The matrix **G** contains genetic variances and covariances of the traits being predicted. In this case, the breeding value for only one trait, height, is being predicted, so **G** is simply a 1 x 1 matrix containing the additive genetic variance for height.

$$\mathbf{G} = [\sigma^2_A] = [4\sigma^2_f] = [0.5320]$$

Economic weights: For this first case, the aggregate genetic worth (w) of 7.5 year-old *P. caribaea* is composed of a single trait, height. Thus, the vector of economic weights (**a**) is a 1 x 1 vector, or a scalar. For simplicity, let **a** = [1].

Application of selection index formulae: With all of the needed vectors and matrices constructed, the predicted genetic worth of each tree in the progeny test can be calculated in either of two ways: 1) by first calculating the index coefficients using Eq. 9.6 and then applying these coefficients to the data (i.e. the **y** vectors) for each tree in the progeny test, or 2) using Eq. 9.7 to calculate the predicted genetic worth of each tree in one instead of two steps. It is computationally easier to first calculate the index coefficients and then apply these coefficients to the data for each tree. If there are 30 families, 6 blocks, and 6 trees per block of each family (a total of 1080 trees) the index will be calculated 1080 times, but the coefficients need to be calculated only once. The selection index coefficients for this example are presented in Table 10.2 and discussed in detail below.

Example 10.2: As an example of the one step method of calculating genetic worth from Eq. 9.7, consider the **y** and $\boldsymbol{\alpha}$ vectors for a particular tree from Example 10.1:

$$\mathbf{y} = \begin{bmatrix} 10.0 \\ 9.2 \\ 10.5 \end{bmatrix} \qquad \boldsymbol{\alpha} = \begin{bmatrix} 9.9 \\ 9.9 \\ 9.3 \end{bmatrix}$$

This particular tree, y_{i243}, is the third tree from the second block and from family 4. Its 7.5 year height is 0.1 m above the plot mean of the 6-tree plot that the tree is located in (i.e. 10.0m - 9.9m), and that particular plot is 0.7m shorter than the overall block mean for block 2 (i.e. 9.2m - 9.9m). The family mean for family 4 is 1.2m above the overall test mean for all 1080 trees (i.e. 10.5m - 9.3m). Given these hypothetical data, the predicted genetic worth for tree y_{i243} is $I = \hat{w} = \mathbf{a'C'V^{-1}(y - \boldsymbol{\alpha})} = 0.56$m. Thus, this tree's breeding value is predicted to be 0.56m above average. Given the mean of the 7.5-year-old test of 9.3m, this is 0.56/9.3 = 6% above average.

To accomplish the stated goal of selecting the top 5% of the individuals from this progeny test, the genetic worth would first be calculated for every tree, and then the trees ranked on this basis. Then the top 0.05 • 1080 = 54 trees would be selected. Of course, many of these could come from the same family and would be related. This is discussed later in this chapter.

To apply the selection index requires the **y** and $\boldsymbol{\alpha}$ vectors for each and every tree in the progeny test. However, many of the formulae in Chapter 9 do not require use of **y** and $\boldsymbol{\alpha}$. That is, we do not need to have the observations on a specific tree or family to calculate such things as the variance of predictions or the error variance of predictions. Since we are dealing with trees in a single test, **V**, **C**, and **a** apply to every tree in the test, thus the following can be calculated:

$\mathbf{b} = \mathbf{V^{-1}Ca}$	(Eq. 9.6),
$Corr(w,\hat{w}) = [\mathbf{a'C'V^{-1}Ca/a'Ga}]^{1/2}$	(Eq. 9.10), and
$\Delta I = i_s \cdot [\mathbf{b'Vb}]^{1/2} = i_s \cdot [\mathbf{a'C'V^{-1}Ca}]^{1/2}$	(Eq. 9.11).

To illustrate some principles of selection index, Eqs. 9.6, 9.10, and 9.11 are used to compare four variations of Case 1: a) using all three observation types (individual, plot mean, family mean) as in the discussion to this point, b) using the individual and the family mean, c) using only the family mean, d) using only the individual observation. For Cases 1b, 1c, and 1d, if a particular observation is not being used, all rows and columns pertaining to that observation are eliminated from **y**, **α**, **V**, and **C**. For example, in Case 1b where the plot mean is eliminated, the second row and second column of **V** is deleted, leaving a 2 x 2 **V** matrix. The estimated coefficients, the correlation between the true and predicted genetic worth, and the gain in the index for Cases 1a - 1d are presented in Table 10.2.

Results for Case 1: For Case 1a, which utilizes all three observation types, the **b'** vector of index coefficients is **b'** = [0.1878 -0.0966 0.3911], and genetic worth is predicted using the formula (Eq. 9.7)

$$\begin{aligned}\hat{w} &= \mathbf{b}'(\mathbf{y} - \boldsymbol{\alpha}) \\ &= 0.1878(y_{ijkl}-\alpha_1) + -0.0966(\bar{y}_{ijk\cdot}-\alpha_2) + 0.3911(\bar{y}_{i\cdot k\cdot}-\alpha_3).\end{aligned}$$

Table 10.2 - Estimated **b** coefficients, correlation between the index and the aggregate genetic worth (Corr(w,$\hat{w}$)), and gain in index units (ΔI)[a] for Cases 1 and 2: selection in open-pollinated *P. caribaea* progeny tests for the single trait of height growth.

Case	Index Coefficients: ind. obs. (y_{ijkl})	plot mean ($\bar{y}_{ijk\cdot}$)	test family mean ($\bar{y}_{i\cdot k\cdot}$)	overall family mean ($\bar{y}_{\cdot\cdot k\cdot}$)	Corr(w,$\hat{w}$)	ΔI: index units	(%)
1a	0.1878	-0.0966	0.3911	.	0.5074	0.7635	(8.2)
1b	0.1631	.	0.3192	.	0.4995	0.7517	(8.1)
1c	0.1987	.	.	.	0.4458	0.6708	(7.2)
1d	.	.	0.4823	.	0.3614	0.5438	(5.9)
2a	0.1878	-0.0966	-0.0242	0.6590	0.5606	0.8436	(9.1)
2b	0.1878	-0.1003	.	0.6385	0.5606	0.8435	(9.1)
2c	0.1631	.	-0.0961	0.6590	0.5535	0.8329	(9.0)
2d	0.1588	.	.	0.5672	0.5519	0.8304	(8.9)
2e	.	.	.	0.7260	0.4319	0.6499	(7.8)

[a]Assuming selection of the top 5% of the individuals in the test. Since selection is on the single trait of height measured in meters, gain in index units is equivalent to gain in height in units of meters.

The coefficient on the plot means is a balance between correcting for between-plot environmental effects and genetic effects, i.e. the plot heritability (Burdon 1982). As plot heritability increases, or the relative contribution of environmental error to plot means decreases, the coefficient on plot means becomes less negative. In Case 1a, the kth family mean has a similar genetic covariance with breeding value of the $ijkl^{th}$ individual as the ijk^{th} plot mean (0.144 and 0.199 from the **C** matrix), but the family mean is substantially more precise. When both family and plot means are used as observations in **y**, the family mean estimates the family effect (f_k) and therefore gets a positive weight, while the plot mean estimates the environmental effect that a particular tree in that row-plot experiences, and therefore gets a negative weight. Heuristically, there is no genetic information contained in the plot mean that is not contained in the family mean; however, because trees within the same row-plot have an environmental covariance as well as a genetic covariance, the plot mean contains information about the quality of the micro-site within a block that a particular tree in that plot experienced.

In Case 1a, the correlation between w and $\hat{w}$ is 0.5074. Assuming that the top 5% of the trees in the test are selected (i=2.063 from Appendix 2), the gain in index units (ΔI) is 0.7635 (Table 10.2). Since aggregate genetic worth in this case is composed only of one trait, height growth in meters, ΔI is equivalent to gain in height in meters. The mean for height in this test is 9.3m (Table 10.1), so the expected gain from selecting the top 5% of the individuals on the basis of their index values in Case 1a is 0.7635m/9.3m = 8.2%.

Case 1b investigates the impact of omitting the plot mean from the selection index. That is, only the individual tree measurements and family mean were used as observations and so the **y** and $\boldsymbol{\alpha}$ vectors were 2 x 1 with the second element of the original vectors omitted. The **C** and **V** matrices were altered accordingly and Eqs. 9.6, 9.10, and 9.11 used to calculate the estimates in Table 10.2.

Omitting the plot mean in Case 1b decreased Corr(w,$\hat{w}$) slightly to 0.4995, and ΔI to 0.7517 (Table 10.2). This is equivalent to 8.1% gain in height, compared to the gain in Case 1a of 8.2%. The greater the amount of environmental error in plot means, i.e. as σ^2_p increases, the greater is the benefit of incorporating plot means in the index.

Finally, Cases 1c and 1d examine selection using only the individual observation, and only the family mean, respectively. Selection based on family means alone will yield 5.9% gain in height, while selection based on individual phenotypic values alone yields an expected gain of 7.2%. The marginal gain from adding both plot and family mean information to the selection index with individual observations is only 1.0%. This is primarily because: 1) the selection is occurring in a half-sib test (thus 3/4 of the additive variance is occurring within families, while only 1/4 of the additive variance occurs between families), and 2) height has a moderate individual tree heritability ($h^2 = 0.20$ from Table 10.1) making individual tree selection fairly effective.

Estimating second moments with ANOVA from a single test: For all the calculations in Case 1, we used the second moments in Table 10.1, which included values for σ^2_f and σ^2_{fe}. However, if the second moments are estimated by conducting an analysis of variance (ANOVA) on a single test, it is not possible to determine σ^2_{fe}. A standard ANOVA for the ith open-pollinated row-plot progeny test yields family, plot, and error variance component estimates (Table 10.3).

The variance component σ^2_{Fi} is the additive genetic variance <u>on a single site</u> and, in terms of the linear model in Eq. 2.23,

$$\sigma^2_{Fi} = \sigma^2_f + \sigma^2_{fe}. \qquad \textbf{Eq. 10.21}$$

When genetic variance among families is estimated based on data from only one test or location it is biased upwards by the amount of genotype x environment interaction variance (Comstock and Moll 1963). This occurs because the effect of the kth family <u>on the ith site</u> (F_{ik}) is the sum of the average family effect and the family by environment effect,

$$F_{ik} = f_k + fe_{ik}. \qquad \textbf{Eq. 10.22}$$

Assume that an ANOVA from a single site is used to estimate the second moments for a selection index application. The true variance components are those in Table 10.1, but from an ANOVA from one site the estimate of family variance will be

$$\begin{aligned}\sigma^2_{Fi} &= \sigma^2_f + \sigma^2_{fe} \\ &= 0.1330 + 0.0443 \\ &= 0.1773.\end{aligned}$$

Table 10.3 - Sources of variance and expected mean squares for an analysis of variance of a row-plot progeny test on a single site.

Source	EMS
Block	- - - - - - - - - - - - -
Family	$\sigma^2_w + k_2\,\sigma^2_p + k_3\,\sigma^2_{Fi}$
Plot	$\sigma^2_w + k_1\,\sigma^2_p$
Error	σ^2_w

Thus from a single site ANOVA, the estimate of additive genetic variance will be 4(0.1773) = 0.7092. Consequently the matrix **G** = [0.7092]. The elements of **V** will not be affected, and thus **V** remains as before. However, all elements in **C** are biased upward from the 'true' values if the estimates from a single site ANOVA are used. The new **C** is:

$$\mathbf{C} = \begin{bmatrix} 0.7092 \\ 0.2660 \\ 0.1921 \end{bmatrix}$$

This inflation of the elements of **C** has several impacts. If more than one trait is included in the genetic worth functions, some will experience more site by family interaction than others. So, the estimated coefficients change, and thus the ranks of individuals on the basis of their index values can change. Another impact of the inflated estimates in **C** is inflated gains estimates. For example, using the above **C** in Eq. 9.10 and 9.11, Corr(w,$\hat{w}$) is calculated as 0.5858, and ΔI is predicted as 1.1078 which is a 10.9% gain in height. This compares to a Corr(w,$\hat{w}$) of 0.5074, ΔI of 0.7635, and a 8.2% gain in height in Case 1a. The 10.9% gain in height can only be expected if progeny of the selected individuals are to be planted back on the specific i^{th} test site and experience the same set of environmental and cultural conditions as experienced by the progeny test. Only under these restrictive conditions does fe_{ik} contribute to the gain. Under any other conditions, only f_k effects contribute to gain, yielding the 8.2% gain calculated in Case 1a.

The implications of this for operational application of selection index are clear. Parameter estimates should be made across a number of tests in order to estimate genotype x environment variance. Indices based on parameter estimates obtained from an ANOVA on a single test can 1) alter ranks, and 2) overestimate expected gain. If data only exist from a single site and it is therefore not possible to estimate parameters across several tests, it may be wise to make assumptions about the relative sizes of the family (σ^2_f) and family x environment (σ^2_{fe}) variances in the family variance component estimate from the single test (σ^2_{Fi}).

Case 2: Multiple Half-Sib Tests, Single Trait

In this example there are three open-pollinated row-plot progeny tests of *P. caribaea* with first and second moments from Table 10.1. Aggregate genetic worth is still defined as genetic worth for a single trait, height growth at 7.5 years. An additional observation is now added to the **y** vector: the overall family mean averaged across all three tests which can be written as

$$\begin{aligned}\bar{y}_{\cdot\cdot k\cdot} &= \sum_i^t \sum_j^b \sum_l^n (\mu + E_i + B_{ij} + f_k + fe_{ik} + p_{ijk} + w_{ijkl})/bn \\ &= \mu + f_k + fe_{\cdot k}/t + p_{\cdot\cdot k}/tb + w_{\cdot\cdot k\cdot}/tbn. \qquad \text{Eq. 10.23}\end{aligned}$$

The expectation of the family mean across all tests is

$$\begin{aligned}E(\bar{y}_{\cdot\cdot k\cdot}) &= E(\mu + f_k + fe_{\cdot k}/t + p_{\cdot\cdot k}/tb + w_{\cdot\cdot k\cdot}/tbn) \\ &= \mu \qquad \text{Eq. 10.24}\end{aligned}$$

since (assuming balanced data): 1) across a whole test the B_{ij} effects sum to zero, 2) across all environments the E_i effects sum to zero, and 3) the expected value of all other effects is zero as defined in Eq. 2.23. The variance of the overall family mean is

$$\begin{aligned}Var(\bar{y}_{\cdot\cdot k\cdot}) &= Var(\mu + f_k + fe_{\cdot k}/t + p_{\cdot\cdot k}/tb + w_{\cdot\cdot k\cdot}/tbn) \\ &= \sigma^2_f + \sigma^2_{fe}/t + \sigma^2_p/tb + \sigma^2_w/tbn \qquad \text{Eq. 10.25}\end{aligned}$$

For three tests, t=3, and using the parameters in Table 10.1, $Var(\bar{y}_{\cdot\cdot k\cdot}) = 0.1883$.

To fill in off-diagonal elements of **V** (which now has dimension 4 x 4), the covariances between the overall family mean and the other 3 observations in **y** are needed. Proceeding in turn, the covariance between the overall family mean and an individual observation is equal to the variance of the overall family mean:

$$\begin{aligned}\mathrm{Cov}(y_{ijkl}, \bar{y}_{\cdot\cdot k\cdot}) &= \mathrm{Cov}(\mu + E_i + B_{ij} + f_k + fe_{ik} + p_{ijk} + w_{ijkl},\\ &\qquad \mu + f_k + fe_{\cdot k}/t + p_{\cdot\cdot k}/tb + w_{\cdot\cdot k\cdot}/tbn)\\ &= \mathrm{Cov}(f_k, f_k) + \mathrm{Cov}(fe_{ik}, fe_{ik})/t + \mathrm{Cov}(p_{ijk}, p_{ijk})/tb\\ &\quad + \mathrm{Cov}(w_{ijkl}, w_{ijkl})/tbn\\ &= \sigma^2_f + \sigma^2_{fe}/t + \sigma^2_p/tb + \sigma^2_w/tbn \qquad \textbf{Eq. 10.26}\\ &= \mathrm{Var}(\bar{y}_{\cdot\cdot k\cdot}).\end{aligned}$$

In an extension of the derivation used in Eq. 10.15, the covariance between both the plot mean and the family mean in one test with the overall family mean is equal to the variance of the overall family mean:

$$\mathrm{Cov}(\bar{y}_{ijk\cdot}, \bar{y}_{\cdot\cdot k\cdot}) = \mathrm{Cov}(\bar{y}_{i\cdot k\cdot}, \bar{y}_{\cdot\cdot k\cdot}) = \mathrm{Var}(\bar{y}_{\cdot\cdot k\cdot}) \qquad \textbf{Eq. 10.27}$$

The covariance between the overall family mean and g_{ijkl} is needed for the last element of **C**:

$$\begin{aligned}\mathrm{Cov}(\bar{y}_{\cdot\cdot k\cdot}, g_{ijkl}) &= \mathrm{Cov}(\mu + f_k + fe_{\cdot k}/t + p_{\cdot\cdot k}/tb + w_{\cdot\cdot k\cdot}/tbn, g_{ijkl})\\ &= \mathrm{Cov}(f_k, g_{ijkl}) + \mathrm{Cov}(w_{\cdot\cdot k\cdot}/tbn, g_{ijkl})\\ &= \mathrm{Cov}(f_k, f_k + d_{ijkl}) + \mathrm{Cov}(e_{i\cdot k\cdot}/tbn + d_{i\cdot k\cdot}/tbn, f_k + d_{ijkl})\\ &= \mathrm{Cov}(f_k, f_k) + (1/tbn)\,\mathrm{Cov}(d_{ijkl}, d_{ijkl})\\ &= \sigma^2_f + (3/tbn)\,\sigma^2_f \qquad \textbf{Eq. 10.28}\\ &= (1/4)\,\sigma^2_A + (3/4tbn)\,\sigma^2_A\\ &= (1/4)\,\sigma^2_A\,(1 + 3/4tbn).\end{aligned}$$

Using Eqs. 10.23 - 10.28 to expand the vectors and matrices from Case 1 (for a single test) to Case 2 (multiple tests), **y**, **α**, **V**, **C**, **G**, and **a** for Case 2 are:

$$\mathbf{y} = \begin{bmatrix} \text{ind. obs.} \\ \text{plot mean} \\ \text{test family mean} \\ \text{overall family mean} \end{bmatrix} = \begin{bmatrix} \underline{y}_{ijkl} \\ \underline{y}_{ijk\cdot} \\ \underline{y}_{i\cdot k\cdot} \\ \underline{y}_{\cdot\cdot k\cdot} \end{bmatrix}$$

$$\boldsymbol{\alpha} = \begin{bmatrix} E(\text{ind. obs.}) \\ E(\text{plot mean}) \\ E(\text{family mean}) \\ E(\text{overall family mean}) \end{bmatrix} = \begin{bmatrix} E(\underline{y}_{ijkl}) \\ E(\underline{y}_{ijk\cdot}) \\ E(\underline{y}_{i\cdot k\cdot}) \\ E(\underline{y}_{\cdot\cdot k\cdot}) \end{bmatrix} = \begin{bmatrix} \mu + E_i + B_{ij} \\ \mu + E_i + B_{ij} \\ \mu + E_i \\ \mu \end{bmatrix}$$

$$\mathbf{V} = \begin{bmatrix} 2.6773 & 0.9065 & 0.2988 & 0.1883 \\ 0.9065 & 0.9065 & 0.2988 & 0.1883 \\ 0.2988 & 0.2988 & 0.2988 & 0.1883 \\ 0.1883 & 0.1883 & 0.1883 & 0.1883 \end{bmatrix} \qquad \mathbf{C} = \begin{bmatrix} 0.5320 \\ 0.1995 \\ 0.1441 \\ 0.1367 \end{bmatrix}$$

$$\mathbf{G} = [0.5320] \qquad \mathbf{a} = [1]$$

The **y** and $\boldsymbol{\alpha}$ vectors have one element added because the overall family mean ($\bar{y}_{\cdot\cdot k\cdot}$) is included in addition to the individual observation, plot mean and test family mean. Thus, **V** has a row and column of variances and covariances between $\bar{y}_{\cdot\cdot k\cdot}$ and the other observations, while **C** has an additional element which is the covariance between $\bar{y}_{\cdot\cdot k\cdot}$ and the breeding value for height. Since genetic worth (w) is still comprised only of genetic worth for height, **G** and **a** remain unchanged from Case 1. As before, rows and columns of the above complete matrices for Case 2 are deleted in a series of case studies to demonstrate certain results. The cases examined are:

2a) using all four observation types,
2b) individual, plot mean, and overall family mean,
2c) individual, test family mean, and overall family mean,
2d) individual, and overall family mean,
2e) only overall family mean.

Results for Case 2: Index coefficients (**b**), the correlation of the index and genetic worth (Corr(w,ŵ)), and the expected gain from selection of the top 5% of the individuals using the index are presented in Table 10.2 for Cases 2a - 2e. Comparing Cases 2a (with all four observation types and 3 tests) to Case 1a (which assumes just a single test and uses all three possible types of observations), Corr(w,$\hat{w}$) increased from 0.5074 to 0.5606 due to the additional information on the extra two test sites. Assuming selection of the top 5% of the trees, ΔI increased from 0.7635 in Case 1a to 0.8436 in Case 2a (Table 10.2), or equivalently, height gain increased from 8.2 to 9.1%. The **b'** vector for Case 2a is

$$\mathbf{b}' = [\ 0.1878 \quad -0.0966 \quad -0.0242 \quad 0.6590\ ].$$

The coefficients on the individual observation and the plot mean remain the same as in Case 1a (Table 10.2); however, in 2a the test family mean has a negative coefficient. Heuristically, in Case 2a, the overall family mean ($\bar{y}_{\cdot\cdot k\cdot}$) is providing information about the family effect (f_k) and gets a positive weight. The test family mean ($\bar{y}_{i\cdot k\cdot}$) and the plot mean ($\bar{y}_{ijk\cdot}$) get negative weights, and simultaneously account for the micro-site environmental effects associated with the ijklth tree.

Cases 2b and 2c utilize only three of the observation types, deleting the test family mean, and the plot mean, respectively. Although the differences are very slight, it is interesting that a better index (i.e., higher Corr(w,$\hat{w}$)) is achieved in Case 2b where the plot mean is retained in the index, as opposed to Case 2c where the plot mean is deleted and the test family mean retained (Table 10.2). In both Cases 2b and 2c the overall family mean estimates f_k, and the individual tree observation provides a measure of d_{ijkl}, the additive genetic effects associated with the within-family deviation. The plot mean or test family mean functions as a measure of the random environmental effects (p_{ijk} and w_{ijkl}) associated with the ijklth tree. From Eqs. 10.3 and 10.5, note that the plot mean contains the effects p_{ijk}, and w_{ijkl}/n, while the family mean contains p_{ijk}/n and w_{ijkl}/bn. The plot mean is slightly more correlated with the environmental effects associated with the ijklth tree than the family mean, since the p_{ijk} and w_{ijkl} effects are divided by smaller numbers than in the family mean. Therefore using the plot means allows for a better adjustment for those effects, and results in a better index than when using the family mean.

Cases 2d and 2e simply illustrate that more efficient indices result if multiple tests are used. For example, in Case 1b the individual observation and the test family mean are used in the index, while in Case 2d the individual observation and the overall family mean are used. Gain in the index is 0.7517 (8.1% in height) in Case 1b, and 0.8304 (8.9% in height) in Case 2d. Similarly, in Case 1d a test family mean alone is used to do selection, while in Case 2e the overall family mean alone is used. The gain in the index (ΔI) goes from 0.5438 (5.9% gain in height) in Case 1d to 0.6499 (7.0% in height) in Case 2e (Table 10.2). These results occur because the variance of the overall family mean is lower than the variance of the test family mean due to 1) a larger number of trees in the mean, and 2) family x environment variance contributes less to the variance of the overall mean than to the variance of the test family mean (compare Eqs. 10.12 and 10.25).

Thus, in both Cases 1 and 2, relatively little is gained above using just two pieces of information (a 2 x 1 **y** vector): the individual measurement and the highest order family mean. In Case 1, this is the individual and the test family mean ($\bar{y}_{i \cdot k \cdot}$), while in Case 2 it is the individual and the overall family mean ($\bar{y}_{\cdot \cdot k \cdot}$). While these conclusions depend on the second moments in Table 10.2, we believe for open-pollinated tests that, in general, selection indices consisting of the individual tree measurement and highest order family mean will be nearly as efficient as more complex indices involving 3 or more levels of observations.

Case 3: Single Full-Sib Test, Single Trait

Often breeders wish to apply selection index within full-sib tests. To demonstrate an application of selection index in full-sib tests, we assume there is a single 7.5 year-old *P. caribaea* cross-pollinated (CP) progeny test. Assume that there are four unrelated parents (A, B, C ,D) crossed in a partial half-diallel mating design, thereby creating 6 full-sib families in the test (AxB, AxC, AxD, BxC, BxD, CxD). These 6 full-sib families are planted in a single test environment in a randomized complete block design. The linear model for full-sib data described in Chapter 6 (Eq. 6.1) is repeated as Eq. 10.29, and adopted for the observed data in Case 3. Aggregate genetic worth is still defined as genetic worth for the single trait of height growth, and the necessary first and second moments are known (Table 10.4).

Determining the y and α vectors: To predict the genetic worth of the mth tree in the jth plot of full-sib family AxB, there are a number of different observations which could be utilized in the **y** vector, e.g. the individual tree observation ($\bar{y}_{ijABm}$), the jth plot mean of the AxB full-sib family ($\bar{y}_{ijAB\cdot}$), and the test mean for full-sib family AxB ($\bar{y}_{i \cdot AB \cdot}$). We might also use the half-sib family (or plot) means of all the progeny of parent A and parent B. Additionally we might choose to use the full-sib family or plot means of AxC, AxD, BxC and BxD because these are related to, and therefore they provide

Table 10.4 - First and second moments for height growth in 7.5 year-old cross-pollinated *P. caribaea* progeny tests.[a,b]

Trait	Mean	h^2	σ^2_f	σ^2_{fe}	σ^2_d	σ^2_{de}	σ^2_p	σ^2_w
Height (m)	9.3	0.20	0.1330	0.0443	0.0665	0.0222	0.3750	1.8590

[a]Data based on Dean et al. (1986), Test D.
[b]Symbols are defined with Eq. 6.1 in Chapter 6.

information about the mth tree in the jth plot of the AxB full-sib family. If we do use those family means it is also of benefit to include information on the CxD full-sib family even though neither C nor D is directly related to $ijABm^{th}$ tree. This occurs because, heuristically, more information on parents C and D allows for better adjustment for their effects when using AxC, AxD, BxC, and BxD to predict the breeding value of a tree in the AxB full-sib family.

For this particular example, we have chosen to let the **y** vector for predicting genetic worth of a tree in the AxB family contain the individual observation, the AxB full-sib family mean, four full-sib family means with one parent either A or B, and the CxB full-sib family mean. All of these observations provide information either directly or indirectly on the breeding value of the $ijABm^{th}$ tree, and **y** is

$$\mathbf{y} = \begin{bmatrix} \underline{y}_{ijABm} \\ \underline{y}_{i\cdot AB\cdot} \\ \underline{y}_{i\cdot AC\cdot} \\ \underline{y}_{i\cdot AD\cdot} \\ \underline{y}_{i\cdot BC\cdot} \\ \underline{y}_{i\cdot BD\cdot} \\ y_{i\cdot CD\cdot} \end{bmatrix} .$$

Plot means are omitted for simplicity, but in this example $\bar{y}_{ijAB\cdot}$ could be included and would substantially improve the index if there is a large amount of plot variance.

Using the model in Eq. 6.1, an individual tree observation is modeled as:

$$y_{ijklm} = \mu + E_i + B_{ij} + f_k + fe_{ik} + h_l + he_{il} + fh_{kl} + fhe_{ikl} + p_{ijkl} + w_{ijklm}, \quad \textbf{Eq. 10.29}$$

and has an expected value of

$$E(y_{ijklm}) = E(\mu + E_i + B_{ij} + f_k + fe_{ik} + h_l + he_{il} + fh_{kl} + fhe_{ikl} + p_{ijkl} + w_{ijklm}).$$

$$= \mu + E_i + B_{ij}. \quad \textbf{Eq. 10.30}$$

A full-sib family mean from a single test is

$$\overline{y}_{i\cdot kl\cdot} = \sum_j^b \sum_m^n (\mu + E_i + f_k + fe_{ik} + h_l + he_{il} + fh_{kl} + fhe_{ikl} + p_{ijkl} + w_{ijkl\cdot})/bn,$$

$$= \mu + E_i + f_k + fe_{ik} + h_l + he_{il} + fh_{kl} + fhe_{ikl} + p_{i\cdot kl}/b + w_{i\cdot kl\cdot}/bn. \quad \textbf{Eq. 10.31}$$

This has an expected value of

$$E(\overline{y}_{i\cdot kl\cdot}) = E(\mu + E_i + f_k + fe_{ik} + h_l + he_{il} + fh_{kl} + fhe_{ikl} + p_{i\cdot kl}/b + w_{i\cdot kl\cdot}/bn) \quad \textbf{Eq. 10.32}$$

$$= \mu + E_i\ .$$

Since the **y** vector contains individual observations and full-sib family means, estimates of the expected values of these observations are needed to enter into the $\boldsymbol{\alpha}$ vector of Eqs. 9.6 and 9.7. As with data from open-pollinated tests, the overall block mean ($\overline{y}_{ij\cdots}$) has the same expected value as any tree in that block. That is, $E(\overline{y}_{ij\cdots}) = \mu + E_i + B_{ij}$, which matches that of Eq. 10.30 and can therefore be used to estimate the expected value of the first element of **y**.

Similarly, the overall test mean ($\bar{y}_{i\cdots\cdot}$) has an expected value equal to the expected value for a full-sib family mean across the full test. Therefore, $\boldsymbol{\alpha}$ can be constructed as:

$$\boldsymbol{\alpha} = \begin{bmatrix} E(y_{ijABm}) \\ E(\bar{y}_{i\cdot AB\cdot}) \\ E(\bar{y}_{i\cdot AC\cdot}) \\ E(\bar{y}_{i\cdot AD\cdot}) \\ E(\bar{y}_{i\cdot BC\cdot}) \\ E(\bar{y}_{i\cdot BD\cdot}) \\ E(\bar{y}_{i\cdot CD\cdot}) \end{bmatrix} = \begin{bmatrix} E(\bar{y}_{ij\cdots}) \\ E(\bar{y}_{i\cdots\cdot}) \\ E(\bar{y}_{i\cdots\cdot}) \\ E(\bar{y}_{i\cdots\cdot}) \\ E(\bar{y}_{i\cdots\cdot}) \\ E(\bar{y}_{i\cdots\cdot}) \\ E(\bar{y}_{i\cdots\cdot}) \end{bmatrix} \approx \begin{bmatrix} \bar{y}_{ij\cdots} \\ \bar{y}_{i\cdots\cdot} \\ \bar{y}_{i\cdots\cdot} \\ \bar{y}_{i\cdots\cdot} \\ \bar{y}_{i\cdots\cdot} \\ \bar{y}_{i\cdots\cdot} \\ \bar{y}_{i\cdots\cdot} \end{bmatrix}$$

Variances and covariances of observations: Variances and covariances of the observations in **y** are needed to enter into the 7 x 7 **V** matrix. Using Eq. 10.29 and the parameters in Table 10.4, the variance of individual observations (element v_{11} in **V**) is

$$\begin{aligned} \mathrm{Var}(y_{ijklm}) &= \mathrm{Var}(\mu + E_i + B_{ij} + f_k + fe_{ik} + h_l + he_{il} + fh_{kl} + fhe_{ikl} \\ &\qquad + p_{ijkl} + w_{ijklm}) \\ &= \sigma^2_f + \sigma^2_{fe} + \sigma^2_f + \sigma^2_{fe} + \sigma^2_d + \sigma^2_{de} + \sigma^2_p + \sigma^2_w \\ &= 2\,\sigma^2_f + 2\,\sigma^2_{fe} + \sigma^2_d + \sigma^2_{de} + \sigma^2_p + \sigma^2_w. \end{aligned} \qquad \textbf{Eq. 10.33}$$

The variance of a full-sib family mean (from Eq. 10.31) is

$$\begin{aligned} \mathrm{Var}(\bar{y}_{i\cdot kl\cdot}) &= \mathrm{Var}(\mu + E_i + f_k + fe_{ik} + h_l + he_{il} + fh_{kl} + fhe_{ikl} \\ &\qquad + p_{i\cdot kl}/b + w_{i\cdot kl\cdot}/bn) \\ &= \sigma^2_f + \sigma^2_{fe} + \sigma^2_f + \sigma^2_{fe} + \sigma^2_d + \sigma^2_{de} + \sigma^2_p/b + \sigma^2_w/bn \\ &= 2\,\sigma^2_f + 2\,\sigma^2_{fe} + \sigma^2_d + \sigma^2_{de} + \sigma^2_p/b + \sigma^2_w/bn. \end{aligned} \qquad \textbf{Eq. 10.34}$$

There are also five types of covariances among the observations in **y** needed for the off-diagonal elements of **V**. The covariance between an individual observation and its full-sib family mean is (in the derivations of Eqs. 10.35 - 10.37, fixed effects are ignored as they contribute nothing to the covariances):

$$\begin{aligned}Cov(y_{ijklm},\bar{y}_{i\cdot kl\cdot}) &= Cov(f_k + fe_{ik} + h_l + he_{il} + fh_{kl} + fhe_{ikl} + p_{ijkl} + w_{ijklm},\\ &\qquad f_k + fe_{ik} + h_l + he_{il} + fh_{kl} + fhe_{ikl} + p_{i\cdot kl}/b + w_{i\cdot kl\cdot}/bn)\\ &= Cov(f_k,f_k) + Cov(fe_{ik},fe_{ik}) + Cov(h_l,h_l) + Cov(he_{il},he_{il})\\ &\quad + Cov(fh_{kl},fh_{kl}) + Cov(fhe_{ikl},fhe_{ikl})\\ &\quad + Cov(p_{ijkl},p_{i\cdot kl}/b) + Cov(w_{ijklm},w_{i\cdot kl\cdot}/bn)\\ &= 2\,\sigma^2_f + 2\,\sigma^2_{fe} + \sigma^2_d + \sigma^2_{de} + \sigma^2_p/b + \sigma^2_w/bn. \qquad \textbf{Eq. 10.35}\\ &= Var(\bar{y}_{i\cdot kl\cdot}).\end{aligned}$$

The covariance between an individual observation in the kl[th] family and the full-sib family mean of family kl' (l' ≠ l, therefore only one parent is common) is

$$\begin{aligned}Cov(y_{ijklm},\bar{y}_{i\cdot kl'\cdot}) &= Cov(f_k + fe_{ik} + h_l + he_{il} + fh_{kl} + fhe_{ikl} + p_{ijkl} + w_{ijklm},\\ &\qquad f_k + fe_{ik} + h_{l'} + he_{il'} + fh_{kl'} + fhe_{ikl'} + p_{i\cdot kl'}/b + w_{i\cdot kl'\cdot}/bn)\\ &= Cov(f_k,f_k) + Cov(fe_{ik},fe_{ik})\\ &= \sigma^2_f + \sigma^2_{fe} \qquad \textbf{Eq. 10.36}\\ &= Cov(y_{ijklm},\bar{y}_{i\cdot k'l\cdot})\end{aligned}$$

Note that this equation also applies for the covariance between an individual observation in the kl[th] family and the full-sib family mean of family k'l (k' ≠ k).

The covariance between two full-sib family means with one common parent is

$$\begin{aligned}Cov(\bar{y}_{i\cdot kl\cdot},\bar{y}_{i\cdot kl'\cdot}) &= Cov(f_k + fe_{ik} + h_l + he_{il} + fh_{kl} + fhe_{ikl} + p_{i\cdot kl}/b + w_{i\cdot kl\cdot}/bn,\\ &\qquad f_k + fe_{ik} + h_{l'} + he_{il'} + fh_{kl'} + fhe_{ikl'} + p_{i\cdot kl'}/b + w_{i\cdot kl'\cdot}/bn)\\ &= Cov(f_k,f_k) + Cov(fe_{ik},fe_{ik})\\ &= \sigma^2_f + \sigma^2_{fe} \qquad \textbf{Eq. 10.37}\\ &= Cov(\bar{y}_{i\cdot kl\cdot},\bar{y}_{i\cdot kl'\cdot})\end{aligned}$$

Lastly, the covariance between an individual observation in the klth family, or the klth family mean and the k'l'th family mean is zero, since no effects are in common,

$$Cov(y_{ijklm},\bar{y}_{i\cdot k'l'\cdot}) = Cov(\bar{y}_{i\cdot kl\cdot\cdot},\bar{y}_{i\cdot k'l'\cdot}) = 0 \qquad \textbf{Eq. 10.38}$$

With the **y** vector presented above, and using Eqs. 10.33 - 10.38, and the parameters presented in Table 10.4, **V** is

$$\mathbf{V} = \begin{bmatrix} 2.6773 & 0.5574 & 0.1773 & 0.1773 & 0.1773 & 0.1773 & 0 \\ 0.5574 & 0.5574 & 0.1773 & 0.1773 & 0.1773 & 0.1773 & 0 \\ 0.1773 & 0.1773 & 0.5574 & 0.1773 & 0.1773 & 0 & 0.1773 \\ 0.1773 & 0.1773 & 0.1773 & 0.5574 & 0 & 0.1773 & 0.1773 \\ 0.1773 & 0.1773 & 0.1773 & 0 & 0.5574 & 0.1773 & 0.1773 \\ 0.1773 & 0.1773 & 0 & 0.1773 & 0.1773 & 0.5574 & 0.1773 \\ 0 & 0 & 0.1773 & 0.1773 & 0.1773 & 0.1773 & 0.5574 \end{bmatrix}$$

Determining traits in w and constructing C: With aggregate genetic worth still defined as genetic worth for the single trait of height at 7.5 years, we are trying to predict a breeding value (g_{ijklm}) for a particular tree (y_{ijklm}), using as data the phenotypic measurement on that individual and all full-sib family means. To derive expressions for the elements of **C** (the covariances between the observations in **y** and the breeding value for height), we must again utilize genetic theory to define the breeding value g_{ijkl} in terms of the linear model in Eq. 6.1.

The breeding value of an individual in the klth full-sib family (g_{ijklm}) can be defined as f_k, the average effect of the kth female parent, plus h_l, the average effect of the lth male parent, plus some within-family deviation due to Mendelian sampling.

$$g_{ijklm} = f_k + h_l + d_{ijklm} \qquad \textbf{Eq. 10.39}$$

where g_{ijklm} = the breeding value of the mth individual in the klth family in the jth block of the ith test,

f_k = random effect of the kth female, $E(f_k)=0$, $Var(f_k) = \sigma^2_f = ¼\sigma^2_A$,

h_l = random effect of the lth male, $E(h_l)=0$, $Var(h_l) = \sigma^2_f = ¼\sigma^2_A$, and

d_{ijklm} = random additive genetic effect of the mth individual in the jth block of the kl[th] family in the ith test, $E(d_{ijklm})=0$, $Var(d_{ijklm}) = 2\sigma^2_f = 1/2\ \sigma^2_A$.

Then

$$Var(g_{ijklm}) = Var(f_k) + Var(h_l) + Var(d_{ijklm}) = \sigma^2_A. \qquad \textbf{Eq. 10.40}$$

The term d_{ijklm} accounts for the part of the breeding value of the ijklm[th] individual which cannot be accounted for by $f_k + h_l$, the average effects of the kth and l[th] parents. In the linear model in Eq. 6.1, the term w_{ijklm} contains d_{ijklm} (the within-family deviation), as well as other random error effects, and can be expanded as follows:

$$w_{ijklm} = d_{ijklm} + e_{ijklm} \qquad \textbf{Eq. 10.41}$$

where e_{ijklm} = random non-additive genetic error and experimental error of mth individual in the ijkl[th] plot, $E(e_{ijklm})=0$, $Var(e_{ijklm}) = \sigma^2_e - 2\sigma^2_f = \sigma^2_e - 1/2\ \sigma^2_A$.

With Eqs. 10.39 - 10.41, we can now determine covariances between the observations in **y** and g_{ijklm}. The covariance between an individual tree observation and g_{ijklm} (the breeding value of that tree) is the first element of **C**:

$$\begin{aligned} Cov(y_{ijklm}, g_{ijklm}) &= Cov(f_k + fe_{ik} + h_l + he_{il} + fh_{kl} + fhe_{ikl} + p_{ijkl} + w_{ijklm}, g_{ijklm}) \\ &= Cov(f_k, g_{ijklm}) + Cov(h_l, g_{ijklm}) + Cov(w_{ijklm}, g_{ijklm}) \\ &= Cov(f_k, f_k + h_l + d_{ijklm}) + Cov(h_l, f_k + h_l + d_{ijklm}) \qquad \textbf{Eq. 10.42} \\ &\quad + Cov(e_{ijklm} + d_{ijklm}, f_k + h_l + d_{ijklm}) \\ &= Cov(f_k, f_k) + Cov(h_l, h_l) + Cov(d_{ijklm}, d_{ijklm}) \\ &= \sigma^2_f + \sigma^2_f + 2\sigma^2_f \\ &= \sigma^2_A. \end{aligned}$$

The covariance between the klth full-sib family mean and the breeding value of an individual from that family (g_{ijklm}) is the second element of **C**:

$$\begin{aligned}
Cov(\overline{y}_{i\cdot kl\cdot}, g_{ijklm}) &= Cov(f_k + fe_{ik} + h_l + he_{il} + fh_{kl} + fhe_{ikl} + p_{i\cdot kl}/b + w_{i\cdot kl\cdot}/bn, g_{ijklm}) \\
&= Cov(f_k, g_{ijklm}) + Cov(h_l, g_{ijklm}) + Cov(w_{i\cdot kl\cdot}/bn, g_{ijklm}) \\
&= Cov(f_k, f_k + h_l + d_{ijklm}) + Cov(h_l, f_k + h_l + d_{ijklm}) \\
&\quad + Cov(e_{i\cdot kl\cdot}/bn + d_{i\cdot kl\cdot}/bn, f_k + h_l + d_{ijklm}) \qquad \textbf{Eq. 10.43} \\
&= Cov(f_k, f_k) + Cov(h_l, h_l) + Cov(d_{ijklm}, d_{ijklm})/bn \\
&= \sigma^2_f + \sigma^2_f + 2\sigma^2_f/bn \\
&= (1/2)\ \sigma^2_A + (1/2bn)\ \sigma^2_A \\
&= (1/2)\ \sigma^2_A\ (1 + 1/bn).
\end{aligned}$$

The covariance between the breeding value of an individual and a full-sib family mean with one parent in common (i.e. between g_{ijklm} and the kl'th full-sib family mean, or between g_{ijklm} and the k'lth family mean) is used for elements 3, 4, 5, and 6 of **C**:

$$\begin{aligned}
Cov(\overline{y}_{i\cdot kl'\cdot}, g_{ijklm}) &= Cov(f_k + fe_{ik} + h_{l'} + he_{il'} + fh_{kl'} + fhe_{ikl'} + p_{i\cdot kl'}/b + w_{i\cdot kl'\cdot}/bn, g_{ijklm}) \\
&= Cov(f_k, g_{ijklm}) \\
&= Cov(f_k, f_k + h_l + d_{ijklm}) \qquad \textbf{Eq. 10.44} \\
&= Cov(f_k, f_k) = \sigma^2_f \\
&= (1/4)\ \sigma^2_A \\
&= Cov(\overline{y}_{i\cdot k'l\cdot}, g_{ijklm})
\end{aligned}$$

Using the parameters found in Table 10.4 and Eqs. 10.42 - 10.44, **C** for this example is:

$$\mathbf{C} = \begin{bmatrix} \mathrm{Cov}(y_{ijABm}, g_{iABlm}) \\ \mathrm{Cov}(\bar{y}_{i\cdot AB\cdot}, g_{iABlm}) \\ \mathrm{Cov}(\bar{y}_{i\cdot AC\cdot}, g_{iABlm}) \\ \mathrm{Cov}(\bar{y}_{i\cdot AD\cdot}, g_{iABlm}) \\ \mathrm{Cov}(\bar{y}_{i\cdot BC\cdot}, g_{iABlm}) \\ \mathrm{Cov}(\bar{y}_{i\cdot BD\cdot}, g_{iABlm}) \\ \mathrm{Cov}(\bar{y}_{i\cdot CD\cdot}, g_{iABlm}) \end{bmatrix} = \begin{bmatrix} 0.5320 \\ 0.2734 \\ 0.1330 \\ 0.1330 \\ 0.1330 \\ 0.1330 \\ 0 \end{bmatrix}$$

Economic weights: Since the aggregate worth is a single trait, the vector of economic weights remains as before, **a** = [1].

Defining G: Since the only trait in the aggregate genetic worth function is height, **G** remains as before, **G** = [0.5320].

Results for Case 3: Using the **y**, $\boldsymbol{\alpha}$, **V**, **C**, **G** and **a** vectors and matrices, and deleting rows and columns when necessary, we examine five cases:

a) using the individual observation, the full-sib family mean with two common parents, the 4 full-sib family means with one common parent, and the one full-sib family mean with no common parents (i.e., the individual and all full-sib family means),
b) the individual and the two-common-parent mean,
c) the two-common-parent mean, the 4 one-common-parent means, and the one full-sib family mean with no common parents (i.e., all full-sib family means, but no individual measurement),
d) the two-common-parent mean, and
e) the individual observation.

For Case 3a, the index equation with the associated coefficients (Table 10.5) is

$$\hat{w} = 0.1220\,(y_{ijABm}-\alpha_1) + 0.2414\,(\bar{y}_{i\cdot AB\cdot}-\alpha_2) + 0.0999\,(\bar{y}_{i\cdot AC\cdot}-\alpha_3)$$
$$+ 0.0999\,(\bar{y}_{i\cdot AD\cdot}-\alpha_4) + 0.0999\,(\bar{y}_{i\cdot BC\cdot}-\alpha_5) + 0.0999\,(\bar{y}_{i\cdot BD}-\alpha_6) + -0.1271\,(\bar{y}_{i\cdot CD\cdot}-\alpha_7)$$

Table 10.5 - Estimated **b** coefficients, correlation between the index and the aggregate genetic worth (Corr(w,$\hat{w}$)), and gain in index units (ΔI)[a] for Case 3: selection in cross-pollinated *P. caribaea* progeny tests for the single trait of height growth.

Case	Index Coefficients				Corr(w,$\hat{w}$)	ΔI	
	ind. obs.	full-sib means				index units	(%)
	(y_{ijklm})	($\bar{y}_{i \cdot kl \cdot}$)	($\bar{y}_{i \cdot kl' \cdot}$) ($\bar{y}_{i \cdot k'l \cdot}$)	($\bar{y}_{i \cdot k'l' \cdot}$)			
3a	0.1220	0.2414	0.0999	-0.1271	0.5882	0.8850	(9.5)
3b	0.1220	0.3685	.	.	0.5580	0.8396	(9.0)
3c	.	0.3634	0.0999	-0.1271	0.5354	0.8056	(8.7)
3d	.	0.4905	.	.	0.5021	0.7555	(8.1)
3e	0.1987	.	.	.	0.4458	0.6708	(7.2)

[a]Assuming selection of the top 5% of the individuals in the test. Since selection is on the single trait of height measured in meters, gain in index units is equivalent to gain in height in units of meters.

Note that the coefficients on the four family means with one-common parent are all the same since they are all equally precise, i.e. have the same variance and have the same covariance with the breeding value. The coefficient on the AxB family, which has both parents in common with the individual tree of interest, is approximately twice as large (.2193 vs 0.0979). Although the variance associated with the AxB mean is the same as all the other family means, the covariance with the breeding value is over twice as large (0.2734 vs. 0.1330, see **C** above) due, primarily, to the fact that there are two parental effects (f_k and h_l) contributing to the covariance instead of one. The small additional amount of covariance (σ^2_A/2bn) occurs because the measurement of the ijABmth tree is contained in the AxB family mean (see Eq. 10.43). Finally there is a negative coefficient on the CxD family mean, the family which has no direct relationship to the ijABmth tree, and therefore no covariance with its breeding value. In effect, if the AxC, AxD, BxC, and BxD family means are above average, the $\hat{w}$

calculated by the index will be tend to above average. The superiority of those four families could be caused by superiority of parents A and B, or by superiority of parents C and D. The negative coefficient on the CxD family mean adjusts $\hat{w}$ downward if CxD is above average, and upward if CxD is below average.

In Case 3a, with the index using the individual tree observation and all six full-sib family means, selection of the top 5% of the individuals in the test according to the predicted index values ($\hat{w}$) yields a gain in the index (ΔI) of 0.8850 (Table 10.5). This is equivalent to a gain in height of 9.5% (100 x 0.8850/9.3). Marginally less efficient indices result when some observations are not used. In Case 3b, the index used only the individual tree observation and the full-sib mean of the family with two-common parents. Expected gain in this case is 0.8396 (Table 10.5), which is equivalent to a 9.0% gain in height. In Case 3c, only the six full-sib means are used, and expected gain is 0.8056, equivalent to a 8.7% gain in height. And in Case 3d, using only the single full-sib mean with two common parents, expected gain is 0.7555, a 8.1% increase in height. Lastly, in Case 3e, selection is based on individual phenotype alone, and expected gain is 0.6708 (Table 10.5), a 7.2% gain in height. This is identical to Case 1c (Table 10.2), since phenotypic variances and covariances with the breeding value are the same. Case 1 dealt with open-pollinated tests, and Case 3 with control-pollinated tests, and we can compare the two cases to examine the effectiveness of index selection of individual trees in the two types of test. The marginal increase in gain from using half-sib family means and individual observations (compared to individual observations alone) can be seen in Cases 1c and 1b, where gain in height went from 7.2% to 8.1% (Table 10.2). The marginal increase in gain from using full-sib family means and individual observations (compared to individual observations alone) can be seen in Cases 3e and 3b, where gain in height went from 7.2% to 9.0% (Table 10.5). This occurs because in OP tests, only 1/4 of the additive variance is between families, and 3/4 within families, while in CP tests, 1/2 of the additive variance is between families, and 1/2 within.

Case 4: Single Half-sib Test, Multiple Traits

Very often breeders are interested in more than a single trait, i.e. aggregate genetic worth is defined as some combination of a number of traits, and they wish to construct a selection index for multiple traits in order to take advantage of the attractive properties of an index outlined in Chapter 9. To demonstrate how such an index can be calculated, consider a single open-pollinated 7.5 year-old progeny test of

P. caribaea with measurements for three traits: height, stem diameter, and branch diameter. Branch diameter is measured on a 1 to 4 scale with a measurement of 1 indicating thick branches, an undesirable trait, while 4 indicates thin branches, a desirable trait. Since these are OP data, we assume the linear model as defined in Eq. 2.23, and assume estimates of the necessary second moments are available for all three traits (Table 10.1, adapted from Dean et al. 1986).

Determining the y and $\boldsymbol{\alpha}$ vectors: For this example we use a simplified **y** vector which includes measurements for the three traits on the individual tree, and family means for the three traits, yielding a total of six observations in **y** (i.e. **y** is a 6 x 1 vector for each candidate). Plot means are not included in this example to simplify the matrices, and because, for this particular set of second moments, plot variance for all three traits is relatively small and therefore including plot means does not significantly improve the index. We denote y_{ijkl} as h_{ijkl} when the trait is height, s_{ijkl} when the trait is stem diameter, and b_{ijkl} when the trait is branch diameter. As derived in Eqs. 10.8 and 10.9, the block mean is used to estimate the expected values of individual observations and the test mean to estimate the expected value of family means for $\boldsymbol{\alpha}$. Thus **y** and $\boldsymbol{\alpha}$ are

$$\mathbf{y} = \begin{bmatrix} h_{ijkl} \\ s_{ijkl} \\ b_{ijkl} \\ \bar{h}_{i \cdot k \cdot} \\ \bar{s}_{i \cdot k \cdot} \\ \bar{b}_{i \cdot k \cdot} \end{bmatrix} \qquad \boldsymbol{\alpha} = \begin{bmatrix} E(h_{ijkl}) \\ E(s_{ijkl}) \\ E(b_{ijkl}) \\ E(\bar{h}_{i \cdot k \cdot}) \\ E(\bar{s}_{i \cdot k \cdot}) \\ E(\bar{b}_{i \cdot k \cdot}) \end{bmatrix} = \begin{bmatrix} E(\bar{h}_{ij \cdot \cdot}) \\ E(\bar{s}_{ij \cdot \cdot}) \\ E(\bar{b}_{ij \cdot \cdot}) \\ E(\bar{h}_{i \cdot \cdot \cdot}) \\ E(\bar{s}_{i \cdot \cdot \cdot}) \\ E(\bar{b}_{i \cdot \cdot \cdot}) \end{bmatrix} \approx \begin{bmatrix} \bar{h}_{ij \cdot \cdot} \\ \bar{s}_{ij \cdot \cdot} \\ \bar{b}_{ij \cdot \cdot} \\ \bar{h}_{i \cdot \cdot \cdot} \\ \bar{s}_{i \cdot \cdot \cdot} \\ \bar{b}_{i \cdot \cdot \cdot} \end{bmatrix}$$

Variances and covariances of observations: We have derived the variance of individual observations in Eq. 10.10, the variance of family means in a single test in Eq. 10.12, and the covariance between individual observations and family means in Eq. 10.14. However, with multiple traits, a number of additional covariances are needed to fill in the entire 6 x 6 **V** matrix.

Let $y_{ijkl} = u_{ijkl}$ when the trait is 'u', and v_{ijkl} when the trait is 'v'. Then the covariance between measurements of u and v on an individual tree is

$$\begin{aligned} Cov(u_{ijkl}, v_{ijkl}) = Cov(&\mu^u + E_i^u + B_{ij}^u + f_k^u + fe_{ik}^u + p_{ijk}^u + w_{ijkl}^u, \\ &\mu^v + E_i^v + B_{ij}^v + f_k^v + fe_{ik}^v + p_{ijk}^v + w_{ijkl}^v) \end{aligned}$$

$$= \text{Cov}(f^u_k,f^v_k) + \text{Cov}(fe^u_{ik},fe^v_{ik}) + \text{Cov}(p^u_{ijk},p^v_{ijk}) + \text{Cov}(w^u_{ijkl},w^v_{ijkl})$$

$$= \sigma_{f,uv} + \sigma_{fe,uv} + \sigma_{p,uv} + \sigma_{w,uv} \qquad \textbf{Eq. 10.45}$$

where superscripts indicate effects associated with either trait u or trait v, and $\sigma_{f,uv}$ indicates the covariance component between traits u and v associated with family effects, and similarly for covariance components associated with other effects.

The covariance between a measurement of trait u on an individual, and a family mean for trait v is

$$\begin{aligned}\text{Cov}(u_{ijkl},\bar{v}_{i\cdot k\cdot}) &= \text{Cov}(\mu^u + E^u_i + B^u_{ij} + f^u_k + fe^u_{ik} + p^u_{ijk} + w^u_{ijkl},\\ &\qquad \mu^v + E^v_i + B^v_{ij} + f^v_k + fe^v_{ik} + p^v_{i\cdot k}/b + w^v_{i\cdot k\cdot}/bn)\\ &= \text{Cov}(f^u_k,f^v_k) + \text{Cov}(fe^u_{ik},fe^v_{ik}) + \text{Cov}(p^u_{ijk},p^v_{i\cdot k}/b) +\\ &\quad\ \text{Cov}(w^u_{ijkl},w^v_{i\cdot k\cdot}/bn) \qquad \textbf{Eq. 10.46}\\ &= \sigma_{f,uv} + \sigma_{fe,uv} + \sigma_{p,uv}/b + \sigma_{w,uv}/bn.\end{aligned}$$

The covariance between family means for traits u and v is

$$\begin{aligned}\text{Cov}(\bar{u}_{i\cdot k\cdot},\bar{v}_{i\cdot k\cdot}) &= \text{Cov}(\mu^u + E^u_i + B^u_{ij} + f^u_k + fe^u_{ik} + p^u_{i\cdot k}/b + w^u_{i\cdot k\cdot}/bn,\\ &\qquad \mu^v + E^v_i + B^v_{ij} + f^v_k + fe^v_{ik} + p^v_{i\cdot k}/b + w^v_{i\cdot k\cdot}/bn)\\ &= \text{Cov}(f^u_k,f^v_k) + \text{Cov}(fe^u_{ik},fe^v_{ik}) + \text{Cov}(p^u_{i\cdot k}/b,p^v_{i\cdot k}/b) +\\ &\quad\ \text{Cov}(w^u_{i\cdot k\cdot}/bn,w^v_{i\cdot k\cdot}/bn) \qquad \textbf{Eq. 10.47}\\ &= \sigma_{f,uv} + \sigma_{fe,uv} + \sigma_{p,uv}/b + \sigma_{w,uv}/bn.\end{aligned}$$

Using Eqs. 10.10, 10.12, 10.14, and 10.45 - 10.47, with the parameters in Table 10.1, **V** is

$$\mathbf{V} = \begin{bmatrix} 2.6773 & 3.6471 & -0.3053 & 0.2988 & 0.4238 & -0.0327\\ 3.6471 & 7.5723 & -0.6762 & 0.4238 & 0.9601 & -0.0784\\ -0.3053 & -0.6762 & 0.2071 & -0.0327 & -0.0784 & 0.0168\\ 0.2988 & 0.4238 & -0.0327 & 0.2988 & 0.4238 & -0.0327\\ 0.4238 & 0.9601 & -0.0784 & 0.4238 & 0.9601 & -0.0784\\ -0.0327 & -0.0784 & 0.0168 & -0.0327 & -0.0784 & 0.0168 \end{bmatrix}$$

Determining traits in w and constructing C: The aggregate genetic worth is comprised of the three traits: height growth, stem diameter, and branch diameter. Therefore, to construct **C**, covariances are needed between the six types of observations used in **y** and the breeding values for the three traits in the aggregate worth function. These covariances are needed to fill in the elements of the 6 x 3 **C** matrix ($m = 6$ = number of observations in **y**, $q = 3$ = number of traits in index).

For the covariance between an individual observation, or a family mean for a particular trait, and the breeding value for that trait, we utilize Eqs. 10.18 and 10.20; however, we must derive the covariances between measurements of both types on one trait, say u, and the breeding value for another trait, say v.

The covariance between the measurement on an individual tree for one trait and the breeding value for another trait is derived below (using Eqs. 10.16 and 10.17):

$$\begin{aligned}
\mathrm{Cov}(y^u_{ijkl}, g^v_{ijkl}) &= \mathrm{Cov}(\mu^u + E^u_i + B^u_{ij} + f^u_k + fe^u_{ik} + p^u_{ijk} + w^u_{ijkl},\ f^v_k + d^v_{ijkl}) \\
&= \mathrm{Cov}(f^u_k, f^v_k) + \mathrm{Cov}(e^u_{ijkl} + d^u_{ijkl},\ f^v_k + d^v_{ijkl}) \\
&= \mathrm{Cov}(f^u_k, f^v_k) + \mathrm{Cov}(d^u_{ijkl},\ d^v_{ijkl}) \\
&= \sigma_{f,uv} + 3\ \sigma_{f,uv} \qquad \textbf{Eq. 10.48} \\
&= \sigma_{A,uv}
\end{aligned}$$

The covariance between the family mean for one trait, and the breeding value of an individual in that family for another trait is

$$\begin{aligned}
\mathrm{Cov}(\bar{y}^u_{i\cdot k\cdot}, g^v_{ijkl}) &= \mathrm{Cov}(\mu^u + E^u_i + B^u_{ij} + f^u_k + fe^u_{ik} + p^u_{i\cdot k}/b + w^u_{i\cdot k\cdot}/bn,\ f^v_k + d^v_{ijkl}) \\
&= \mathrm{Cov}(f^u_k, f^v_k) + \mathrm{Cov}(e^u_{i\cdot k\cdot}/bn + d^u_{i\cdot k\cdot}/bn,\ f^v_k + d^v_{ijkl}) \\
&= \mathrm{Cov}(f^u_k, f^v_k) + \mathrm{Cov}(d^u_{ijkl},\ d^v_{ijkl})/bn \\
&= \sigma_{f,uv} + (3/bn)\ \sigma_{f,uv} \qquad \textbf{Eq. 10.49} \\
&= (1/4)\ \sigma_{A,uv} + (3/4bn)\ \sigma_{A,uv} \\
&= (1/4)\ \sigma_{A,uv}\ (1 + 3/bn)
\end{aligned}$$

We are using six observations (an individual observation and a family mean for each of three traits) to predict three breeding values which are aggregated into a single value measuring genetic worth ($\hat{w}$). Therefore the matrix **C** of covariances between the observations and the breeding values to be predicted has dimension 6 x 3. With Eqs. 10.48 and 10.49, and the parameter estimates in Table 10.1, **C** is constructed as

$$
\mathbf{C} = \begin{bmatrix} 0.5320 & 0.7772 & -0.0564 \\ 0.7772 & 1.8668 & -0.1432 \\ -0.0564 & -0.1432 & 0.0212 \\ 0.1441 & 0.2105 & -0.0153 \\ 0.2105 & 0.5056 & -0.0388 \\ -0.0153 & -0.0388 & 0.0057 \end{bmatrix}
$$

Defining G: Since 3 breeding values are being predicted as part of the genetic worth function, **G** is a 3 x 3 matrix of the variances and covariances of those breeding values. From the parameter estimates in Table 10.1, **G** is

$$
\mathbf{G} = \begin{bmatrix} 0.5320 & 0.7772 & -0.0564 \\ 0.7772 & 1.8668 & -0.1432 \\ -0.0564 & -0.1432 & 0.0212 \end{bmatrix}
$$

Economic weights: In Cases 1 through 3, aggregate genetic worth was composed of a single trait, and therefore the economic weight on that trait was simply one (**a** = [1]). In this Case, there are three traits and **a** is a 3 x 1 vector with elements expressing the relative economic values of a change in one measurement unit of each of the three traits. Since economic weights are relatively difficult to determine, it is often advisable to try a number of different economic weights to examine the sensitivity of the index and the four **a** vectors used in this example are:

a) $\mathbf{a}' = [1 \quad 1 \quad 1]$,
b) $\mathbf{a}' = [10 \quad 1 \quad 1]$,
c) $\mathbf{a}' = [1 \quad 10 \quad 1]$, and
d) $\mathbf{a}' = [1 \quad 1 \quad 10]$.

Results for Case 4: Once the necessary vectors and matrices (**y**, $\boldsymbol{\alpha}$, **V**, **C**, **G**, and **a**) are set up for a multiple trait, the equations presented in Chapter 9 are applied in the usual way; thus, the actual coefficients and calculations are not presented. Rather, we illustrate the effects of different sets of economic weights, and an approach to conducting a type of sensitivity analysis. Calculations which are of particular interest include the correlation between the predicted aggregate genetic worth ($\hat{w}$ = I) and the true genetic worth (w),

$$\text{Corr}(w,\hat{w}) = [\mathbf{a'C'V^{-1}Ca/a'Ga}]^{½} \qquad \text{(Eq. 9.10).}$$

and the predicted genetic gain in each of the traits in w ($\hat{\mathbf{g}}_s$). When selection is based on the index values,

$$\begin{aligned}\hat{\mathbf{g}}_s &= i_s \cdot [\mathbf{C'V^{-1}Ca}/\text{Var(I)}^{½}] \qquad \text{(Eq. 9.12)}\\ &= i_s \cdot [\mathbf{C'V^{-1}Ca/(a'C'V^{-1}Ca)}^{½}].\end{aligned}$$

As discussed in Chapter 9, $\hat{\mathbf{g}}_s$ is a 3 x 1 vector since there are three traits (height, stem diameter, and branch diameter) in w.

One approach to assigning economic weights is simply to assume that they are of equal importance, i.e. that the value of a change in one measurement unit in one trait is equal to the value of a change in one measurement unit in any other trait. This approach was utilized in Case 4a where **a'** = [1 1 1]. Given these economic weights, the correlation between the predicted and true aggregate genetic worth is 0.5311 (Table 10.6). If the top 5% of the trees in the test are selected on the basis of the index calculated in Case 4a, the following gains are expected: a gain of 0.676 m (7.3%) in height growth, 1.503 cm (9.3%) in stem diameter growth, and -0.1073 branch diameter (-4.0%), indicating slightly thicker branches (Table 10.6). The percent gains in the three traits are correlated to their heritabilities. Stem diameter has a heritability of 0.25 and a gain of 9.3% is expected, height has a heritability of 0.20 and an expected gain of 7.3%, and branch diameter has the lowest heritability (0.10) (and is negatively correlated with the other two traits), leading to a negative gain (-4.0%) (Tables 10.1 and 10.6).

Cases 4b, 4c and 4d place a weight of 10 on height, stem diameter and branch diameter, respectively, with weights of 1 on the remaining two traits. Note that the only change in these cases from Case 4a is that the vector of economic weights, **a**, has changed. All other vectors and matrices remain the same, and therefore once they have been determined it is a relatively simple matter to try different sets of economic weights to examine their effects.

The most efficient index, in terms of the correlation between the predicted and true genetic worth (Corr(w,ŵ) = 0.5471), is Case 4c when the maximum economic weight is placed on diameter, the most highly heritable trait. Similarly the least efficient index (Corr(w,ŵ) = 0.4438) is Case 4d, when the maximum economic weight is placed on the least heritable trait, branch diameter (Table 10.6).

Table 10.6 - Vector of economic weights (**a'**) for the traits of height growth, stem diameter, and branch diameter, correlation between the index and aggregate genetic worth (Corr(w,ŵ), and genetic gain for each of the three traits ($\hat{\mathbf{g}}_s$') for Case 4: selection in an open-pollinated *P. caribaea* progeny test.

Case	Economic weights (**a'**)	Corr(w,ŵ)	Genetic Gain[a] ($\hat{\mathbf{g}}_s$') Height m (%)	Stem Diam. cm (%)	Branch Diam. 4 unit scale[b] (%)
4a	[1 1 1]	0.531	0.676 (7.3)	1.503 (9.2)	-0.107 (-4.0)
4b	[10 1 1]	0.507	0.751 (8.1)	1.281 (7.8)	-0.090 (-3.3)
4c	[1 10 1]	0.547	0.591 (6.4)	1.548 (9.4)	-0.114 (-4.2)
4d	[1 1 10]	0.444	0.635 (6.8)	1.146 (7.0)	-0.046 (-1.7)

[a]Assuming selection of the top 5% of the individuals in the test according to index value.

[b]Branch diameter is measured on a 4 unit scale, where 1 is thick branches, and 4 is thin branches.

Remember that Cases 4a-4d define 4 different sets of genetic worth functions. This was done because of uncertainty in how much weight to place on each of the three traits. If the true genetic worth is Case 4c (where most weight is on stem diameter which has the highest heritability), then the correlation of true and predicted genetic worth is the highest. This is not necessarily a reason to choose the weights in 4c if they are not optimal in an economic sense.

The maximum gain in each of the three traits occurs when that trait receives the maximum economic weight, 8.1% for height in Case 4b, 9.4% for stem diameter in Case 4c, and (the least negative) -1.7% for branch diameter in Case 4d (Table 10.6). The genetic progress in a particular trait is a balance between the economic weight placed on the trait, its heritability, and its correlation with the other traits. In Case 4d, even with an economic weight 10 times that on height and stem diameter, genetic progress in branch diameter was negative (which is undesirable since thick branches = 1, and thin branches = 4) because the heritability of the trait was so low and it was negatively correlated with the other two traits (see Table 10.1).

A final item of interest can be seen by comparing Cases 4a and 4d. Going from Case 4a, where equal weights were placed on all three traits, to Case 4d, where branch diameter received a weight of 10, the expected gain in height went from 0.6762 m to 0.6345 m (Table 10.6). Gain in height in Case 4d was 94% of the gain expected in Case 4a. For stem diameter, gain went from 1.5027 cm in Case 4a to 1.1459 cm in Case 4d, thus gain in Case 4d was only 76% of that expected in Case 4a. Although both height and stem diameter are negatively correlated with branch diameter, the genetic correlation of stem and branch diameters is more negative than that between height and branch diameter (-0.72 vs -0.53, from Table 10.1). Thus, when the trait of branch diameter receives the highest economic weight, genetic progress in stem diameter is decreased more than in height growth.

SELECTION INDEX WITH UNBALANCED DATA

One of the assumptions of all four previous case studies in this chapter was that all the data were balanced, i.e. that every individual had exactly the same quality and quantity of data; thus, a single index equation could be developed which would apply to all candidates. Although this assumption might be reasonable in some circumstances, there may often be situations where violations of this assumption are serious enough to warrant a somewhat different approach to the selection of individuals. In this section, two situations are discussed which are common in many forestry applications and require departures from the traditional selection index approach.

Different Classes of Data

In many programs which are breeding and testing hundreds (or possibly thousands) of individuals it is simply impossible to maintain balanced progeny tests. Generally, some families are included in more progeny tests than others, or some families are included in a different series of tests than others. Unless all tests or series of tests are identical, it is not possible to develop a single index equation to apply to all candidates. However, if the progeny testing has been relatively well organized, it may be possible to identify certain subsets or classes of families which do meet the classical assumption of balanced data. In this case, a different index for each subset or class of families can be developed which is appropriate for the type of data available for that class.

Example 10.3: Suppose you have five control-pollinated progeny tests of *P. caribaea* which are designated PC1 - PC5, and you wish select individuals from these tests using the selection index approach. Each test includes 50 full-sib families as follows: tests PC1 and PC2 include the same 50 full-sib families, while tests PC3 and PC4 include a different 50 families. Finally 25 of the families included in PC3 and PC4 are also tested in PC5, along with 25 additional families which are not included in any other test. Thus, there are four sub-sets of families which can be identified, and the tests they are included in are: 1) PC1 and PC2, 2) PC3 and PC4, 3) PC3, PC4, and PC5, 4) PC5. If these tests have had good survival so the data are fairly balanced, it would be appropriate to develop an index for each of the four classes and calculate index values for all candidates in each class. As long as the same definition of genetic worth (w) is used in all four classes, then all four sets of predicted genetic worths are directly comparable. So all candidates can now be ranked according to their index value, and selections made ignoring which class a candidate is in.

Heterogeneous Second Moments

In Case 2, the selection index application across three tests utilized the second moments in Table 10.1. One way in which second moments might be estimated for an operational application of selection index would be to conduct a analysis of variance across the three sites, and assume that the calculated variance component estimates apply to all test sites. This assumption may be inappropriate if the tests have widely different means, and therefore variances, or if sites differ in their uniformity so that heritability is high in one test, and low in others. Closely related to the topic of heterogeneous second moments is the topic of target environments, discussed extensively in Chapter 7. If the trait of interest, say volume growth, is expressed differently in different environments (for this is what heterogeneous second moments implies), then we can no longer speak of predicting a breeding value for volume growth, rather we must predict volume growth in a specific target environment which has certain characteristics. Those characteristics may include the mean for volume growth on that site, the environmental factors like soil fertility, altitude, water availability, age, or environmental variability. Thus, if the second moments are heterogeneous, care must be taken in specifying both the genetic worth function (for a specific target environment), and in constructing the **V** and **C** matrices to treat a trait expressed in different environments as different traits.

Selection of Individuals With BLP

As different classes of data are used to develop different indices, and we allow for the data to have heterogeneous second moments, then we are essentially using Best Linear Prediction to predict the breeding value (or aggregate genetic worth of an individual forward selection. Classical selection index is simply BLP with the restriction that the data is completely balanced, and has homogeneous second moments. When these assumptions are relaxed, the result is the selection of individual trees using a BLP approach, which is no different from BLP as described throughout this book. All principles, procedures caveats, etc. which apply to BLP in general also apply to the case when prediction are being made for forward selections.

APPROXIMATION TO SELECTION INDEX

In an operational breeding program where selections are being made there are generally two objectives considered important: 1) to make maximum genetic gain while 2) maintaining genetic diversity. If N individuals are needed to carry into the next generation and only the objective of maximizing genetic gain is considered, this could be accomplished simply by 1) using a selection index or BLP to rank all individual trees on the basis of their predicted aggregate genetic worth, and then 2) picking the top 1 through N individuals. This approach, however, will generally lead to the selection of many related individuals from just a few families, and thus will not be satisfactory if both objectives (gain and genetic diversity), are considered important. For this reason, even if a breeder uses an index approach to rank all candidates, it will generally be necessary to place some kind of restriction on the number of related individuals carried into the next generation.

Given that it is important to balance the two objectives in most tree improvement programs, an alternate process may be as effective and easier to apply operationally. The process would basically involve the following steps:

1) First predict the breeding values, or aggregate genetic worth of all parents in the nth generation which have progeny being considered for the $(n+1)^{th}$ generation.
2) Consider all possible families available, determine which families should contribute selections, and how many selections should come from each family.
3) Select the best individuals within the best families identified in step 2.

This approach breaks up the selection of the best families, and the best individuals within those families into separate steps. The advantage of this is essentially logistical: instead of predicting breeding values for hundreds of thousands of individual candidate trees, first predict breeding values for the few hundred or few thousand parent trees. Then only calculate breeding values for trees within families where selections will definitely be made. Calculation procedures are reduced, and in addition, keeping track of relatedness is simplified, thus making it easier to accomplish the objective of maintaining a broad genetic base.

SUMMARY AND CONCLUSIONS

The applications of selection index methodology presented in this chapter involved relatively small examples. Even so, the amount of algebra involved in determining expressions for elements of $\boldsymbol{\alpha}$, $\mathbf{V}$ and $\mathbf{C}$ can become formidable. We believe that it is important to take a systematic approach and closely follow the 8 steps outlined in the first section of this chapter. In particular, it is critical to completely specify the linear model for the data, including which effects are fixed and which are random. Once specified, expressions for expected values, variances and covariances are developed by closely adhering to the linear model. After this, the application of the selection index formula becomes essentially computational.

Even for more complex situations than described in this chapter, the computations of selection indices are not tedious. Thus, it is often beneficial to examine index coefficients ($\mathbf{b}$) and rankings based on predicted genetic worths ($\hat{w}$) for a variety of scenarios. This was done for four different vectors of economic weights in Case 4. Vectors of index coefficients can also be calculated for different assumptions involving elements in $\mathbf{V}$ and $\mathbf{C}$ (such as different genetic correlations among the traits). If index coefficients do not change much and rankings are not greatly affected by different assumed second moments, this gives confidence that the selections made will not be greatly influenced by the particular set of assumptions (within the range tested). Conversely, alternative assumptions that greatly alter index coefficients and rankings should be critically examined and point the way to types of research that is needed.

PROBLEMS

The following problems are all are based on Case 3 presented in this chapter. The example involves developing selection indices for a four parent partial half-diallel. Second moments are provided in Table 10.4. Recall that for this example there is only one trait, height, in the genetic worth function. Thus, genetic worth (w) is used synonomously with breeding value (g) below.

Problem 10.1: Suppose you have the following data (all in units of meters): $\bar{y}_{i....} = 9.3$, $\bar{y}_{i1...} = 9.2$, $\bar{y}_{i2...} = 10.2$, $\bar{y}_{i\cdot AB\cdot} = 10.5$, $\bar{y}_{i\cdot AC\cdot} = 9.7$, $\bar{y}_{i\cdot AD\cdot} = 9.4$, $\bar{y}_{i\cdot BC\cdot} = 9.4$, $\bar{y}_{i\cdot BD\cdot} = 8.9$, $\bar{y}_{i\cdot CD\cdot} = 8.3$, $y_{i1AB2} = 11.2$, and $y_{i2AB4} = 8.2$. Note that **V**, **C**, and **G** as presented in the text for Case 3 apply to this problem. Note that there are more than two blocks in the test so that the block means do <u>not</u> average to the test mean.

a) Use the data above and the coefficients presented for Case 3a (Table 10.5) to set up **y** and $\boldsymbol{\alpha}$, and predict a genetic worth (breeding value for height) for the individual trees y_{i1AB2} and y_{i2AB4} using Eq. 9.6.

b) Use Eq. 9.7 and the matrices in Case 3a to predict the genetic worths of these two trees.

c) Use the data above and the coefficients presented for Case 3b (Table 10.5) to set up **y** and $\boldsymbol{\alpha}$ (using only the individual observation and the AB full-sib family mean), and predict a genetic worth (breeding value for height) for the individual trees y_{i1AB2} and y_{i2AB4}.

d) Calculate $\mathrm{Var}(\hat{g})$ for Cases 3a and 3b.

Problem 10.2: Suppose you wish to use the data (in Problem 10.1 above) on the individual observation y_{i1AB2} and all six full-sib family means to simultaneously predict the 4 x 1 vector, **g**, containing breeding values for the individual y_{i1AB2}, parent A, parent B, and the average offspring of the cross AxB. Note that **V** as presented in the text applies to this problem.

a) Set up the appropriate **C** matrix.

b) Set up the appropriate **G** matrix.

c) Use BLP Eq. 4.2 or selection index Eq. 9.7 (with the vector of economic weights, **a**, omitted) to predict $\hat{\mathbf{g}}$ using y_{i1AB2} in **y**. Repeat using y_{i2AB4} in **y**. Are the predicted breeding values for the individuals different? Are the predicted breeding values for parents A, B, and AxB ($\hat{g}_A$, $\hat{g}_B$, and $\hat{g}_{AxB}$, respectively) different?

d) Calculate the matrix of coefficients (**b'** = **C'V**$^{-1}$). Write out the index equation which predicts the breeding value of parent A. What is the coefficient on the individual observation y_{i1AB2}? Why?

e) How is $\hat{g}_{AxB}$ related to $\hat{g}_A$ and $\hat{g}_B$?

Problem 10.3: Because you are concerned about high plot variance (σ^2_p), you now wish to include the plot mean ($\bar{y}_{ijkl\cdot}$) in the selection index outlined in Case 3a. That is, the **y** vector will be a 8 x 1 vector containing the individual observation, the plot mean, and all six full-sib family means. To fill in all elements of **V** and **C**, you will need to derive some additional variances and covariances. For each of the following, 1) derive expressions in terms of variance components defined for the linear model, and 2) calculate the variance using the second moments in Table 10.4:

a) $\mathrm{Var}(\bar{y}_{ijkl\cdot})$,
b) $\mathrm{Cov}(y_{ijklm}, \bar{y}_{ijkl\cdot})$,
c) $\mathrm{Cov}(\bar{y}_{ijkl\cdot}, \bar{y}_{i\cdot kl\cdot})$,
d) $\mathrm{Cov}(\bar{y}_{ijkl\cdot}, \bar{y}_{i\cdot k'l\cdot})$,
e) $\mathrm{Cov}(\bar{y}_{ijkl\cdot}, \bar{y}_{i\cdot k'l'\cdot})$, and
f) $\mathrm{Cov}(\bar{y}_{ijkl\cdot}, g_{ijklm})$.

Now you wish to examine the effect of including the plot mean in the index.

g) Set up the new **V** and **C**.

h) Calculate the coefficients. Is the coefficient on the plot mean positive or negative? Why?

i) Calculate $\mathrm{Corr}(\hat{g},g)$ and ΔI (or genetic gain since we are dealing with a single trait) assuming selection of the top 5% of the individuals (i=2.063 from Appendix 2). How do these compare to $\mathrm{Corr}(\hat{g},g)$ and ΔI if the plot mean is omitted? Would you include the plot mean in the index? Why or why not?

Chapter 11
Best Linear Unbiased Prediction: Introduction

INTRODUCTION

The derivation of Best Linear Prediction (Chapter 4) and selection index theory (Chapter 9) assumed that both the first and second moments of the joint distribution of **g** and **y** were known constants. That is, the means, variances and covariances are assumed known. In practice, these are never known exactly and estimates must be used (as was done in Chapters 4-10). This means that in all applications of BLP, the predicted breeding values are approximations to BLP. To the extent that the estimated first and second moments closely approximate the true moments, the predicted breeding values closely approximate best linear predictions (see Chapter 7).

In previous chapters, estimates of the fixed effects, $\boldsymbol{\alpha}$, have been obtained by simple arithmetic averages (ordinary least squares method) prior to predicting the genetic values. Each fixed effect was often estimated using a small portion of the data such as data from just one block to estimate a block mean in a progeny test. The two step process (first estimating fixed effects, then predicting random, genetic effects) is conceptually and computationally simpler than using all data in one step to obtain estimates of fixed effects and prediction of the random effects. The estimates of $\boldsymbol{\alpha}$ were assumed to be both accurate and precise and treated as known constants throughout the BLP process. We believe that for many types of forestry data sets (i.e. those that contain several genetic check lots and/or many random samples from the same genetic population), these estimates of fixed effects are probably of the necessary accuracy and precision so that the predictions of previous chapters closely approximate BLP.

However, this is <u>not</u> always the case and this chapter examines the situation where ordinary least squares estimates of the fixed effects, $\boldsymbol{\alpha}$, are not satisfactory. In these cases, ordinary least squares estimates of the first moments (i.e. means of fixed factors) should <u>not</u> simply be estimated with means from the data with as we have done in Chapters 4-10. Rather, the estimation of the fixed effects must account for the representation and variances of the genetic material. For example, suppose that in a series of progeny tests, there are no check lots planted and that the families in any given test are <u>not</u> a random sample from the base population of families. This might occur if families were grouped in some manner (such as second generation families in some tests, first generation in other tests) or if only a few families were planted in each test. Under these circumstances, any estimate of the environmental means obtained by simple averages of family means would be both imprecise and biased. As an example of bias consider the average of family means from a test containing only very good families. This average would overestimate the environmental effect of the test because of confounding from the above average contribution of genetic effects.

In cases where "good" (i.e. precise and unbiased) estimates of the first moments can not be obtained in a simple fashion, Best Linear Unbiased Prediction (BLUP) is used to obtain both estimates of the fixed effects and predictions of the breeding values. BLUP is used extensively by animal breeders and was developed by Henderson (1963, 1973, 1977) specifically to deal with the horribly unbalanced data in breeding of dairy cattle. Specifically, in evaluating sires based on the performance of their progeny, the fixed effects are difficult to estimate because 1) subclass cell means are often 90% missing, 2) there are genetic trends in the data (due to progress from selection) and 3) records are subject to culling such that there are more data on better animals (Henderson 1973, 1974, McDaniel 1974, Iloeje and Wilcox 1981, Wilcox and Delorenzo 1983). This led to the development of BLUP which incorporates best linear unbiased estimates of the fixed effects through generalized least squares with best linear unbiased prediction of the random (i.e. genetic) effects.

The purpose of this chapter is to give a brief introduction to the theory and properties of BLUP and to highlight some possible uses of this prediction method in forestry. BLUP is conceptually more complicated than BLP and can become computationally very difficult. We stress concepts here and a complete account of BLUP is beyond the scope of this book. The reader should consult Henderson (1973, 1977, 1984), Kennedy and Sorensen (1988) and Quaas et al. (1984) for derivations and details of implementation. The next chapter provides a few specific examples of BLUP to both illustrate the method and highlight possible applications in forestry.

THEORY OF BEST LINEAR UNBIASED PREDICTION

Definition of the Mixed Linear Model

Throughout all previous chapters, we have implicitly assumed a mixed linear model for the data records composed of fixed effects (such as those due to blocks, test environments, treatment effects, etc.) and unobservable random effects. The random effects are of two types: one type is the genetic effects that are correlated with the genetic effects we are trying to predict (such as family effects in an open-pollinated progeny test are correlated with parental breeding values being predicted), while the other type is associated with experimental error such as plot and within-plot effects (these effects may contain confounded genetic effects such as the portions of the additive and dominance variances contained in the within-plot error).

In the description of best linear unbiased prediction, it is convenient to be more explicit about the form of this mixed linear model and the following model is assumed to apply to the vector of data records (see Henderson 1973, 1977, 1984):

$$\mathbf{y} = \mathbf{X}\boldsymbol{\beta} + \mathbf{Z}\mathbf{u} + \mathbf{e} \qquad \text{Eq. 11.1}$$

where

y is an n x 1 vector of observed data records (i.e. measurements),
$\boldsymbol{\beta}$ is a t x 1 vector of fixed effects such as due to blocks, test environments, fertilizers, genetic groups,
X is an n x t design or incidence matrix containing 0's and 1's that relate the fixed effects in $\boldsymbol{\beta}$ to the elements in **y**,
u is an s x 1 vector of random genetic effects such as due to family effects,
Z is an n x s incidence matrix containing 0's and 1's that relate the random effects in **u** to the elements in **y**, and
e is an n x 1 vector of random effects associated with components of experimental error such as plot effects and within plot tree effects (these may contain confounded genetic effects).

In this model, $\boldsymbol{\beta}$, **X**, **Z**, contain constants, while **y**, **u**, and **e** are vectors of random variables. The means of **u** and **e** are assumed zero, i.e. E(**u**) = E(**e**) = **0**. and thus

$$\begin{aligned} E(\mathbf{y}) &= E(\mathbf{X}\boldsymbol{\beta} + \mathbf{Z}\mathbf{u} + \mathbf{e}) \\ &= E(\mathbf{X}\boldsymbol{\beta}) + E(\mathbf{Z}\mathbf{u}) + E(\mathbf{e}) \\ &= \mathbf{X}\boldsymbol{\beta} \qquad \text{Eq. 11.2} \\ &= \boldsymbol{\alpha}. \end{aligned}$$

So as always, the expected values of the data records are those of the fixed effects, $\boldsymbol{\alpha}$. Using this model, however, the n x 1 vector of fixed effects in $\boldsymbol{\alpha}$ for the n data records are specified in terms of a smaller number (t) of common fixed effects in $\boldsymbol{\beta}$ that are related to $\boldsymbol{\alpha}$ through the incidence matrix, **X** (see Example 11.1).

The following second moments for the random vectors are defined:

Var(**y**) = **V** is the same n x n variance-covariance matrix of previous chapters which contains variances of and covariances among all measured data records,

Var(**u**) = **U** is an s x s matrix containing variances of and covariances among genetic effects in the linear model specified for the records in **y**,

Var(**e**) = **E** is an n x n matrix containing variances of and covariances among experimental error components of the linear model.

The error and genetic components are assumed uncorrelated, i.e. Cov(**u**,**e**) = **0**, and so the variances and covariances of the records can be specified in terms of the random components of the linear model:

$$\begin{aligned} \mathrm{Var}(\mathbf{y}) &= \mathbf{V} \\ &= \mathrm{Var}(\mathbf{X}\boldsymbol{\beta} + \mathbf{Zu} + \mathbf{e}) \\ &= \mathrm{Var}(\mathbf{Zu}) + \mathrm{Var}(\mathbf{e}) && \text{(since } \mathbf{X}\boldsymbol{\beta} \text{ constant)} \\ &= \mathbf{Z}\mathrm{Var}(\mathbf{u})\mathbf{Z}' + \mathrm{Var}(\mathbf{e}) && \text{(using Eq. 2.30)} \\ &= \mathbf{ZUZ}' + \mathbf{E}. \end{aligned} \qquad \textbf{Eq. 11.3}$$

Example 11.1: As a very small example to illustrate the above definitions, consider a full-sib progeny test of Coastal Douglas-fir in which six full-sib families are planted in a randomized complete block design with two blocks and ten trees from each family in each block. This example is used throughout this chapter to illustrate principles of BLUP. The six full-sib families (F_1, F_2, ... F_6) represent a complete 2 x 3 factorial among five parents (P_1, P_2, ... P_5) in which parents 1 and 2 were females and 3, 4, and 5 were males. That is, the following matings were made to form six families:

		Male Parents		
		P_3	P_4	P_5
Female Parents	P_1	F_1	F_2	F_3
	P_2	F_4	F_5	F_6

Suppose measurements are taken on height at ten years of age on each tree. A linear model for each of these 120 records (2 blocks x 6 families x 10 trees) is:

$$y_{jkl} = \mu + B_j + F_k + p_{jk} + w_{jkl}$$

where

μ is an overall mean,
B_j is a fixed effect associated with the jth block,
F_k is a random effect associated with the kth full-sib family,
$E(F_k) = 0$ and $Var(F_k) = \sigma^2_F$,
p_{jk} is a random plot effect associated with the plot of the kth family in the jth block, $E(p_{jk}) = 0$ and $Var(p_{jk}) = \sigma^2_p$, and
w_{jkl} is a random tree effect of the lth tree in the jk^{th} plot,
$E(w_{jkl}) = 0$ and $Var(w_{jkl}) = \sigma^2_w$.

For this example, we suppose that all ten trees are present in each plot and that plot means are used for all analyses. A linear model for each of the 12 plot means (2 blocks x 6 families) is obtained by averaging over the 10 trees in each plot, and shorthand notation, y_{jk}, is used to represent the 12 plot means:

$$y_{jk} = \bar{y}_{jk.} = \mu + B_j + F_k + p_{jk} + \bar{w}_{jk\cdot}$$

where $\bar{w}_{jk\cdot}$ represents the random tree effects averaged over the ten trees in plot jk. With y_{jk} as the experimental unit, **y** is a 12 x 1 vector of plot means, $\boldsymbol{\beta}$ is a 3 x 1 vector of fixed effects containing μ, B_1, and B_2, and **u** is a 5 x 1 vector of full-sib family effects. The 12 x 1 error vector, **e**, contains the plot and averaged tree errors associated with each of the 12 plot means. The vectors and matrices of Equation 11.1 are completely specified below:

$$\underset{\mathbf{y}}{\begin{bmatrix} y_{11}\\ y_{12}\\ y_{13}\\ y_{14}\\ y_{15}\\ y_{16}\\ y_{21}\\ y_{22}\\ y_{23}\\ y_{24}\\ y_{25}\\ y_{26} \end{bmatrix}} = \underset{\mathbf{X}}{\begin{bmatrix} 1&1&0\\ 1&1&0\\ 1&1&0\\ 1&1&0\\ 1&1&0\\ 1&1&0\\ 1&0&1\\ 1&0&1\\ 1&0&1\\ 1&0&1\\ 1&0&1\\ 1&0&1 \end{bmatrix}} \underset{\boldsymbol{\beta}}{\begin{bmatrix} \mu\\ B_1\\ B_2 \end{bmatrix}} + \underset{\mathbf{Z}}{\begin{bmatrix} 1&0&0&0&0&0\\ 0&1&0&0&0&0\\ 0&0&1&0&0&0\\ 0&0&0&1&0&0\\ 0&0&0&0&1&0\\ 0&0&0&0&0&1\\ 1&0&0&0&0&0\\ 0&1&0&0&0&0\\ 0&0&1&0&0&0\\ 0&0&0&1&0&0\\ 0&0&0&0&1&0\\ 0&0&0&0&0&1 \end{bmatrix}} \underset{\mathbf{u}}{\begin{bmatrix} F_1\\ F_2\\ F_3\\ F_4\\ F_5\\ F_6 \end{bmatrix}} + \underset{\mathbf{e}}{\begin{bmatrix} p_{11} + \bar{w}_{11\cdot}\\ p_{12} + \bar{w}_{12\cdot}\\ p_{13} + \bar{w}_{13\cdot}\\ p_{14} + \bar{w}_{14\cdot}\\ p_{15} + \bar{w}_{15\cdot}\\ p_{16} + \bar{w}_{16\cdot}\\ p_{21} + \bar{w}_{21\cdot}\\ p_{22} + \bar{w}_{22\cdot}\\ p_{23} + \bar{w}_{23\cdot}\\ p_{24} + \bar{w}_{24\cdot}\\ p_{25} + \bar{w}_{25\cdot}\\ p_{26} + \bar{w}_{26\cdot} \end{bmatrix}}$$

This is simply a matrix equivalent of the scalar linear model for this experiment and the scalar linear model for any y_{jk} is obtained by multiplying the proper rows of the incidence matrices, **X** and **Z**, by $\boldsymbol{\beta}$ and **u** and then adding on the proper row of **e**. For example for y_{23}, multiply the ninth row of **X** by $\boldsymbol{\beta}$, add the product of the ninth row of **Z** with **u** and add the ninth row of **e**. The result is

$$y_{23} = \mu + B_2 + F_3 + p_{23} + \bar{w}_{23\cdot} \;.$$

The expected values of all records are obtained by Equation 11.2. For the ninth row of E(**y**), $E(y_{23}) = \mu + B_2$. The second moments and use of Equation 11.3 follow directly and are illustrated in Chapter 12.

Theory of Best Linear Unbiased Prediction

The goal is the same as in all previous chapters: to develop a function of the observed n x 1 data vector, **y**, that results in accurate and precise predictions of **g**, a q x 1 vector of non-observable, random genetic values. These genetic values can be, for example, breeding values of parents or breeding values or total genetic values (clonal values) of individuals. The dimension of **g** reflects both the number of traits and the number of genetic entities being predicted. For example, if breeding values are predicted for 3 traits on each of 100 trees, then q = 300.

As with BLP, **y** must be jointly distributed (i.e. correlated) with **g** for there to be useful information in **y** for predicting **g**. Note from Equation 11.1 that this joint distribution happens via the random part of the model for **y**, that is via **u**. Retaining the definitions of Chapter 4, Var(**g**) = **G** is a q x q matrix containing variances and covariances among genetic values being predicted and Cov(**y**,**g**') = **C** is an n x q matrix of covariances between the observations and the genetic values being predicted. We assume E(**g**) = **0** meaning that all genetic entities being predicted come from the same population whose mean is zero.

From Equation 11.2, E(**y**) = **X*β*** and the fixed effects, ***β***, are assumed unknown. This is a critical difference from BLP where they are assumed known in the derivation. Here, ***β*** must be estimated and **g** must be predicted from the same data. In addressing this problem, Henderson (1963, 1984) showed that the Best Linear Unbiased Prediction of **g** is:

$$\hat{\mathbf{g}} = \mathbf{C'V^{-1}}(\mathbf{y} - \mathbf{X}\hat{\boldsymbol{\beta}}) \qquad \textbf{Eq. 11.4}$$

where

$$\hat{\boldsymbol{\beta}} = (\mathbf{X'V^{-1}X})^{-}(\mathbf{X'V^{-1}y}). \qquad \textbf{Eq. 11.5}$$

Remember from Chapter 1 that $(\mathbf{X'V^{-1}X})^{-}$ is the generalized inverse of $\mathbf{X'V^{-1}X}$. To apply Equation 11.4, fixed effects can be first estimated using Equation 11.5 and then substituted into Equation 11.4. There are other approaches that may be more efficient computationally than this method; these are discussed in a later section, Computational Considerations with BLUP.

Equation 11.4 is identical to the BLP Equation 4.2 except estimated fixed effects have been substituted for fixed effects assumed known. The particular estimate, $\hat{\boldsymbol{\beta}}$, of the fixed effects that is specified in Equation 11.5 is called the weighted or generalized least squares estimate of ***β*** and is the Best Linear Unbiased Estimate (BLUE) of ***β*** (Kmenta 1971, Graybill 1976, Henderson 1984 p 17).

Equation 11.4 is called the best linear unbiased prediction of **g** because it results in the minimum error variance of prediction among all possible linear functions of the data that are also unbiased. When Henderson derived BLUP, he began by considering only linear functions of the data. Thus, he sought a function, **B'y** + **a**, (where **B** and **a** are a matrix and vector of constants) such that the choice of **B** and **a** would result in predictions of **g** that were in some sense optimal. Just as with BLP, the optimality property he chose was to minimize was the error variance of the predictions, $E(g_i - \hat{g}_i)^2$. However, he found that with the fixed effects assumed unknown, he could not find a linear function of the data to minimize the error variance. So, he placed an additional restriction on the class of functions under consideration (already the class was restricted to only linear functions). As an additional restriction, Henderson considered only those linear functions of the data that were also unbiased, i.e. $E(\hat{\mathbf{g}}) = E(\mathbf{g}) = \mathbf{0}$. With this additional constraint, he was able to derive Equation 11.4.

The difference between BLP and BLUP is somewhat subtle, but informative. If the fixed effects are assumed known, then a function can be found that minimizes the error variance of prediction among the class of all linear functions (both biased and unbiased predictions are considered); this is BLP. If the fixed effects are considered unknown, then in the more restricted class of linear functions that result in unbiased predictions, Equation 11.4 minimizes the error variance of prediction; this is BLUP. In this context, there may be other linear functions that result in smaller error variances, but biased predictions. If the joint distribution of **g** and **y** is multivariate normal then other optimality properties accrue.

In an analogous manner, $\hat{\boldsymbol{\beta}}$, is termed BLUE for $\boldsymbol{\beta}$ because among the class of linear functions that results in unbiased estimates of $\boldsymbol{\beta}$, Equation 11.5 minimizes the error variance. The ordinary least squares estimate of $\boldsymbol{\beta}$ is $(\mathbf{X'X})^{-}(\mathbf{X'y})$ and is equivalent to the generalized least squares estimate only if all random effects are assumed to have the same variance and be uncorrelated with each other. In this special case, **V** can be written as $\mathbf{I}\sigma^2_e$ (where **I** is the identity matrix) and does not appear in the equation.

Heuristically, the inverse of the **V** matrix in Equation 11.5 serves to weight the data according to the inverse of the variance and to provide a connectedness among observations that are correlated. For example, individuals from the same family growing in two different blocks are correlated because they have one or two common parents (i.e. they are half-sibs or full-sibs, respectively). The performance of individuals in block one therefore provides information about how individuals in block two should perform and vice versa. This provides a connectedness among the blocks and is used to get a better estimate of both block means compared to ordinary least squares. This is illustrated further in Example 11.2.

Example 11.2: With the data balanced as assumed in Example 11.1, block means obtained by averaging over all 6 plot means in a block (ordinary least estimates) would serve as best linear unbiased estimates (BLUE) of $\mu + B_i$. This requires only the data from block one be used to estimate the mean for block one and similarly for block two. However, if there were only 4 of the six families planted in block 1 and 4 planted in block 2, block means obtained by simple averaging of the 4 families in each test would not account for the different genetic representation in each block and would not be BLUE for the fixed effects.

Suppose for example, that of the four families in blocks 1 and 2, two are in both blocks, and 2 in each block are only in that block. If the two families which are only in block one happen to be the very best ones genetically, then a simple average of the four families in block one would overestimate $\mu + B_1$ because of the confounded genetic effects. The inverse of the **V** matrix recognizes the connectedness that the two common families provide between blocks and utilizes this in the estimation of μ, B_1 and B_2 to provide unbiased and more precise estimates compared to OLS.

Properties of BLUE and BLUP

While we are more interested in BLUP of **g** than in BLUE of $\boldsymbol{\beta}$, there are a few properties of the GLS estimate of $\boldsymbol{\beta}$ that are useful when discussing BLUP. These, along with the properties of best linear unbiased predictions are discussed below and illustrated in Chapter 12.

In the derivations below, recall from Chapter 2 that for any matrix **A** and its generalized inverse $\mathbf{A}^-$, $\mathbf{AA^-A} = \mathbf{A}$. Also note that **X**, **V** and **C** are matrices of known constants and that for any random vector **y** and vector of constants **A**, Var(**Ay**) = **A**Var(**y**)**A'** (see Chapter 2 for details). Finally, $(\mathbf{X'V^{-1}X})^-$ is symmetric so its transpose is itself.

Defining a matrix, Q, to use in equations below: The equations below become somewhat cumbersome and to use as shorthand, we define a new matrix, **Q**

$$\mathbf{Q} = \mathbf{V}^{-1} - \mathbf{V}^{-1}\mathbf{X}(\mathbf{X'V}^{-1}\mathbf{X})^-\mathbf{X'V}^{-1}. \qquad \textbf{Eq. 11.6}$$

While **Q** appears formidable at first, it is simply an n x n matrix of constants (since **V** and **X** are constants). Once calculated, it can be used repeatedly to simplify subsequent calculations. As an example, note that the BLUP Equation 11.4 can be rewritten as $\hat{\mathbf{g}} = \mathbf{C'Qy}$; this expression reemphasizes that the prediction is just a linear combination of **y** (the linear transformation matrix is **C'Q**). That Equation 11.4 can be written in this fashion is seen by evaluating the expression:

$$\begin{aligned}
\mathbf{C'Qy} &= \mathbf{C'[V^{-1} - V^{-1}X(X'V^{-1}X)^{-}X'V^{-1}]y} && \text{(from Eq. 11.6)}\\
&= \mathbf{C'V^{-1}y - C'V^{-1}X(X'V^{-1}X)^{-}X'V^{-1}y}\\
&= \mathbf{C'V^{-1}y - C'V^{-1}X\hat{\beta}} && \text{(from Eq. 11.5)}\\
&= \mathbf{C'V^{-1}(y - X\hat{\beta})}.
\end{aligned}$$ **Eq. 11.7**

Two useful properties of **Q** are:

1) **Q** is symmetric (i.e. $\mathbf{Q'} = \mathbf{Q}$) **Eq. 11.8**
2) $\mathbf{QVQ} = \mathbf{Q}$. **Eq. 11.9**

The first property of **Q** derives because **V** and **X'X** are symmetric; the second can be shown by substituting the complete expression for **Q** from Equation 11.6 and carrying out a fair amount of matrix algebra (see Problem 12.1). Another interesting aspect is that **Q** equals the inverse of the **V** matrix minus $\mathbf{V^{-1}X(X'V^{-1}X)^{-}X'V^{-1}}$. It can be shown (see Problem 11.2) that

$$\mathrm{Var}(\mathbf{V^{-1}X\hat{\beta}}) = \mathbf{V^{-1}X(X'V^{-1}X)^{-}X'V^{-1}}.$$ **Eq. 11.10**

Thus, this last portion of **Q** is always positive (since it is a variance). Further, this last portion is related to the precision of the estimates of the fixed effects. If the fixed effects are precisely estimated, then $\mathrm{Var}(\mathbf{V^{-1}X\hat{\beta}})$, which is equivalent to the last portion of **Q**, will be smaller and less will be subtracted off from $\mathbf{V^{-1}}$.

The properties below assume only that **g** and **y** are jointly distributed (form of the distribution is unknown) and that the second moments of the distribution are known exactly.

1. The GLS estimate of $\boldsymbol{\beta}$ in Equation 11.5 is unbiased: This was a condition of derivation since $\hat{\boldsymbol{\beta}}$ is the minimum variance estimate of $\boldsymbol{\beta}$ in the class of linear functions of the data that are also <u>unbiased</u>. The fact is demonstrated in the derivation.

Derivation:

$$\begin{aligned}
E(\hat{\beta}) &= E[(\mathbf{X'V^{-1}X})^{-}(\mathbf{X'V^{-1}y})] && \text{(from Eq. 11.5)}\\
&= (\mathbf{X'V^{-1}X})^{-}(\mathbf{X'V^{-1}}) \cdot E(\mathbf{y}) && \text{(since X, V constant)}\\
&= (\mathbf{X'V^{-1}X})^{-}(\mathbf{X'V^{-1}}) \cdot \mathbf{X}\boldsymbol{\beta} && \text{(from Eq. 11.2)}\\
&= (\mathbf{X'V^{-1}X})^{-}(\mathbf{X'V^{-1}X})\boldsymbol{\beta}\\
&= \boldsymbol{\beta} && \text{(property of G-inverse)}
\end{aligned}$$

2. The variance of $\hat{\boldsymbol{\beta}}$ can be estimated: The sampling (error) variance associated with the GLS estimator of $\boldsymbol{\beta}$ in Eq 11.5 is minimum among the class of linear unbiased estimates of $\boldsymbol{\beta}$ and is derived below.

Derivation:

$$\begin{aligned} \text{Var}(\hat{\boldsymbol{\beta}}) &= \text{Var}[(\mathbf{X'V^{-1}X})^{-}(\mathbf{X'V^{-1}y})] && \text{(from Eq. 11.5)} \\ &= (\mathbf{X'V^{-1}X})^{-}(\mathbf{X'V^{-1}}) \cdot \text{Var}(\mathbf{y}) \cdot (\mathbf{V^{-1}X})(\mathbf{X'V^{-1}X})^{-} \\ &= (\mathbf{X'V^{-1}X})^{-}(\mathbf{X'V^{-1}}) \cdot \mathbf{V} \cdot (\mathbf{V^{-1}X})(\mathbf{X'V^{-1}X})^{-} \\ &= (\mathbf{X'V^{-1}X})^{-}(\mathbf{X'V^{-1}X})(\mathbf{X'V^{-1}X})^{-} && \text{(since } \mathbf{VV^{-1}} = \mathbf{I}) \\ &= (\mathbf{X'V^{-1}X})^{-} && \text{(property of G-inverse).} \end{aligned}$$

3. The predictions are unbiased: This was a condition of the derivation which must hold.

Derivation:

$$\begin{aligned} E(\hat{\mathbf{g}}) &= E[\mathbf{C'V^{-1}}(\mathbf{y} - \mathbf{X}\hat{\boldsymbol{\beta}})] && \text{(substituting from Eq. 11.4)} \\ &= \mathbf{C'V^{-1}}\,[E(\mathbf{y} - \mathbf{X}\hat{\boldsymbol{\beta}})] && \text{(since } \mathbf{C}, \mathbf{V} \text{ are constants)} \\ &= \mathbf{C'V^{-1}}\,[E(\mathbf{y}) - \mathbf{X}E(\hat{\boldsymbol{\beta}})] && \text{(since } \mathbf{X} \text{ is constant)} \\ &= \mathbf{C'V^{-1}}\,[E(\mathbf{y}) - \mathbf{X}\boldsymbol{\beta}] && \text{(see Property 1)} \\ &= \mathbf{C'V^{-1}}\,[\mathbf{X}\boldsymbol{\beta} - \mathbf{X}\boldsymbol{\beta}] && \text{(From Eq. 11.2)} \\ &= \mathbf{0} = E(\mathbf{g}). \end{aligned}$$

4. The correlation between the true and predicted genetic values is maximum: In the class of linear functions that yield unbiased predictions, BLUP maximizes the correlation between g_i and $\hat{g}_i$ (Henderson 1977, 1984).

5. When multivariate normal, the probability of selecting the better candidate is maximum within the class of linear functions with mean zero: If **g** and **y** are distributed jointly with a multivariate normal distribution, then in the class of linear functions with mean predictions of zero, BLUP maximizes the probability of the correct ranking of the elements of **g** (Henderson 1973) and hence maximizes the probability of selecting the better candidate from any pair (Henderson 1977).

6. The variance among predictions can be estimated: The variance among predictions expected upon repeated sampling (i.e. repeated runs of the experiment) is $\text{Var}(\hat{\mathbf{g}}) = \mathbf{C'QC}$. This q x q matrix contains variances among predictions on the diagonal and covariances among predictions on the off-diagonal. For example, the first element estimates the variance for the prediction of the first element of **g**. This is an estimate of the variability or spread among lots of predictions all made with data of the same quality of data as for g_1.

While **C'QC** is most convenient for computing Var($\hat{\mathbf{g}}$), the last three lines in the derivation below are more informative. In the last line, it can be seen that two sources contribute to the variance among predictions in BLUP. Var($\hat{\mathbf{g}}$) is written as $\mathbf{C'V^{-1}C}$ - Var($\mathbf{C'V^{-1}X\hat{\beta}}$). The first part of this expression ($\mathbf{C'V^{-1}C}$) is the formula for variance among predictions using best linear prediction (see Eq. 4.4). As discussed in Chapter 4, the variance among predictions in BLP is larger for predictions based on high quality and large quantities of data. When the data in the **y** vector are poor for predicting **g**, the predictions in BLP are "shrunk" back toward the mean (i.e. less spread among the predictions).

In BLUP, Var($\mathbf{C'V^{-1}X\hat{\beta}}$) also contributes to the variance among predictions by reducing the spread in predictions if the fixed effects are poorly estimated. In BLP the fixed effects are assumed known, but in BLUP they are estimated. Thus, in BLUP Var($\hat{\mathbf{g}}$) will be large if the data are of high quality and quantity for predicting **g** and if the fixed effects are precisely estimated. Note that if the fixed effects are perfectly estimated (i.e. known) then Var($\hat{\boldsymbol{\beta}}$) = 0 and Var($\hat{\mathbf{g}}$) for BLUP equals that for BLP. The last 5 lines represent alternative computational formulae:

Derivation:

$$\begin{aligned}
\text{Var}(\hat{\mathbf{g}}) &= \text{Var}[\mathbf{C'V^{-1}(y - X\hat{\beta})}] && \text{(from Eq. 11.4)} \\
&= \text{Var}(\mathbf{C'Qy}) && \text{(from Eq. 11.7)} \\
&= \mathbf{C'Q} \cdot \text{Var}(\mathbf{y}) \cdot \mathbf{Q'C} && \text{(since } \mathbf{C}, \mathbf{Q} \text{ are constants)} \\
&= \mathbf{C'QVQC} && \text{(Eq. 11.9 \& definition of } \mathbf{V}) \\
&= \mathbf{C'QC} && \textbf{Eq. 11.11} \\
&= \mathbf{C'[V^{-1} - V^{-1}X(X'V^{-1}X)^{-}X'V^{-1}]C} && \text{(substitute for } \mathbf{Q}\text{, Eq. 11.6)} \\
&= \mathbf{C'V^{-1}C - C'V^{-1}X(X'V^{-1}X)^{-}X'V^{-1}C} && \\
&= \mathbf{C'V^{-1}C - C'V^{-1}X[\text{Var}(\hat{\beta})]X'V^{-1}C} && \text{(from Property 2 above)} \\
&= \mathbf{C'V^{-1}C} - \text{Var}(\mathbf{C'V^{-1}X\hat{\beta}}) && \text{(from Eq. 2.30).}
\end{aligned}$$

7. The covariance between true and predicted genetic values can be estimated: Cov($\hat{\mathbf{g}}$,g') is a q x q matrix of covariances with the covariances between the true and predicted genetic value for a given genetic value, i.e. Cov($\hat{g}_i, g_i$) on the diagonal. The off-diagonal elements are covariances between true and predicted genetic values for different genetic values, i.e. Cov($\hat{g}_i, g_{i'}$). The diagonal elements are useful for calculating correlations between true and predicted genetic values (see Property 9 below). As with BLP, it is also true for BLUP that $\text{Cov}(\hat{\mathbf{g}},\mathbf{g}') = \text{Var}(\hat{\mathbf{g}})$.

Derivation:

$$
\begin{aligned}
\text{Cov}(\hat{\mathbf{g}},\mathbf{g}') &= \text{Cov}(\mathbf{C}'\mathbf{Q}\mathbf{y},\mathbf{g}') && \text{(from Eq. 11.7)} \\
&= \mathbf{C}'\mathbf{Q}[\text{Cov}(\mathbf{y},\mathbf{g}')] && \text{(since } \mathbf{C}, \mathbf{Q} \text{ are constants)} \\
&= \mathbf{C}'\mathbf{Q}\mathbf{C} && \text{(by definition of } \mathbf{C}\text{)} \\
&= \text{Var}(\hat{\mathbf{g}}) && \text{(from Property 6).}
\end{aligned}
$$

8. The error variance of predictions can be estimated: $\text{Var}(\hat{\mathbf{g}} - \mathbf{g})$ is a q x q matrix of error variances and covariances expected upon repeated sampling. The diagonal elements are estimated error variances associated with the prediction of a given genetic value, i.e. $\text{Var}(\hat{g}_i - g_i)$, and are useful in assessing the precision of the predictions. From the discussion of $\text{Var}(\hat{\mathbf{g}})$ in Property 6, the error variance of predictions from BLUP can be written as $\mathbf{G} - \mathbf{C}'\mathbf{V}^{-1}\mathbf{C} + \text{Var}(\mathbf{C}'\mathbf{V}^{-1}\mathbf{X}\hat{\boldsymbol{\beta}})$. This equals the error variance of predictions from BLP (i.e. $\mathbf{G} - \mathbf{C}'\mathbf{V}^{-1}\mathbf{C}$, Eq. 4.6) plus a term associated with the precision with which the fixed effects are estimated. If the fixed effects are poorly estimated, the error variance of predictions is larger.

Derivation:

$$
\begin{aligned}
\text{Var}(\hat{\mathbf{g}} - \mathbf{g}) &= \text{Var}(\hat{\mathbf{g}}) + \text{Var}(\mathbf{g}) - 2\text{Cov}(\hat{\mathbf{g}},\mathbf{g}) \\
&= \text{Var}(\hat{\mathbf{g}}) + \text{Var}(\mathbf{g}) - 2\text{Var}(\hat{\mathbf{g}}) && \text{(from Property 7)} \\
&= \text{Var}(\mathbf{g}) - \text{Var}(\hat{\mathbf{g}}) \\
&= \mathbf{G} - \mathbf{C}'\mathbf{Q}\mathbf{C}. && \textbf{Eq. 11.12}
\end{aligned}
$$

9. The correlation between true and predicted genetic values can be estimated: Our favorite measure of precision is to estimate how closely the predictions of genetic values correlate with the true genetic values. The derivation for this and the formula to use are identical to those for BLP and are not derived here (see Eq. 4.7). For the ith genetic value in **g**, $\text{Corr}(\hat{g}_i,g_i) = \{\text{Var}(\hat{g}_i)/\text{Var}(g_i)\}^{1/2}$. To apply this formula, find the ith diagonal elements from $\mathbf{C}'\mathbf{Q}\mathbf{C}$ (for the numerator) and **G** (for the denominator).

10. BLUP of a linear transformation of the genetic values is that function of the predictions: If the prediction desired is a linear function of the genetic values, i.e. $\mathbf{Lg}$, the best linear unbiased prediction of that function is the same linear function of the predicted genetic values, i.e. $\mathbf{L}\hat{\mathbf{g}}$. The proof of this for BLUP follows along the exact lines as for BLP presented as Property 5 in Chapter 4 and is not repeated here. The importance of this property for both BLP and BLUP is that the q x 1 vector of individual genetic values can first be predicted and then the predicted genetic values aggregated and combined via any linear function to result in the BLUP of that linear function. If **L** contains economic weights, then the aggregated genetic values will be BLUPs of genetic worth (i.e. w of Chapter 9).

TOPICS IN BEST LINEAR UNBIASED PREDICTIONS

In the next several sections, we introduce topics dealing with different aspects of applying BLUP. These discussions are neither definitive nor exhaustive. While the concepts behind BLUP are straightforward, its application can become quite complicated. In the case of breeding dairy cattle, the data sets to which BLUP is applied tend to have the dubious distinction of suffering from all possible types of complications at once: 1) the data sets are huge (often containing hundreds of thousands of records) making computations difficult, 2) the data sets are horribly unbalanced (often over 90% of the subclasses have no observations), 3) the animals often represent many overlapping generations so that the effects of selection and relatedness must be accounted for, and 4) the data records have been subject to culling such that more data is available on better animals (Henderson 1973, 1974, McDaniel 1974, Iloeje and Wilcox 1981, Wilcox and DeLorenzo 1983).

These problems complicate the application of BLUP and much of the dairy literature relates to dealing with these issues. In particular, because the data sets are so large and interconnected, all data observations must be included in the analysis. This makes computational efficiency a necessity and much effort has been devoted to this aspect of BLUP application (e.g. inverting the relationship matrix, see Henderson 1976; the reduced animal model, see Quass et al. 1984 and Henderson 1988). In forestry and many crop breeding programs, we believe BLUP may have utility, but with only some, not all, of the complications. For example, in forest tree breeding, data within a test site are generally balanced (or nearly so) at the subclass or plot level (i.e. few if any missing plots). Also, it may often be possible to divide a large data set into many smaller, unconnected ones so that analysis presents fewer computational problems.

Thus, problems and concerns with BLUP applications in forestry will likely be different than in dairy science. As BLUP is new to forestry and most crop breeding, more work is needed on all of the topics discussed below. The discussions below represent only current impressions.

Estimating Second Moments

As with BLP, BLUP assumes the second moments of the joint distribution of **g** and **y** are known; hence, **C** and **V** are treated as matrices of constants (not random variables) in the derivations above. In practice, **C** and **V** must be estimated either from the data being used for calculating predictions or from external data sets. Thus, all applications of BLUP are approximate. Use of more accurate and precise estimates of the elements of **C** and **V** result in predictions that are more closely BLUP. In general, selection decisions are probably less affected by errors in the second moments than are estimates of genetic progress (Sales and Hill 1976). That is, the rank order of the predictions is less sensitive than the absolute values of the predictions. Nevertheless, more research is needed on the effects of estimating **C** and **V** on the predictions in forestry.

The theory of both BLP and BLUP allow for complete generality in the **C** and **V** matrices and when properly specified, both BLP and BLUP correctly weight and scale the data (see Gianola 1986). We have often specified heterogeneous second moments (e.g. Chapters 7 and 8) both to account for the fact that data were of different ages and from different test locations. For example, the variance among family means from a well-tended, precise test will not be the same as from an imprecise progeny test of the same age; nor will the correlations of these tests with the breeding values being predicted be the same.

Because of the large data sets in dairy cattle breeding, the assumptions of homogeneous genetic and error variances are often made to allow computations to be tractable. A common model in sire evaluation (Henderson 1973, 1977, 1984, 1988, Thompson 1979, Quaas and Pollak 1981) is to write $\mathbf{U} = \mathbf{A}\sigma^2_s$ and $\mathbf{E} = \mathbf{I}\sigma^2_e$ where **A** is the numerator relationship matrix specifying the genetic relations among the animals, **I** is the identity matrix and these matrices are multiplied by scalars representing the sire (genetic) and error variances assumed to apply to all observations in **y**. Henderson (1976) found an easy way to specify the inverse of **A** and make BLUP computationally much more feasible.

When second moments are assumed homogeneous, but really are not, prediction errors (and therefore selection errors) will result. For example, if error variances are assumed homogeneous but in fact differ dramatically among observations, then too many selections will be made from environments with large experimental error. This is because the observations from these environments will tend to have large variability (i.e. very high and low) and these observations are not properly scaled to reflect that the variability is caused by experimental error and not genetic differences. The dairy cattle breeders are finding that the assumption of homogeneous variances is resulting in too many selections from herds with higher error variance.

When data sets are large or even of moderate size, computational considerations become quite important and the assumption of homogeneous second moments may be required in order to make the computations feasible. If so, one possibility may be to standardize the data to variance = 1 prior to analysis (see Hill 1984). Then the assumption of homogeneous variances would be met. For example, if the goal is predicting breeding values of parents based on performance of their progeny growing in different test locations, the family means from each test location might be used as the unit of observation in $\mathbf{y}$ and standardized to have variance = 1 from each test. Thus, all tests would have the same variance. Care would need to be exercised that all elements in $\mathbf{C}$ and $\mathbf{V}$ apply to the standardized values.

Predictions for Genetic Values from Different Populations

It is sometimes the case that the genetic values being predicted do not come from the same population and thus, we may not want to assume that $E(\mathbf{g}) = \mathbf{0}$. In forestry, progeny from both first and second generation selections may be mixed to different degrees at various progeny test locations. If no check lots are planted in the tests, environmental effects associated with the test locations will be confounded with the "generation effects". For example, if only second generation selections occur in a specific test, the test location effect will be biased upward and the genetic predictions biased downward if the arithmetic test mean is used to estimate the test location fixed effect. In this instance, the generations might be considered different populations with selections (i.e. parents) nested within populations.

Another example in forestry occurs in provenance testing where several (say 30) different geographical seed sources and many different parents within each seed source are included in a testing program. The goals may be to both find the best seed source for different planting environments and to predict breeding values for parents within each seed source. It is often inconvenient, impossible or inefficient to plant all provenances and or all families at all locations and so a subset is planted at each location. In this instance, the arithmetic mean at each location may be influenced (biased) both by the specific provenances and families at that location. It is common to analyze each test separately or to group test locations for combined analysis that have the same provenances and families; however, there may be more efficient ways of analysis that would include all data at once and increase the precision of the estimates and predictions. In this example, provenances would be considered different populations with families nested within.

One approach to such an analysis is to consider the different populations (sometimes called "groups" in the dairy literature) as fixed effects to be estimated along with predictions of genetic values. In the above examples, generations and provenances would be considered fixed, while parents would be random variables whose breeding values were to be predicted. This approach has been used and discussed in the dairy literature (Henderson 1973, 1975b, 1984, Thompson 1979, Quaas and Pollak 1981). If group or population effects are considered fixed then they can be estimated as part of $\hat{\boldsymbol{\beta}}$ where $\boldsymbol{\beta}' = [\boldsymbol{\beta}'_p \mid \boldsymbol{\beta}'_o]$. Here, the vector of fixed effects is partitioned into a set of effects dealing with the populations, $\boldsymbol{\beta}_p$, and a set containing all other fixed effects (such as locations, blocks, treatments, etc.).

Henderson (1973, 1977, 1984) shows that for any linear, estimable functions of the fixed effects, say $\mathbf{P}\boldsymbol{\beta}$,

$$\hat{\mathbf{w}} = \mathbf{P}\hat{\boldsymbol{\beta}} + \mathbf{L}\hat{\mathbf{g}} \qquad \textbf{Eq. 11.13}$$

is BLUP for w where $\hat{\boldsymbol{\beta}}$ is from Eq. 11.5 and $\hat{\mathbf{g}}$ is from Eq. 11.4. That is, new linear functions of the best linear unbiased estimates of the fixed effects and the best linear unbiased predictions of the genetic values will result in BLUP of w. Variances, covariances and other attributes of these predictions are presented in the cited references.

A portion of **P**, say $\mathbf{P}_p$, could be written that would be an incidence matrix of 1's and 0's relating the data to the groups (i.e. populations) they are in; all other fixed effects could be ignored. For example, if the ith observation in **y** is in group 2, then the ith row of $\mathbf{P}_p$ has a 1 in the second column and zeros elsewhere. In this case, to predict the merit of a parent from a given population, write

$$\hat{\mathbf{w}} = \mathbf{P}_p\hat{\boldsymbol{\beta}}_p + \hat{\mathbf{g}}. \qquad \textbf{Eq. 11.14}$$

In this case, note that the genetic value predictions (breeding value predictions in our example) are predicted as deviations from population means because the population effects are treated as fixed and are therefore part of the $\mathbf{X}\hat{\boldsymbol{\beta}}$ used to express data records as deviations in Equation 11.4.

If population (e.g. provenance) by location interaction terms were included as part of the estimated fixed effects, it would seem possible to predict performances of parents from particular provenances when planted at specific locations. By writing a **P** that picked up main and interaction effects for all provenances at a certain location, it should be possible to predict (and assess the reliability of the predictions) which provenance does best at which location.

There may be other uses for this general concept of predicting linear combinations of the fixed and random effects.

Effects of Inbreeding and Coancestry

If the genotypes being evaluated are not all unrelated to each other then incorporating the relationships among relatives reduces the error variance of prediction. This is a direct result of considering genetic values as random variables to be predicted instead of fixed effects to be estimated. Then relatives have positive genetic covariances that are part of the **V** matrix; an example would be that sibs with one parent in common would have a greater correlation of relationship than individuals with no parents in common but a smaller coefficient of relationship than sibs with both parents in common. The usefulness of this type of information has been demonstrated in Chapters 6 and 10 when discussing full-sib and other types of progeny tests.

It will happen after some generations of selection that other types of relationships occur in addition to classical sibling ones. For example in tree improvement programs, two or more second generation selections are sometimes made from the same full-sib family in a first-generation progeny test. When these selections are progeny tested in the second generation, offspring from one selection are related to and therefore provide information about all selections from that full-sib family. Incorporating this information into the **V** matrix decreases the error variance of predicted breeding values for all selections (Henderson 1975c).

It will also happen that inbred offspring result from breeding related selections. For example in forestry, Lindgren (1986) has suggested that all top selections each generation be selfed or crossed to relatives as a method of increasing genetic progress. If an additive genetic model is assumed (no dominance or epistasis), then BLUP accounts for the effects of inbreeding through the relationships and variances in the **V** matrix (Henderson 1975a, 1984, Kennedy and Sorensen 1988). For example, one effect of inbreeding is to increase variance among individuals that are inbred by a factor of $1 + \bar{F}$ (where $\bar{F}$ is the average inbreeding coefficient of the parents) above that of non-inbred parents (see Kennedy and Sorensen 1988). Thus, variances on the diagonal of the **V** matrix will be larger for inbred parents. Likewise, covariances (off diagonal elements) will also change. For an additive genetic model then, BLUP will account for the effects of inbreeding if the variances and covariances are properly specified in **V**.

If dominance exists (as it certainly does in most outcrossing species), then inbred offspring will suffer from inbreeding depression (Falconer 1982). It is not clear to us how BLUP as usually applied in animal breeding programs (i.e. completely additive model) accounts for an average depression of inbred offspring. It seems that if inbreeding depression is ignored, breeding values for parents with inbred offspring will be underpredicted. One approach might be to use a covariance adjustment for average inbreeding depression on all data records in **y** prior to computing BLUP.

In many animal breeding programs, genetic relationships are specified in what is called the numerator relationship matrix (often written **A**). This matrix contains 1 + F_i for the ith diagonal element of **V** and appropriate coefficients of genetic relationships as off diagonal elements (see Thompson 1979, Henderson 1984, Kennedy and Sorensen 1988). The value of such a matrix in large BLUP problems is that the inverse of **A** can be readily specified (Henderson 1976). So, to simplify computing, dairy cattle breeders make several assumptions concerning homogeneous genetic and error variances which allow **V** to be specified as sums and products of simpler matrices so $\mathbf{V}^{-1}$ is calculated more easily. The relationship matrix is used in this process (see sections on Computing BLUP and on Estimating Second Moments). While the numerator relationship matrix may have much utility in forestry and other breeding programs, it is critical to examine all assumptions implied by the linear model. When the assumption of homogeneous genetic and error variances for all families, individuals and test locations is violated, the utility of the numerator relationship matrix is seemingly lessened. In this case, elements of **V** should be directly specified (not indirectly specified by $\mathbf{A}\sigma^2_s$) but if the data set is large, **V** may be computationally difficult to invert. The trade offs between computational tractability and more correct specification of the model will need to be balanced.

Effects of Selection on BLUP

When the genotypes being evaluated represent several generations of selection, BLUP and BLP account for the effects of selection (and can therefore be used to estimate genetic progress) as long as all genotypes (and their relationships) back to the original unselected population are included. This statement holds as long effects of selection on genetic variances and covariances are small or accounted for. Kennedy and Sorensen (1988) provide an excellent example demonstrating that the predictions for the unselected population average zero while those from later generations show the influence of selection. If all genotypes or relationships are not included (perhaps not known or perhaps for computational feasibility), then one option is to group the genotypes into different populations as described above and estimate the groups as fixed effects.

Computing Best Linear Unbiased Predictions

As mentioned several times, computational feasibility and efficiency can become important considerations even for moderate sized data sets. This has led dairy cattle breeders to simplified linear models (see the section Estimating Second Moments in this chapter) and to using what are called Mixed Model Equations (Henderson 1973, 1975b, 1977, 1984, Quaas et al. 1984, Kennedy and Sorensen 1988). A description of mixed model equations is beyond the scope of this book. We only point out that they are used because they reduce the dimensions of the matrices that need to be inverted and yet give equivalent results as using Equations 11.4 and 11.5. They also depend upon the specific linear models that are assumed for **y** and for linking **g** and **y**. When applying a mixed model equation from the dairy literature, the assumptions implicit in the specific linear models should be evaluated for appropriateness.

It appears that there will always be trade-offs in the application of BLUP between the computational problems associated with specification of a detailed more general linear model (such as allowing for heterogeneous second moments) and the more tractable computations associated with specification of a simpler linear model. Some items involved in the decision are judgement, suitable approximations (e.g. breaking up large data sets into smaller ones even if a small amount of connectedness is lost), standardization (so that homogeneous second moments is a better assumption), choice of units of observation (e.g. using family means as opposed to individuals, see Chapter 7) and computational capabilities.

BLP vs BLUP

The sole difference between BLP and BLUP is the assumption in BLP that fixed effects associated with the measured data are known without error. This is never the case and therefore all uses of BLP are approximations. However as pointed out above, applications of both BLP and BLUP are also approximate because the assumed-known second moments must be estimated. The choice between prediction methodologies must then be based on which assumptions are most suitable given computational capabilities.

If simple estimates of the fixed effects (such as arithmetic averages) can be found that are accurate and precise and if use of these estimates reduces the computational complexity so that a more general linear model can be employed for the data, then BLP is more suitable. For example, using test means to estimate environmental effects in Chapter 8 allowed predictions of breeding values of unrelated parents to be calculated one at a time.

On the other hand, there are other instances where use of BLP would not greatly simplify calculations (i.e. where the size of **V** is not reduced) or where generalized least squares estimates are clearly needed to obtain unbiased, precise estimates of the fixed effects. Remember that in concept, BLUP is just BLP using GLS estimates of the fixed effects.

SUMMARY AND CONCLUSIONS

Neither BLP nor BLUP are perfect prediction methodologies for all situations. If **C** and **V** are only poorly estimated, the predictions from these approaches may be quite imprecise and biased. Also, for binomial traits the predictions may be further from having the optimality properties associated with multivariate normality. Nevertheless, we believe the concepts of BLUP are straightforward and that BLUP is quite elegant in the way it handles data. As tree improvement programs mature and collect data across several generations of selection and at several ages, BLUP should find more use as a method of predicting genetic values. It could also be more widely used in other breeding programs.

To use either BLP or BLUP, the linear model and all of its assumptions must be specified. This has the advantages of 1) allowing the effects of violating these assumptions to be tested through simulation studies and 2) allowing new data, better estimates and more general models to be utilized as soon as possible. For example, poor estimates for **C** and **V** really mean that breeding value prediction, and hence selection, will be a crap shoot however accomplished because both the heritabilities and relationships among traits and the quality of various types of data are not well understood. Use of BLP or BLUP uses the best estimates available at the time and can be recalculated any time better estimates or more data become available.

Unfortunately, computational complexities often make it difficult to actually apply BLUP and may often even obscure the elegance with which all data are utilized to both estimate fixed effects and predict genetic values. This serves as a challenge to the analyst to find an appropriate linear model such that the data are modeled as well as possible and yet the problem is computationally feasible.

PROBLEMS

Problem 11.1: Show, by conducting the necessary matrix algebra, that **QVQ** = **Q** where **Q** is defined in Equation 11.6.

Problem 11.2: Derive Equation 11.10.

Problem 11.3: List and briefly describe types of problems and situations in forestry or in other breeding programs where BLUP would provide much better predictions of genetic values than BLP.

Chapter 12
Best Linear Unbiased Prediction: Applications

INTRODUCTION

Best linear unbiased prediction (BLUP) can be thought of as best linear prediction (BLP) with one difference: the vector of fixed effects ($\boldsymbol{\alpha}$) associated with the vector of observations (**y**) is assumed known in BLP, but estimated in BLUP. Many of the steps required to accomplish BLUP and characteristics of how BLUP handles data are similar to those discussed for BLP in Chapters 4 - 8; they are therefore not discussed in detail in this chapter. The purpose of this chapter is to apply BLUP for two simple cases, and to illustrate some interesting aspects of prediction that are unique to BLUP, and therefore not covered in earlier chapters. Finally, we present some examples where use of BLUP might accrue significant advantages over BLP.

Application of BLUP follows these general steps:

1) Define a linear model appropriate for the data.
2) Decide what kinds of observations will make up the n x 1 **y** vector.
3) Determine the phenotypic variances and covariances of the observations in **y**, and construct **V**.
4) Calculate the genetic covariances between the observations in **y** and the genetic values being predicted (**g**), and construct **C**.
5) Determine the genetic variances and covariances for all traits being predicted, and construct **G**.
6) Construct **X**, the n x t design matrix, where t is the number of fixed effects.

When these steps have been accomplished, the equations in Chapter 11 can be used to 1) make generalized least squares estimates of the fixed effects (Eq. 11.5), 2) make best linear unbiased predictions of the random effects (Eq. 11.4), and 3) estimate precisions associated with these estimates and predictions (Eq. 11.12). We illustrate the application of BLUP with two simple cases.

CASE 1: DISCONNECTED DIALLELS IN THREE TESTS

Assume that nine loblolly pine parents (P1 - P9) have been mated in three three-parent disconnected half-diallels (omitting selfs), designated A, B, and C. Thus, there are nine full-sib families as in Fig 12.1. Offspring from the three diallels are planted in three progeny tests in the following design: test 1 contains diallels A and B, test 2 contains B and C, and test 3 contains C and A. The three tests are planted

DIALLEL A	P2	P3
P1	A1	A2
P2		A3

DIALLEL B	P5	P6
P4	B1	B2
P5		B3

DIALLEL C	P8	P9
P7	C1	C2
P8		C3

Figure 12.1 - Mating design for Case 1: three 3-parent diallels.

in randomized complete block designs, with 6 blocks and 6 trees per plot. Each test contains only six full-sib crosses and no checklots were planted. All 3 tests have been measured for height at 12 years and estimates of the second moments are available (Table 6.1). The goal of the analysis is to predict the breeding values of the nine parents.

Linear Model and the y Vector

Assume that the linear model defined in Eq. 6.1 is appropriate for the progeny test data, and that full-sib family means from a single test (as defined in Eq. 6.2) are the units of observation. The **y** vector is:

$$\mathbf{y} = \begin{bmatrix} \bar{y}_{1\cdot12\cdot} \\ \bar{y}_{1\cdot13\cdot} \\ \bar{y}_{1\cdot23\cdot} \\ \bar{y}_{1\cdot45\cdot} \\ \bar{y}_{1\cdot46\cdot} \\ \bar{y}_{1\cdot56\cdot} \\ \bar{y}_{2\cdot45\cdot} \\ \bar{y}_{2\cdot46\cdot} \\ \bar{y}_{2\cdot56\cdot} \\ \bar{y}_{2\cdot78\cdot} \\ \bar{y}_{2\cdot79\cdot} \\ \bar{y}_{2\cdot89\cdot} \\ \bar{y}_{3\cdot78\cdot} \\ \bar{y}_{3\cdot79\cdot} \\ \bar{y}_{3\cdot89\cdot} \\ \bar{y}_{3\cdot12\cdot} \\ \bar{y}_{3\cdot13\cdot} \\ \bar{y}_{3\cdot23\cdot} \end{bmatrix} = \begin{bmatrix} 38.8 \\ 33.3 \\ 37.1 \\ 25.6 \\ 35.2 \\ 33.5 \\ 24.6 \\ 27.8 \\ 25.6 \\ 28.1 \\ 25.7 \\ 29.9 \\ 25.3 \\ 33.9 \\ 31.2 \\ 34.6 \\ 36.0 \\ 33.1 \end{bmatrix}$$

Obs. #	Diallel
1	A
2	A
3	A
4	B
5	B
6	B
7	B
8	B
9	B
10	C
11	C
12	C
13	C
14	C
15	C
16	A
17	A
18	A

There are 6 full-sib means from each of three tests for a total of 18 observations in the 18 x 1 **y** vector. The labels of observation number and diallel are for convenience and later reference.

Constructing V, C, G

For full-sib data in general, and for the second moments given in Table 6.1 in particular, the construction of **V**, **C**, and **G** have been described in detail in Chapter 6. As construction of these matrices proceeds in exactly the same manner for both BLP and BLUP, their construction is not detailed here. For this example, the final form of **V** is

(Obs. #)

$$
\begin{array}{cccccccl}
 & 1,2,3 & 4,5,6 & 7,8,9 & 10,11,12 & 13,14,15 & 16,17,18 & \text{(Obs \#)} \\
 & \mathbf{V1} & \mathbf{Z} & \mathbf{Z} & \mathbf{Z} & \mathbf{Z} & \mathbf{V2} & 1,2,3 \\
 & \mathbf{Z} & \mathbf{V1} & \mathbf{V2} & \mathbf{Z} & \mathbf{Z} & \mathbf{Z} & 4,5,6 \\
\mathbf{V} = & \mathbf{Z} & \mathbf{V2} & \mathbf{V1} & \mathbf{Z} & \mathbf{Z} & \mathbf{Z} & 7,8,9 \\
 & \mathbf{Z} & \mathbf{Z} & \mathbf{Z} & \mathbf{V1} & \mathbf{V2} & \mathbf{Z} & 10,11,12 \\
 & \mathbf{Z} & \mathbf{Z} & \mathbf{Z} & \mathbf{V2} & \mathbf{V1} & \mathbf{Z} & 13,14,15 \\
 & \mathbf{V2} & \mathbf{Z} & \mathbf{Z} & \mathbf{Z} & \mathbf{Z} & \mathbf{V1} & 16,17,18
\end{array}
$$

where the sub-matrices **V1**, **V2**, and **Z** are below:

$$
\mathbf{V1} = \begin{bmatrix} 8.51 & 2.70 & 2.70 \\ 2.70 & 8.51 & 2.70 \\ 2.70 & 2.70 & 8.51 \end{bmatrix} \qquad
\mathbf{V2} = \begin{bmatrix} 4.50 & 1.80 & 1.80 \\ 1.80 & 4.50 & 1.80 \\ 1.80 & 1.80 & 4.50 \end{bmatrix} \qquad
\mathbf{Z} = \begin{bmatrix} 0 & 0 & 0 \\ 0 & 0 & 0 \\ 0 & 0 & 0 \end{bmatrix}
$$

This is a short-hand method of writing **V**; **V** is expanded by writing out each of the sub-matrices **V1**, **V2**, and **Z** in full in **V** above. The first **V1** sub-matrix (in the upper left of **V**) relates to the three full-sib means from diallel A planted in test 1. The diagonal elements in **V1** represent variances of full-sib family means for height at 12 years (this is 8.51 using the parameters from Table 6.1 in Eq. 6.4). The off-diagonal elements in **V1** are covariances between two full-sib family means with one common parent and planted at the same test site (this is 2.70 using these parameters in Eq. 6.7). Note that the 6 **V1**s occurring in **V** represent the variances and covariances for a block of observations from a diallel in one test.

For **V2**, consider that the **V2** in the upper right corner of **V** represents covariances of the first three elements of **y** with elements 16, 17, and 18. These are the covariances between the observations from diallel A in test 1 and the observations from diallel A in test 2. There are two kinds of covariances involved with the covariances of a block of observations from a given diallel in one test, and a block of observations on the same diallel in another test. First, the diagonal elements of **V2** are covariances between full-sib family means with both parents in common planted on different sites (4.50 using parameters from Table 6.1 in Eq. 6.5). For example, the covariance between observations 1 ($\bar{y}_{1 \cdot 12 \cdot}$) and 16 ($\bar{y}_{3 \cdot 12 \cdot}$) is the first diagonal element of the upper right **V2**, i.e., $Cov(\bar{y}_{1 \cdot 12 \cdot}, \bar{y}_{3 \cdot 12 \cdot})$. These observations are full-sib family means with both parents 1 and 2 in common, planted in tests 1 and 3. Second, the off-diagonal elements in **V2** are covariances of full-sib family means with one common parent planted in different tests (1.80 using parameters from Table 6.1 in Eq. 6.6).

The sub-matrix **Z** represents covariances between a block of observations in one diallel and a block of observations in a different disconnected diallel. Disconnected means that there are no parents in common between the two diallels. Thus, information on any cross in one diallel provides no information about a cross in a different diallel, and all possible covariances are zero.

The **C** matrix contains covariances between the observations in **y** and the genetic values being predicted (in **g**). For this example, there are 18 observations in **y** and 9 breeding values being predicted in **g**, so **C** is the 18 x 9 matrix shown on the top of the next page. Whenever the parental breeding value being predicted (labeled across the top of **C**) matches a parent in one of the observations (labeled down the rows in **C**), this element in **C** is a covariance between a breeding value of that parent and that full-sib mean ($Cov(\bar{y}_{i \cdot kl \cdot}, g_k)$ = 3.60 using parameters from Table 6.1 in Eq. 6.9). Thus, **C** is:

$$
\begin{array}{c}
\text{Parental BV} \\
\begin{array}{ccccccccc} 1 & 2 & 3 & 4 & 5 & 6 & 7 & 8 & 9 \end{array}
\end{array}
$$

$$
\mathbf{C} = \begin{bmatrix}
3.6 & 3.6 & 0 & 0 & 0 & 0 & 0 & 0 & 0 \\
3.6 & 0 & 3.6 & 0 & 0 & 0 & 0 & 0 & 0 \\
0 & 3.6 & 3.6 & 0 & 0 & 0 & 0 & 0 & 0 \\
0 & 0 & 0 & 3.6 & 3.6 & 0 & 0 & 0 & 0 \\
0 & 0 & 0 & 3.6 & 0 & 3.6 & 0 & 0 & 0 \\
0 & 0 & 0 & 0 & 3.6 & 3.6 & 0 & 0 & 0 \\
0 & 0 & 0 & 3.6 & 3.6 & 0 & 0 & 0 & 0 \\
0 & 0 & 0 & 3.6 & 0 & 3.6 & 0 & 0 & 0 \\
0 & 0 & 0 & 0 & 3.6 & 3.6 & 0 & 0 & 0 \\
0 & 0 & 0 & 0 & 0 & 0 & 3.6 & 3.6 & 0 \\
0 & 0 & 0 & 0 & 0 & 0 & 3.6 & 0 & 3.6 \\
0 & 0 & 0 & 0 & 0 & 0 & 0 & 3.6 & 3.6 \\
0 & 0 & 0 & 0 & 0 & 0 & 3.6 & 3.6 & 0 \\
0 & 0 & 0 & 0 & 0 & 0 & 3.6 & 0 & 3.6 \\
0 & 0 & 0 & 0 & 0 & 0 & 0 & 3.6 & 3.6 \\
3.6 & 3.6 & 0 & 0 & 0 & 0 & 0 & 0 & 0 \\
3.6 & 0 & 3.6 & 0 & 0 & 0 & 0 & 0 & 0 \\
0 & 3.6 & 3.6 & 0 & 0 & 0 & 0 & 0 & 0
\end{bmatrix}
\begin{array}{c}
\text{(Obs. \#)} \\ 1 \\ 2 \\ 3 \\ 4 \\ 5 \\ 6 \\ 7 \\ 8 \\ 9 \\ 10 \\ 11 \\ 12 \\ 13 \\ 14 \\ 15 \\ 16 \\ 17 \\ 18
\end{array}
$$

The **G** matrix contains variances and covariances among the 9 breeding values in **g** and is thus:

$$
\mathbf{G} = \begin{bmatrix}
7.2 & 0 & 0 & 0 & 0 & 0 & 0 & 0 & 0 \\
0 & 7.2 & 0 & 0 & 0 & 0 & 0 & 0 & 0 \\
0 & 0 & 7.2 & 0 & 0 & 0 & 0 & 0 & 0 \\
0 & 0 & 0 & 7.2 & 0 & 0 & 0 & 0 & 0 \\
0 & 0 & 0 & 0 & 7.2 & 0 & 0 & 0 & 0 \\
0 & 0 & 0 & 0 & 0 & 7.2 & 0 & 0 & 0 \\
0 & 0 & 0 & 0 & 0 & 0 & 7.2 & 0 & 0 \\
0 & 0 & 0 & 0 & 0 & 0 & 0 & 7.2 & 0 \\
0 & 0 & 0 & 0 & 0 & 0 & 0 & 0 & 7.2
\end{bmatrix}
$$

The non-zero elements are variances of the breeding values ($Var(g_k)$), and the off-diagonal elements are covariances between breeding values ($Cov(g_k,g_{k'})$). Since the parents are assumed to be unrelated, all of these latter covariances are zero.

Estimating the Fixed Effects

Recall that in BLUP breeding values are predicted with Eq. 11.4 as

$$\hat{\mathbf{g}} = \mathbf{C'V}^{-1}(\mathbf{y} - \mathbf{X}\hat{\boldsymbol{\beta}})$$

where $\hat{\boldsymbol{\beta}}$ is the t x 1 vector of estimated fixed effects, **X** is an n x t design matrix relating the fixed effects (elements of $\hat{\boldsymbol{\beta}}$) to the **y** vector of records. The vector $\boldsymbol{\beta}$ is estimated by Eq. 11.5 as $\hat{\boldsymbol{\beta}} = (\mathbf{X'V^{-1}X})^{-}(\mathbf{X'V^{-1}y})$.

There is a problem in actually solving for $\hat{\boldsymbol{\beta}}$ in that $(\mathbf{X'V^{-1}X})$ is singular (i.e. does not have a unique inverse). There are many (actually an infinite number) of generalized inverses of $(\mathbf{X'V^{-1}X})$ and therefore, many solutions to $\hat{\boldsymbol{\beta}}$; thus $\hat{\boldsymbol{\beta}}$ is not unique. However, some functions (called estimable functions) of $\hat{\boldsymbol{\beta}}$ are unique regardless of the solution to $\hat{\boldsymbol{\beta}}$. It can be shown that $\mathbf{X}\hat{\boldsymbol{\beta}}$ is estimable and results in unique BLUE estimates of the fixed effects regardless of the solution of $\hat{\boldsymbol{\beta}}$ used. That is, the estimates of the fixed effects needed in the BLUP equation ($\mathbf{X}\hat{\boldsymbol{\beta}}$ in Eq. 11.4) are the same for quite different estimates of $\boldsymbol{\beta}$. While a complete discussion of estimability is beyond the scope of this book (see Graybill 1976, Henderson 1984, Milliken and Johnson 1984), three examples are presented below to demonstrate these concepts.

Three approaches to estimating fixed effects ('over-specified' effects approach, 'sum-to-zero' approach, and 'means' model approach) are used to show that for 3 three different sets of **X** and $\boldsymbol{\beta}$ in the diallel problem, the function $\mathbf{X}\hat{\boldsymbol{\beta}}$ is the same for all three. We first present the three approaches and then show the relationships among the solutions. The goal of all three approaches is to develop unique estimates of $\mathbf{X}\hat{\boldsymbol{\beta}}$, which is an 18 x 1 vector containing the estimated fixed effects associated with the 18 observations. The three different sets of **X** and $\boldsymbol{\beta}$ are superscripted with different letters below to distinguish the three approaches.

Over-specified effects approach: The first approach is called the 'over-specified' effects or G-inverse approach. Since the data in **y** are family means in a single test ($\bar{y}_{i\cdot kl\cdot}$ calculated according to Eq. 6.2), there are two types of fixed effects associated with each observation: the general mean (μ) and one environmental (i.e. test) effect (E_i). Since there are three different test sites, there are a total of four fixed effects (μ, E_1, E_2, E_3). However for this problem, estimates of these four parameters are not needed; three other estimates (which are linear functions of these 4 parameters) are needed (i.e. $\mu + E_1$, $\mu + E_2$, and $\mu + E_3$). For these data records which are full-sib family means, the expected value of any record is (from Eq. 6.3):

$$E(\bar{y}_{i\cdot kl\cdot}) = \mu + E_i \ .$$

All full-sib family means from a given test have the same expected value. Since there are three tests, there are three different expected values, i.e. $\mu + E_1$, $\mu + E_2$, and $\mu + E_3$. This is what $\mathbf{X}\hat{\boldsymbol{\beta}}$ must estimate in order to express the data in **y** as deviations from their fixed effects in the BLUP equation (Eq. 11.4). All the observations in **y** have one of these three values as an expected value. The design matrix, $\mathbf{X}^*$, and the parameter vector, $\hat{\boldsymbol{\beta}}^*$, for this approach are shown in Figure 12.2.

The matrix $\mathbf{X}^*$ has 18 rows, corresponding to the 18 observations, and four columns. The first column represents the fixed effect μ, and the remaining three columns represent the fixed E_i effect for each test. The entire first column of $\mathbf{X}^*$ contains 1's, indicating that every observation contains the fixed effect μ in its expected value. The elements of the 2nd, 3rd and 4th columns of $\mathbf{X}^*$ are either 0 or 1 depending on which test the observation came from. The first element of **y** is $\bar{y}_{1\cdot 12\cdot}$,

---

$$
\begin{array}{cccc}
E(\mathbf{y}) & = & \mathbf{X}^* & \boldsymbol{\beta}^*
\end{array}
$$

$$
E\begin{bmatrix}
\bar{y}_{1\cdot 12\cdot} \\ \bar{y}_{1\cdot 13\cdot} \\ \bar{y}_{1\cdot 23\cdot} \\ \bar{y}_{1\cdot 45\cdot} \\ \bar{y}_{1\cdot 46\cdot} \\ \bar{y}_{1\cdot 56\cdot} \\ \bar{y}_{2\cdot 45\cdot} \\ \bar{y}_{2\cdot 46\cdot} \\ \bar{y}_{2\cdot 56\cdot} \\ \bar{y}_{2\cdot 78\cdot} \\ \bar{y}_{2\cdot 79\cdot} \\ \bar{y}_{2\cdot 89\cdot} \\ \bar{y}_{3\cdot 78\cdot} \\ \bar{y}_{3\cdot 79\cdot} \\ \bar{y}_{3\cdot 89\cdot} \\ \bar{y}_{3\cdot 12\cdot} \\ \bar{y}_{3\cdot 13\cdot} \\ \bar{y}_{3\cdot 23\cdot}
\end{bmatrix}
=
\begin{bmatrix}
1 & 1 & 0 & 0 \\
1 & 1 & 0 & 0 \\
1 & 1 & 0 & 0 \\
1 & 1 & 0 & 0 \\
1 & 1 & 0 & 0 \\
1 & 1 & 0 & 0 \\
1 & 0 & 1 & 0 \\
1 & 0 & 1 & 0 \\
1 & 0 & 1 & 0 \\
1 & 0 & 1 & 0 \\
1 & 0 & 1 & 0 \\
1 & 0 & 1 & 0 \\
1 & 0 & 0 & 1 \\
1 & 0 & 0 & 1 \\
1 & 0 & 0 & 1 \\
1 & 0 & 0 & 1 \\
1 & 0 & 0 & 1 \\
1 & 0 & 0 & 1
\end{bmatrix}
\begin{bmatrix}
\mu \\ E_1 \\ E_2 \\ E_3
\end{bmatrix}
$$

Figure 12.2 - Fixed effects portion of the linear model with design matrix and parameter vector for the 'over specified' effects or G-inverse approach.

an observation from test 1. Thus, the line in $\mathbf{X}^*$ pertaining to that observation (the first row of $\mathbf{X}$) will have 1 in the first column for the mean μ, and 1 in the second column for the effect E_1, and zeros in the third and fourth columns for the effects E_2 and E_3. If we premultiply $\boldsymbol{\beta}^*$ by the first row of $\mathbf{X}^*$ to get the expected value of $\overline{y}_{1\cdot 12\cdot}$,

$$\begin{aligned} E(\overline{y}_{1\cdot 12\cdot}) &= 1\ (\mu) + 1\ (E_1) + 0\ (E_2) + 0\ (E_3) \\ &= \mu + E_1. \end{aligned}$$

This approach is called the 'over-specified' effects or G-inverse approach because $\mathbf{X}^*$ is of less than full column rank. That is, there are more effects (4: μ, E_1, E_2, E_3) than can be estimated (only 3). This means that a unique inverse of $\mathbf{X}^{*\prime}\mathbf{V}^{-1}\mathbf{X}^*$ does not exist (the matrix is singular); however, there are an infinite number of generalized inverses. Since $\hat{\boldsymbol{\beta}}^*$ is estimated as $(\mathbf{X}^{*\prime}\mathbf{V}^{-1}\mathbf{X}^*)^-(\mathbf{X}^{*\prime}\mathbf{V}^{-1}\mathbf{y})$, it follows that a unique solution for $\hat{\boldsymbol{\beta}}^*$ does not exist. There are many solutions, since there are many generalized inverses of $(\mathbf{X}^{*\prime}\mathbf{V}^{-1}\mathbf{X}^*)$. However, $\mathbf{X}\hat{\boldsymbol{\beta}}$ estimates $\mu + E_i$ for every element of $\mathbf{y}$ regardless of the choice of generalized inverse because $\mathbf{X}\hat{\boldsymbol{\beta}}$ is an estimable function. This is demonstrated after the next two approaches are defined.

Sum-to-zero approach: Another approach to solving for $\hat{\boldsymbol{\beta}}$ is to place certain restrictions on the parameters so that the model is not over-specified. One of these types of restrictions is the 'sum-to-zero' restriction, so called because the effects are restricted to sum to zero. In the above case for example, $E_1 + E_2 + E_3 = 0$, and $E_3 = - E_2 - E_3$. Using the 'sum-to-zero' restriction, the new design matrix $\mathbf{X}^z$ and parameter vector $\boldsymbol{\beta}^z$ are as shown in Figure 12.3. The matrix $\mathbf{X}^z$ is of full column rank, and $\mathbf{X}^{z\prime}\mathbf{V}^{-1}\mathbf{X}^z$ has a unique inverse; thus there is a unique solution for $\boldsymbol{\beta}^z$. As before, $\mathbf{X}^z\hat{\boldsymbol{\beta}}^z$ calculates the expected value for any observation in $\mathbf{y}$. For example, for the last observation in $\mathbf{y}$,

$$\begin{aligned} E(\overline{y}_{3\cdot 23\cdot}) &= 1\ (\mu) + -1\ (E_1) + -1\ (E_2) \\ &= \mu - E_1 - E_2 \\ &= \mu + E_3. \end{aligned}$$

Note that this utilizes the restriction that $E_3 = - E_1 - E_2$.

Means model: Another possible approach to developing estimates for the fixed effects is called the 'means model'. Applying this approach to this example, there are three elements of $\boldsymbol{\beta}^m$. These are $\mu + E_i$ for each of three tests, and $\mathbf{X}^m$ and $\boldsymbol{\beta}^m$ are shown in Figure 12.4. The matrix $\mathbf{X}^m$ is of full rank, and $\mathbf{X}^{m\prime}\mathbf{V}^{-1}\mathbf{X}^m$ has a unique inverse; thus, there is a unique solution for $\boldsymbol{\beta}^m$. As before, the function $\mathbf{X}^m\hat{\boldsymbol{\beta}}^m$ calculates the expected value for every observation in **y**.

Comparison of different approaches -- parameter estimates: There are many parameterizations of the fixed effects portion of the model that can be specified in any BLUP application. We have presented three: 'over-specified' effects approach, 'sum-to-zero' approach, and the 'means' model. Using any of these three parameterizations (or any other legitimate one) yields identical 1) estimates of the fixed effects ($\mathbf{X}\hat{\boldsymbol{\beta}} = \hat{\boldsymbol{\alpha}}$), 2) breeding value predictions ($\hat{\mathbf{g}}$), 3) variances of the predictions (Var($\hat{\mathbf{g}}$)), 4) error variances of the predictions (Var($\hat{\mathbf{g}}$-**g**)), and 5) correlations between the true and predicted breeding values (Corr($\hat{\mathbf{g}}$,**g**). For many BLUP applications where the fixed

---

$$
\begin{array}{cccc}
E(\mathbf{y}) & = & \mathbf{X}^z & \boldsymbol{\beta}^z \\
E\begin{bmatrix} \bar{y}_{1\cdot12\cdot} \\ \bar{y}_{1\cdot13\cdot} \\ \bar{y}_{1\cdot23\cdot} \\ \bar{y}_{1\cdot45\cdot} \\ \bar{y}_{1\cdot46\cdot} \\ \bar{y}_{1\cdot56\cdot} \\ \bar{y}_{2\cdot45\cdot} \\ \bar{y}_{2\cdot46\cdot} \\ \bar{y}_{2\cdot56\cdot} \\ \bar{y}_{2\cdot78\cdot} \\ \bar{y}_{2\cdot79\cdot} \\ \bar{y}_{2\cdot89\cdot} \\ \bar{y}_{3\cdot78\cdot} \\ \bar{y}_{3\cdot79\cdot} \\ \bar{y}_{3\cdot89\cdot} \\ \bar{y}_{3\cdot12\cdot} \\ \bar{y}_{3\cdot13\cdot} \\ \bar{y}_{3\cdot23\cdot} \end{bmatrix} & = & \begin{bmatrix} 1 & 1 & 0 \\ 1 & 1 & 0 \\ 1 & 1 & 0 \\ 1 & 1 & 0 \\ 1 & 1 & 0 \\ 1 & 1 & 0 \\ 1 & 0 & 1 \\ 1 & 0 & 1 \\ 1 & 0 & 1 \\ 1 & 0 & 1 \\ 1 & 0 & 1 \\ 1 & 0 & 1 \\ 1 & -1 & -1 \\ 1 & -1 & -1 \\ 1 & -1 & -1 \\ 1 & -1 & -1 \\ 1 & -1 & -1 \\ 1 & -1 & -1 \end{bmatrix} & \begin{bmatrix} \mu \\ E_1 \\ E_2 \end{bmatrix}
\end{array}
$$

Figure 12.3 - Fixed effects portion of the linear model with design matrix and parameter vector for the 'sum to zero' approach.

effects are all nuisance effects (environments, blocks, etc.), the investigator has no interest in the specific estimates of the fixed effects ($\hat{\boldsymbol{\beta}}$), or their variances (Var($\hat{\boldsymbol{\beta}}$)). Since $\mathbf{X}\hat{\boldsymbol{\beta}}$ is the same regardless of the approach, any of these three approaches (or any other legitimate solution to $\hat{\boldsymbol{\beta}}$) is appropriate.

When the fixed effects include genetic effects, or other factors that are of interest, the interpretation of $\hat{\boldsymbol{\beta}}$ and Var($\hat{\boldsymbol{\beta}}$) depends upon the specific approach used. Remember, $\hat{\boldsymbol{\beta}}$ will vary depending on the approach; $\mathbf{X}\hat{\boldsymbol{\beta}}$ is the same for all approaches since it forms estimable functions of the parameters.

Thus, using Eq. 11.5 to estimate $\hat{\boldsymbol{\beta}}$ for each of the three approaches, three different vectors result (Figure 12.5). The $\boldsymbol{\beta}^*$ vector is one of an infinite number of possible solutions for $\hat{\boldsymbol{\beta}}$ using the 'over-specified' effects or G-inverse approach. This particular one was solved for using a Moore-Penrose generalized inverse of $\mathbf{X}^{*\prime}\mathbf{V}^{-1}\mathbf{X}^*$ (Graybill 1976).

$$
\begin{array}{ccccc}
\mathrm{E}(\mathbf{y}) & = & \mathbf{X}^{\mathrm{m}} & & \boldsymbol{\beta}^{\mathrm{m}}
\end{array}
$$

$$
\mathrm{E}\begin{bmatrix}
\bar{y}_{1\cdot 12\cdot} \\ \bar{y}_{1\cdot 13\cdot} \\ \bar{y}_{1\cdot 23\cdot} \\ \bar{y}_{1\cdot 45\cdot} \\ \bar{y}_{1\cdot 46\cdot} \\ \bar{y}_{1\cdot 56\cdot} \\ \bar{y}_{2\cdot 45\cdot} \\ \bar{y}_{2\cdot 46\cdot} \\ \bar{y}_{2\cdot 56\cdot} \\ \bar{y}_{2\cdot 78\cdot} \\ \bar{y}_{2\cdot 79\cdot} \\ \bar{y}_{2\cdot 89\cdot} \\ \bar{y}_{3\cdot 78\cdot} \\ \bar{y}_{3\cdot 79\cdot} \\ \bar{y}_{3\cdot 89\cdot} \\ \bar{y}_{3\cdot 12\cdot} \\ \bar{y}_{3\cdot 13\cdot} \\ \bar{y}_{3\cdot 23\cdot}
\end{bmatrix}
=
\begin{bmatrix}
1 & 0 & 0 \\ 1 & 0 & 0 \\ 1 & 0 & 0 \\ 1 & 0 & 0 \\ 1 & 0 & 0 \\ 1 & 0 & 0 \\ 0 & 1 & 0 \\ 0 & 1 & 0 \\ 0 & 1 & 0 \\ 0 & 1 & 0 \\ 0 & 1 & 0 \\ 0 & 1 & 0 \\ 0 & 0 & 1 \\ 0 & 0 & 1 \\ 0 & 0 & 1 \\ 0 & 0 & 1 \\ 0 & 0 & 1 \\ 0 & 0 & 1
\end{bmatrix}
\begin{bmatrix}
\mu + E_1 \\ \mu + E_2 \\ \mu + E_3
\end{bmatrix}
$$

Figure 12.4 - Fixed effects portion of the linear model with design matrix and parameter vector for the means model approach.

At first glance, the $\hat{\boldsymbol{\beta}}$ vectors seem quite different, but a little algebra shows there are certain linear functions of the parameter estimates that give identical values for each of the three $\hat{\boldsymbol{\beta}}$ vectors. For example, calculate an estimate of $\mu + E_3$ for each of the three models:

$$\begin{aligned}
(\mu + E_3)^* &= (\mu^* + E_3^*) & &= 23.304 + 8.272 &&= 31.576 \\
(\mu + E_3)^z &= (\mu^z + E_3^z) &= (\mu^z - E_1^z - E_2^z) &= (31.072 - 2.559 + 3.062) &&= 31.576 \\
(\mu + E_3)^m &= (\mu^m + E_3^m) &= & &&= 31.576
\end{aligned}$$

Parameters or functions of parameters that give identical estimates regardless of the parameterization of the model are called **estimable functions of the parameters** (Milliken and Johnson 1984). The function $\mu + E_3$ is an estimable function, as are $\mu + E_1$, and $\mu + E_2$. Thus, the three elements of $\boldsymbol{\beta}^m$ are unique, estimable functions of those parameters. This is demonstrated by writing an $\mathbf{L}^*$ and an $\mathbf{L}^z$ such that both $\mathbf{L}^*\hat{\boldsymbol{\beta}}^*$ and $\mathbf{L}^z\hat{\boldsymbol{\beta}}^z$ equal $\hat{\boldsymbol{\beta}}^m$, i.e. a 3 x 1 vector estimating $\mu + E_i$ for each of the three tests. These are shown on the top of the next page.

$$\boldsymbol{\beta}^* = \begin{bmatrix} \mu^* \\ E_1^* \\ E_2^* \\ E_3^* \end{bmatrix} = \begin{bmatrix} 23.304 \\ 10.327 \\ 4.706 \\ 8.272 \end{bmatrix} \qquad \boldsymbol{\beta}^z = \begin{bmatrix} \mu \\ E_1^z \\ E_2^z \end{bmatrix} = \begin{bmatrix} 31.072 \\ 2.559 \\ -3.062 \end{bmatrix}$$

$$\boldsymbol{\beta}^m = \begin{bmatrix} \mu + E_1 \\ \mu + E_2 \\ \mu + E_3 \end{bmatrix} = \begin{bmatrix} 33.631 \\ 28.010 \\ 31.576 \end{bmatrix}$$

Figure 12.5 - Vectors of parameter estimates using three different approaches to estimating the fixed effects: 'over-specified' effects ($\boldsymbol{\beta}^*$); 'sum to zero' ($\boldsymbol{\beta}^z$); and means model ($\boldsymbol{\beta}^m$).

$$\mathbf{L}^* \quad \hat{\boldsymbol{\beta}}^* \quad = \quad \mathbf{L}^z \quad \hat{\boldsymbol{\beta}}^z \quad = \quad \hat{\boldsymbol{\beta}}^m$$

$$\begin{bmatrix} 1\,1\,0\,0 \\ 1\,0\,1\,0 \\ 1\,0\,0\,1 \end{bmatrix} \begin{bmatrix} \mu^* \\ E_1^* \\ E_2^* \\ E_3^* \end{bmatrix} = \begin{bmatrix} 1 & 1 & 0 \\ 1 & 0 & 1 \\ 1 & -1 & -1 \end{bmatrix} \begin{bmatrix} \mu^z \\ E_1^z \\ E_2^z \end{bmatrix} = \begin{bmatrix} (\mu + E_1)^m \\ (\mu + E_2)^m \\ (\mu + E_3)^m \end{bmatrix}$$

$$\begin{bmatrix} 1\,1\,0\,0 \\ 1\,0\,1\,0 \\ 1\,0\,0\,1 \end{bmatrix} \begin{bmatrix} 23.30 \\ 10.33 \\ 4.71 \\ 8.27 \end{bmatrix} = \begin{bmatrix} 1 & 1 & 0 \\ 1 & 0 & 1 \\ 1 & -1 & -1 \end{bmatrix} \begin{bmatrix} 31.07 \\ 2.56 \\ -3.06 \end{bmatrix} = \begin{bmatrix} 33.63 \\ 28.01 \\ 31.58 \end{bmatrix}$$

This 3 x 1 vector of fixed effect estimates contains the only 3 unique estimates needed in this example. The 18 x 1 vector of fixed effect estimates ($\mathbf{X}\hat{\boldsymbol{\beta}}$) needed to apply BLUP with Eq. 11.4 contains each of these 3 estimates 6 times. It is left to the reader (Problem 12.4) to show that $\mathbf{X}^*\hat{\boldsymbol{\beta}}^*$, $\mathbf{X}^z\hat{\boldsymbol{\beta}}^z$, and $\mathbf{X}^m\hat{\boldsymbol{\beta}}^m$ all result in the same 18 x 1 vector of estimated fixed effects.

Comparison of different approaches -- variances of estimates: Each of the estimates of $\boldsymbol{\beta}$ has a variance associated with it, which can be estimated from the equation in Property 2 in Chapter 11. For example, $\text{Var}(\hat{\boldsymbol{\beta}}^z) = (\mathbf{X}^{z\prime}\mathbf{V}^{-1}\mathbf{X}^z)^-$ which is

$$\text{Var}(\hat{\boldsymbol{\beta}}^z) = \begin{bmatrix} 1.223 & 0 & 0 \\ 0 & 0.791 & -0.396 \\ 0 & -0.396 & 0.791 \end{bmatrix}$$

The diagonal elements of $\text{Var}(\hat{\boldsymbol{\beta}}^z)$ are variances associated with the parameter estimates $\hat{\mu}^z$, $\hat{E}_1^z$, and $\hat{E}_2^z$, and the off-diagonal elements are the covariances between those estimates. Knowing that in the 'set-to-zero' approach the sum of the effects is restricted to equal zero (i.e. $\hat{E}_3^z = -\hat{E}_1^z - \hat{E}_2^z$), we can calculate $\text{Var}(\hat{E}_3^z)$ as

$$\begin{aligned} \text{Var}(\hat{E}_3^z) &= \text{Var}(\hat{E}_1^z) + \text{Var}(\hat{E}_2^z) + 2\,\text{Cov}(\hat{E}_1^z,\hat{E}_2^z) \\ &= 0.791 + 0.791 + 2\,(-0.396) \\ &= 0.791 \qquad \text{(discounting round-off error).} \end{aligned}$$

Thus $\text{Var}(\hat{E}_3^z) = \text{Var}(\hat{E}_1^z) = \text{Var}(\hat{E}_2^z)$, which is reasonable in this case where the data are balanced.

Since $(\mathbf{L}^{*}\hat{\boldsymbol{\beta}}^{*})$, $(\mathbf{L}^{z}\hat{\boldsymbol{\beta}}^{z})$, and $(\hat{\boldsymbol{\beta}}^{m})$ yield the same estimates of the fixed effects (as illustrated above), it follows that the variance associated with those linear functions will be the same (see Problem 12.2):

$$\mathrm{Var}(\mathbf{L}^{*}\hat{\boldsymbol{\beta}}^{*}) = \mathrm{Var}(\mathbf{L}^{z}\hat{\boldsymbol{\beta}}^{z}) = \mathrm{Var}(\hat{\boldsymbol{\beta}}^{m})$$
$$\mathbf{L}^{*}(\mathrm{Var}(\hat{\boldsymbol{\beta}}^{*})\mathbf{L}^{*\prime} = \mathbf{L}^{z}(\mathrm{Var}(\hat{\boldsymbol{\beta}}^{z}))\mathbf{L}^{z\prime} = \mathrm{Var}(\hat{\boldsymbol{\beta}}^{m})$$

It can also be shown (see Problem 12.5) that the 18 x 18 variance matrix associated with the 18 x 1 vector of the fixed effect estimates for use in the BLUP equation (i.e. $\mathbf{X}\hat{\boldsymbol{\beta}}$ in Eq. 11.4) is the same for all three model parameterizations, or specifically, $\mathrm{Var}(\mathbf{X}^{*}\hat{\boldsymbol{\beta}}^{*}) = \mathrm{Var}(\mathbf{X}^{z}\hat{\boldsymbol{\beta}}^{z}) = \mathrm{Var}(\mathbf{X}^{m}\hat{\boldsymbol{\beta}}^{m})$.

Results of Case 1

To demonstrate how BLUP handles data and in particular how it estimates and accounts for fixed effects differently than BLP, it is instructive to calculate certain results using both a BLP and a BLUP approach. Recall that three diallels are planted in three progeny tests in the following design: test 1 contains diallels A and B, test 2 contains B and C, and test 3 contains C and A. Each test contains six full-sib crosses and no genetic checklots. To apply BLP, we would normally estimate $\boldsymbol{\alpha}$, the expected value of **y**, in one of two ways: 1) with checklots, or 2) with the grand mean across all six families in a given test (which is an ordinary least squares approach). The latter approach is reasonable (the estimate is precise and unbiased) only if the test includes a large random sample of parents. In this case there are no checks and only a small sample of parents on each site, so it is preferable to apply BLUP where the generalized least squares estimate $(\mathbf{X}\hat{\boldsymbol{\beta}})$ is used as an estimate of the fixed effects. However, for purposes of comparison with BLUP, we apply BLP by estimating the expected value of a family mean from each test using the grand mean from that test. The BLP assumption is that these values are 'known' elements of $\boldsymbol{\alpha}$. Equations 4.2, 4.4, 4.7, 11.4, and 11.11 are used to calculate $\hat{\mathbf{g}}$, $\mathrm{Var}(\hat{\mathbf{g}})$, and $\mathrm{Corr}(\hat{\mathbf{g}},\mathbf{g})$ for the BLP and BLUP approaches. In this section the fixed effects estimated for BLP are written $\hat{\boldsymbol{\alpha}}$, while those estimated by GLS and used in BLUP are written $\mathbf{X}\hat{\boldsymbol{\beta}}$.

Comparison of BLP and BLUP estimates of fixed effects: For both BLP and BLUP, the expected values of observations in the **y** vector (family means from a single test) are

$$E(\overline{y}_{i\cdot kl\cdot}) = \mu + E_i$$

for each element of **y**. In BLP these expected values are assumed known, and are estimated in this case by calculating the grand mean There are three of these grand test means obtained by averaging the 6 family means in a given test (as shown in the **y** vector). These form the elements of the 18 x 1 $\hat{\boldsymbol{\alpha}}$ vector. In BLUP, $\mathbf{X}\hat{\boldsymbol{\beta}}$, the generalized least squares (GLS) estimates of the expected values, is used. Comparing the BLP and BLUP estimates of the fixed effects, the estimates for $\mu + E_i$ are: for test 1, 33.92 *vs* 33.63; for test 2, 26.95 *vs* 28.01; and for test 3, 32.35 *vs* 31.58 (Fig. 12.6, $\hat{\boldsymbol{\alpha}}$ *vs* $\hat{\boldsymbol{\beta}}^m$). The BLUP approach (actually the GLS approach to estimating fixed effects) uses all the data on all sites to estimate the fixed effect on any one site. Thus, the GLS approach 'adjusts' for the fact that some tests contain parents that, on average, are genetically superior to those in other tests. For example, the grand mean of all 6 families in test 2, containing diallels B and C, is 26.95. This is the estimate of $\mu + E_2$ used for BLP. The GLS estimate from the BLUP approach is 28.01. This indicates that part of the reason for the generally shorter trees found in test 2 is genetic inferiority of the parents in diallels B and C. The ordinary least squares (OLS) estimate of the overall site mean for test 2 (used for BLP) makes no adjustment for this. The GLS approach does because of the linkage of diallel sets across sites. The parents in diallels A, B, and C are compared pairwise on 3 different sites; GLS uses these data to 'sort out' genetic and environmental influences. The information to do this is specified in **V** of the GLS equation (Eq. 11.5). If a group of average parents had been mated in diallels and tested on the site of test 2, the average height growth at 12 years would have been closer to 28 feet. The environment of test 2 would still be the poorest of the three sites, but not quite as bad as it might look at first glance.

Comparison of BLP and BLUP breeding value predictions: The differences between OLS and GLS estimates of the fixed effects lead to differences in breeding value predictions between BLP and BLUP. The average of the breeding value predictions

-- **BLUP** --

$$\hat{\mathbf{g}} = \begin{bmatrix} 2.21 \\ 2.59 \\ 0.85 \\ -2.08 \\ -3.73 \\ 1.68 \\ -1.81 \\ -1.17 \\ 1.45 \end{bmatrix} \qquad \mathrm{Var}(\hat{\mathbf{g}}) = \begin{bmatrix} 2.99 & -.05 & -.05 & -.48 & -.48 & -.48 & -.48 & -.48 & -.48 \\ -.05 & 2.99 & -.05 & -.48 & -.48 & -.48 & -.48 & -.48 & -.48 \\ -.05 & -.05 & 2.99 & -.48 & -.48 & -.48 & -.48 & -.48 & -.48 \\ -.48 & -.48 & -.48 & 2.99 & -.05 & -.05 & -.48 & -.48 & -.48 \\ -.48 & -.48 & -.48 & -.05 & 2.99 & -.05 & -.48 & -.48 & -.48 \\ -.48 & -.48 & -.48 & -.05 & -.05 & 2.99 & -.48 & -.48 & -.48 \\ -.48 & -.48 & -.48 & -.48 & -.48 & -.48 & 2.99 & -.05 & -.05 \\ -.48 & -.48 & -.48 & -.48 & -.48 & -.48 & -.05 & 2.99 & -.05 \\ -.48 & -.48 & -.48 & -.48 & -.48 & -.48 & -.05 & -.05 & 2.99 \end{bmatrix}$$

$$\hat{\boldsymbol{\beta}}^{\mathbf{m}} = \begin{bmatrix} 33.63 \\ 28.01 \\ 31.58 \end{bmatrix} \approx \begin{bmatrix} \mu + E_1 \\ \mu + E_2 \\ \mu + E_3 \end{bmatrix} \qquad \mathrm{Corr}(\hat{g}_i, g_i) = 0.64$$

-- **BLP** --

$$\hat{\mathbf{g}} = \begin{bmatrix} 1.86 \\ 2.24 \\ 0.51 \\ -1.83 \\ -3.48 \\ 1.94 \\ -1.71 \\ -1.08 \\ 1.55 \end{bmatrix} \qquad \mathrm{Var}(\hat{\mathbf{g}}) = \begin{bmatrix} 3.60 & 0.55 & 0.55 & 0 & 0 & 0 & 0 & 0 & 0 \\ 0.55 & 3.60 & 0.55 & 0 & 0 & 0 & 0 & 0 & 0 \\ 0.55 & 3.60 & 0.55 & 0 & 0 & 0 & 0 & 0 & 0 \\ 0 & 0 & 0 & 3.60 & 0.55 & 0.55 & 0 & 0 & 0 \\ 0 & 0 & 0 & 0.55 & 3.60 & 0.55 & 0 & 0 & 0 \\ 0 & 0 & 0 & 0.55 & 0.55 & 3.60 & 0 & 0 & 0 \\ 0 & 0 & 0 & 0 & 0 & 0 & 3.60 & 0.55 & 0.55 \\ 0 & 0 & 0 & 0 & 0 & 0 & 0.55 & 3.60 & 0.55 \\ 0 & 0 & 0 & 0 & 0 & 0 & 0.55 & 0.55 & 3.60 \end{bmatrix}$$

$$\hat{\boldsymbol{\alpha}} = \begin{bmatrix} 33.92 \\ 26.95 \\ 32.35 \end{bmatrix} \approx \begin{bmatrix} \mu + E_1 \\ \mu + E_2 \\ \mu + E_3 \end{bmatrix} \qquad \mathrm{Corr}(\hat{g}_i, g_i) = 0.71$$

Figure 12.6 - Results from BLP and BLUP approaches to predicting breeding values for Case 1. Vectors and matrices are: predicted breeding values ($\hat{\mathbf{g}}$), variances and covariances of predicted breeding values ($\mathrm{Var}(\hat{\mathbf{g}})$), and estimated fixed effects ($\hat{\boldsymbol{\alpha}}$, grand family means, and $\mathbf{X}\hat{\boldsymbol{\beta}}^{\mathbf{m}}$, generalized least squares estimates).

is zero for both the BLP and BLUP approaches because in both instances, 1) we assumed all parents came from the same population and 2) only the data in **y** were used to calculate the expected values of **y** (i.e. there were no checks or any other diallels planted on those test sites, so no additional information could be used to increase the precision of the expected value estimates). The first three values in the $\hat{\mathbf{g}}$ vectors (for both BLP and BLUP in Figure 12.6) are predicted breeding values for 12-year height for parents 1, 2, and 3 in diallel A. The second three values are for parents 4, 5, and 6 in diallel B, and the last three are for parents 7, 8, and 9 in diallel C. The predictions from BLP and BLUP are similar for all parents; however, some parents have higher prediction under the BLP approach while others are higher under the BLUP approach. This depends upon which diallel a parent is included in. All parents in diallel A have higher predictions under BLUP than BLP, while those in diallel B are substantially lower in BLUP than BLP, and those in C are slightly lower in BLUP. Whether a parental breeding value prediction is higher in BLUP or BLP depends on the average genetic superiority of all the parents in that diallel. For both BLUP and BLP, diallel A contains better than average trees (mean BLUP $\hat{g}_i = 1.88$, mean BLP $\hat{g}_i = 1.54$ for the three parents in A), diallel B contains a rather poor group (mean BLUP $\hat{g}_i = -1.38$, mean BLP $\hat{g}_i = -1.12$), and diallel C contains a slightly less than average group (mean BLUP $\hat{g}_i = -0.51$, mean BLP $\hat{g}_i = -0.41$).

The pattern is clear: a good diallel (genetically superior parents on average) has higher predictions under BLUP than BLP, while a poor diallel has lower predictions under BLUP. This occurs because of the difference in the vector of estimated expected values of **y**: ($\hat{\boldsymbol{\alpha}}$ in BLP *vs* $\mathbf{X}\hat{\boldsymbol{\beta}}$ in BLUP). Remember that BLP Eq. 4.2 is identical to BLUP Eq. 11.4 except for the difference in estimates of the fixed effects. In both instances $\mathbf{y}-\hat{\boldsymbol{\alpha}}$ (BLP) or $\mathbf{y}-\mathbf{X}\hat{\boldsymbol{\beta}}$ (BLUP) is used to express the data as deviations from fixed effects. In this example, this expresses full-sib family means as deviations from environmental test means. Consider diallel A: it occurs in tests 1 and 3, and in both those tests, the estimate of the expected value of **y** is higher for BLP than in BLUP (compare elements 1 and 3 in $\hat{\boldsymbol{\alpha}}$ *vs* $\mathbf{X}\hat{\boldsymbol{\beta}}$ in Fig. 12.6). Thus, the deviations from the estimated expected values for families in diallel A ($\mathbf{y}-\hat{\boldsymbol{\alpha}}$, or $\mathbf{y}-\mathbf{X}\hat{\boldsymbol{\beta}}$) are lower in the BLP approach than the BLUP approach. For example, parents 1 and 2 are in diallel A, and the first value in **y** is $\bar{y}_{1\cdot 12\cdot} = 38.8$. The first row of the $\mathbf{y}-\hat{\boldsymbol{\alpha}}$ vector is $(38.8 - 33.92) = 4.78$, while the first row of $\mathbf{y}-\mathbf{X}\hat{\boldsymbol{\beta}}$ is $(38.8 - 33.63) = 5.17$. Thus using the BLP approach, the full-sib family from the cross parent 1 x parent 2 seems to be exhibiting less genetic superiority than under the BLUP approach, and this results in a lower breeding value prediction. Now consider diallel B: it occurs in tests 1 and 2. In test 1, the BLP estimate $\hat{\alpha}_1 = \hat{\mu} + \hat{E}_1$ is slightly too high (33.92 *vs* 33.63), but in test 2 the BLP estimate is substantially too low (26.95 *vs* 28.01). Parents 4 and 5 are in diallel B, $\bar{y}_{2 \cdot 45 \cdot} = 24.6$, and the corresponding row of the $\mathbf{y}$-$\hat{\boldsymbol{\alpha}}$ vector will be (24.6 - 26.95) = -2.35, while the first row of the $\mathbf{y}$-$\mathbf{X}\hat{\boldsymbol{\beta}}$ will be (24.6 - 28.01) = -3.41. Thus using the BLP approach, the full-sib family from the cross parent 4 x parent 5 does not seem quite as inferior as under the BLUP approach, and this results in a higher (i.e. less negative) breeding value prediction.

The differences in predicted breeding values from BLP *vs* BLUP result from the different estimates of the fixed effects. The grand mean of the full-sib families in a given test (used to estimate fixed effects in BLP) is influenced by the genetic quality of the parents. The more precise GLS estimates from BLUP use all the data on all sites to estimate the site means adjusted for the confounding influences of different quality parents on different sites.

Comparison of BLP and BLUP: estimated precision of predictions: Var($\hat{\mathbf{g}}$) matrices calculated with both the BLP and BLUP approaches (Fig. 12.6) are interpreted in a similar manner as was described in Chapter 6. Diagonal elements are variances of the individual predictions (Var($\hat{g}_i$)), and are 3.60 and 2.99 for BLP and BLUP, respectively. The estimated variance of the BLUPs is less than that of the BLPs due to a variance associated with estimating the fixed effects (see Eq. 11.11). Heuristically, since BLP assumes fixed effects are known, BLP makes less conservative predictions than BLUP, i.e. BLUP centers all predictions more closely around the mean. That is, since BLP treats fixed effects as known constants, predictions are spread out more. BLUP 'shrinks' the predictions more closely toward zero to reflect the fact that fixed effects are <u>not</u> known, but rather estimated.

In a similar manner, since Var($\hat{\mathbf{g}}$) = Cov($\hat{\mathbf{g}}$,$\mathbf{g}$) (Property 7, Chapter 11), the estimated correlation between the predicted and true breeding value (Corr($\hat{g}_i$,g_i)) is higher for BLP (0.71 for BLP *vs* 0.64 for BLUP in Fig. 12.6). Comparing the two methods in this manner will always indicate a higher Corr($\hat{\mathbf{g}}$,$\mathbf{g}$) for BLP than BLUP, <u>not</u> because BLP is a superior approach, but rather because of the assumption with BLP that $\boldsymbol{\alpha}$ is known. Since $\boldsymbol{\alpha}$ is never really known (although it may be estimated very precisely), Corr($\hat{\mathbf{g}}$,$\mathbf{g}$) calculated under the BLP approach is always an overestimate, and indicates that predicted breeding values are more correlated with true values than they really are. This upward bias occurs because the error in estimating $\boldsymbol{\alpha}$ is not accounted for, $\boldsymbol{\alpha}$ is assumed known. In practical terms, this bias may be irrelevant depending upon the precision of the estimates of $\boldsymbol{\alpha}$.

Comparison of BLP and BLUP: Covariances among predictions: Off-diagonal elements of Var($\hat{\mathbf{g}}$) are covariances between predicted breeding values. If BLP is used, there is zero covariance between predictions for parents in different diallels, and positive covariance between predictions for parents in the same diallel. There is zero covariance between predictions for parents in different diallels because 1) all parents are unrelated, and 2) the diallels are disconnected, so information on any families or parents in say diallel A provide no information about those in diallel B. Positive covariances between predictions for parents in the same diallel occur despite the fact that the parents are unrelated because information on the full-sib cross of parent 1 x parent 2 provides information about the breeding value of both parents (see Chapter 6 for further discussion).

If BLUP is used, however, the situation is quite different than that for BLP. There is a negative covariance between the predictions for parents in different diallels, and a near-zero covariance between predictions for parents in the same diallels. Heuristically, these negative covariances arise because BLUP 'constrains' the average of the breeding values to be zero, though somewhat indirectly. The observed data is used to estimate the fixed effects, re-express **y** as deviations ($\mathbf{y}$-$\mathbf{X}\hat{\boldsymbol{\beta}}$) which sum to zero, and then predict breeding values from those deviations. Hence, the predicted breeding values sum to zero also. When a sum of variables is constrained to be zero there must be, on the average, a negative covariance between the variables. If some breeding values are predicted above average, some others have to be predicted below average in order for them all to sum to zero.

Thus, in Case 1 under the BLUP approach, breeding value predictions for parents in different diallels have a negative covariance because of this indirect 'sum-to-zero' constraint. Breeding value predictions for parents within a diallel have two opposing contributions to overall covariance, 1) a negative covariance arising from the sum-to-zero constraint, and 2) a positive covariance arising because data on a full-sib family is used to predict the breeding values of both parents (as discussed above for the BLP approach). The result is a near zero covariance.

The difference between BLP and BLUP in the estimated covariances among predictions again arises because the fixed effects are assumed known in BLP. Thus, since BLP assumes $\boldsymbol{\alpha}$ is not estimated from the data, there is no constraint that all breeding values sum to zero (even though in this example they do, since we really did estimate $\boldsymbol{\alpha}$ from the data). BLUP recognizes that the fixed effects are not known, and that all fixed effects are estimated with the data, and this sets up a covariance between all pairs of predictions. BLP simply does not 'recognize' that $\boldsymbol{\alpha}$ is estimated, and thus those covariance estimates are less precise.

CASE 2: SOURCE TRIALS OR PROVENANCE TESTS

Occasionally situations arise where fixed effects include not only nuisance effects such as blocks, locations and treatments, but genetic effects of some interest. Tests including provenances (fixed) and parents nested within provenances (random) may be an example of such a case. Assume there are three distinct provenances (A, B, and C) of loblolly pine, and that open-pollinated seed have been collected from three mother trees in each of the provenances. The open-pollinated (assumed half-sib) families from these mother trees have been planted in row-plot progeny tests with 4 blocks and 6 trees per plot on each of three sites. But only two provenances are tested on each site: test 1 contains provenance A and B, test 2 contains B and C, test 3 contains C and A. The tests are measured for height at age 12, and we wish to predict breeding values for the nine mother trees, which are assumed to be unrelated.

Defining the Linear Model

Since the seedlings planted are open-pollinated, the linear model for half-sib data defined in Eq. 2.23 is used, with the addition of a fixed provenance (or source) effect, T_r. Thus, the model is

$$y_{ijrkl} = \mu + E_i + B_{ij} + T_r + f_{rk} + fe_{rik} + p_{rijk} + w_{rijkl}$$

where T_r = fixed effect of the rth provenance, $r = 1, 2, \ldots P$,

f_{rk} = random effect of the kth family in the rth provenance, $k= 1, 2, \ldots s_r$, $E(f_{rk}) = 0$, $Var(f_{rk}) = \sigma^2_f$, and

all other random effects are similarly modified from Eq. 2.23 to be defined for or within the rth provenance.

While a fixed provenance x location interaction term could be specified, it is omitted for simplicity. Parameter estimates are found in Table 12.1. Since provenances are assumed fixed constants, there is no variance among provenance effects.

The y Vector, V, C, and G

The y vector: Units of observation in **y** are half-sib family means calculated in a single test (i.e. $\bar{y}_{i \cdot rk \cdot}$). The data are ordered in **y** first by test, then by provenance and family:

$$\mathbf{y} = \begin{bmatrix} \bar{y}_{1 \cdot A1 \cdot} \\ \bar{y}_{1 \cdot A2 \cdot} \\ \bar{y}_{1 \cdot A3 \cdot} \\ \bar{y}_{1 \cdot B1 \cdot} \\ \bar{y}_{1 \cdot B2 \cdot} \\ \bar{y}_{1 \cdot B3 \cdot} \\ \bar{y}_{2 \cdot B1 \cdot} \\ \bar{y}_{2 \cdot B2 \cdot} \\ \bar{y}_{2 \cdot B3 \cdot} \\ \bar{y}_{2 \cdot C1 \cdot} \\ \bar{y}_{2 \cdot C2 \cdot} \\ \bar{y}_{2 \cdot C3 \cdot} \\ \bar{y}_{3 \cdot C1 \cdot} \\ \bar{y}_{3 \cdot C2 \cdot} \\ \bar{y}_{3 \cdot C3 \cdot} \\ \bar{y}_{3 \cdot A1 \cdot} \\ \bar{y}_{3 \cdot A2 \cdot} \\ \bar{y}_{3 \cdot A3 \cdot} \end{bmatrix} = \begin{bmatrix} 38.8 \\ 33.3 \\ 37.1 \\ 25.6 \\ 35.2 \\ 33.5 \\ 24.6 \\ 27.8 \\ 25.6 \\ 28.1 \\ 25.7 \\ 29.9 \\ 25.3 \\ 33.9 \\ 31.2 \\ 34.6 \\ 36.0 \\ 33.1 \end{bmatrix} .$$

Note that the fourth observation, $\bar{y}_{1 \cdot B1 \cdot}$, is the family mean for family 1 in provenance B in test 1. All other elements of **y** are similarly designated.

Table 12.1 - Hypothetical values for 12-year-old open-pollinated progeny tests of three loblolly pine provenances used in Case 2.[a]

Trait	$\bar{y}_{i\cdots\cdot}$	σ^2_t	σ^2_F	σ^2_f	σ^2_{fe}	σ^2_p	σ^2_w	$Var(\bar{y}_{i\cdot kl\cdot})$[b]
height (ft)	30	43	2.70	1.80	0.90	6.00	34.3	5.63

[a] Note that σ^2_t = total variance = $\sigma^2_f + \sigma^2_{fe} + \sigma^2_p + \sigma^2_w$ and that $\sigma^2_F = \sigma^2_f + \sigma^2_{fe}$.
[b] $Var(\bar{y}_{i\cdot rk\cdot})$ is the variance of half-sib family means within a provenance assuming six blocks (b=4) and six trees of each family in each block (n=6).

The second moments - V, C, and G: Equations in Chapter 5 showing various second moments assumed that half-sib family means in a single progeny test were the observations in **y**. For this particular case we have modified that linear model for half-sib data by adding a fixed provenance effect. This fixed effect has no impact on any second moment (since it is a fixed constant), and so all derivations presented earlier remain valid. Using the parameters in Table 12.1 in Eqs. 5.4, 5.6, 5.7, and 5.8, estimates of the necessary second moments needed to construct **V**, **C**, and **G** are:

$$Var(\bar{y}_{i\cdot rk\cdot}) = 5.63,$$
$$Cov(\bar{y}_{i\cdot rk\cdot}, \bar{y}_{i'\cdot rk\cdot}) = 1.80,$$
$$Cov(\bar{y}_{i\cdot rk\cdot}, g_{rk}) = 3.60,$$
$$Var(g_{rk}) = 7.20.$$

The final **V** and **C** matrices are shown in Figure 12.7. For this Case, the final form of **G** is identical to that for Case 1 because the additive genetic variance was assumed the same. The 9 x 9 **G** matrix (not shown) is a diagonal matrix with $\sigma^2_f = 7.20$ on the diagonals, and zeros on the off-diagonals since the mother trees are assumed unrelated.

$$
\mathbf{V} = \begin{bmatrix}
5.63 & 0 & 0 & 0 & 0 & 0 & 0 & 0 & 0 & 0 & 0 & 0 & 0 & 0 & 0 & 1.8 & 0 & 0 \\
0 & 5.63 & 0 & 0 & 0 & 0 & 0 & 0 & 0 & 0 & 0 & 0 & 0 & 0 & 0 & 0 & 1.8 & 0 \\
0 & 0 & 5.63 & 0 & 0 & 0 & 0 & 0 & 0 & 0 & 0 & 0 & 0 & 0 & 0 & 0 & 0 & 1.8 \\
0 & 0 & 0 & 5.63 & 0 & 0 & 1.8 & 0 & 0 & 0 & 0 & 0 & 0 & 0 & 0 & 0 & 0 & 0 \\
0 & 0 & 0 & 0 & 5.63 & 0 & 0 & 1.8 & 0 & 0 & 0 & 0 & 0 & 0 & 0 & 0 & 0 & 0 \\
0 & 0 & 0 & 0 & 0 & 5.63 & 0 & 0 & 1.8 & 0 & 0 & 0 & 0 & 0 & 0 & 0 & 0 & 0 \\
0 & 0 & 0 & 1.8 & 0 & 0 & 5.63 & 0 & 0 & 0 & 0 & 0 & 0 & 0 & 0 & 0 & 0 & 0 \\
0 & 0 & 0 & 0 & 1.8 & 0 & 0 & 5.63 & 0 & 0 & 0 & 0 & 0 & 0 & 0 & 0 & 0 & 0 \\
0 & 0 & 0 & 0 & 0 & 1.8 & 0 & 0 & 5.63 & 0 & 0 & 0 & 0 & 0 & 0 & 0 & 0 & 0 \\
0 & 0 & 0 & 0 & 0 & 0 & 0 & 0 & 0 & 5.63 & 0 & 0 & 1.8 & 0 & 0 & 0 & 0 & 0 \\
0 & 0 & 0 & 0 & 0 & 0 & 0 & 0 & 0 & 0 & 5.63 & 0 & 0 & 1.8 & 0 & 0 & 0 & 0 \\
0 & 0 & 0 & 0 & 0 & 0 & 0 & 0 & 0 & 0 & 0 & 5.63 & 0 & 0 & 1.8 & 0 & 0 & 0 \\
0 & 0 & 0 & 0 & 0 & 0 & 0 & 0 & 0 & 1.8 & 0 & 0 & 5.63 & 0 & 0 & 0 & 0 & 0 \\
0 & 0 & 0 & 0 & 0 & 0 & 0 & 0 & 0 & 0 & 1.8 & 0 & 0 & 5.63 & 0 & 0 & 0 & 0 \\
0 & 0 & 0 & 0 & 0 & 0 & 0 & 0 & 0 & 0 & 0 & 1.8 & 0 & 0 & 5.63 & 0 & 0 & 0 \\
1.8 & 0 & 0 & 0 & 0 & 0 & 0 & 0 & 0 & 0 & 0 & 0 & 0 & 0 & 0 & 5.63 & 0 & 0 \\
0 & 1.8 & 0 & 0 & 0 & 0 & 0 & 0 & 0 & 0 & 0 & 0 & 0 & 0 & 0 & 0 & 5.63 & 0 \\
0 & 0 & 1.8 & 0 & 0 & 0 & 0 & 0 & 0 & 0 & 0 & 0 & 0 & 0 & 0 & 0 & 0 & 5.63
\end{bmatrix}
$$

$$
\mathbf{C} = \begin{bmatrix}
3.6 & 0 & 0 & 0 & 0 & 0 & 0 & 0 & 0 \\
0 & 3.6 & 0 & 0 & 0 & 0 & 0 & 0 & 0 \\
0 & 0 & 3.6 & 0 & 0 & 0 & 0 & 0 & 0 \\
0 & 0 & 0 & 3.6 & 0 & 0 & 0 & 0 & 0 \\
0 & 0 & 0 & 0 & 3.6 & 0 & 0 & 0 & 0 \\
0 & 0 & 0 & 0 & 0 & 3.6 & 0 & 0 & 0 \\
0 & 0 & 0 & 3.6 & 0 & 0 & 0 & 0 & 0 \\
0 & 0 & 0 & 0 & 3.6 & 0 & 0 & 0 & 0 \\
0 & 0 & 0 & 0 & 0 & 3.6 & 0 & 0 & 0 \\
0 & 0 & 0 & 0 & 0 & 0 & 3.6 & 0 & 0 \\
0 & 0 & 0 & 0 & 0 & 0 & 0 & 3.6 & 0 \\
0 & 0 & 0 & 0 & 0 & 0 & 0 & 0 & 3.6 \\
0 & 0 & 0 & 0 & 0 & 0 & 3.6 & 0 & 0 \\
0 & 0 & 0 & 0 & 0 & 0 & 0 & 3.6 & 0 \\
0 & 0 & 0 & 0 & 0 & 0 & 0 & 0 & 3.6 \\
3.6 & 0 & 0 & 0 & 0 & 0 & 0 & 0 & 0 \\
0 & 3.6 & 0 & 0 & 0 & 0 & 0 & 0 & 0 \\
0 & 0 & 3.6 & 0 & 0 & 0 & 0 & 0 & 0
\end{bmatrix}
$$

Figure 12.7 - Completely specified **V** and **C** matrices for Case 2. Case 2 involves best linear unbiased prediction of breeding values for a total of nine loblolly pine parents (three parents from each of three provenances) tested in three test sites.

The Design Matrix X

With family means from a single test being used as the observations in **y**, the expected value of any observation is written as $E(\bar{y}_{i \cdot rk \cdot}) = \mu + E_i + T_r$. Thus, there are three types of fixed effects to account for in the design matrix **X**: μ, E_i, and T_r. Using the 'sum-to-zero' restrictions on both the E_i and T_r effects:

$$E_1 + E_2 + E_3 = 0, \text{ and } E_3 = - E_2 - E_3,$$
$$T_1 + T_2 + T_3 = 0, \text{ and } T_3 = - T_2 - T_3.$$

Now the expected value of **y**, the design matrix $\mathbf{X^z}$, and $\boldsymbol{\beta}^z$ are written as:

$$E(\mathbf{y}) \quad = \quad \mathbf{X^z} \quad \boldsymbol{\beta}^z$$

$$E\begin{bmatrix} \bar{y}_{1 \cdot A1 \cdot} \\ y_{1 \cdot A2 \cdot} \\ y_{1 \cdot A3 \cdot} \\ y_{1 \cdot B1 \cdot} \\ y_{1 \cdot B2 \cdot} \\ y_{1 \cdot B3 \cdot} \\ y_{2 \cdot B1 \cdot} \\ y_{2 \cdot B2 \cdot} \\ y_{2 \cdot B3 \cdot} \\ y_{2 \cdot C1 \cdot} \\ y_{2 \cdot C2 \cdot} \\ y_{2 \cdot C3 \cdot} \\ y_{3 \cdot C1 \cdot} \\ y_{3 \cdot C2 \cdot} \\ y_{3 \cdot C3 \cdot} \\ y_{3 \cdot A1 \cdot} \\ y_{3 \cdot A2 \cdot} \\ y_{3 \cdot A3 \cdot} \end{bmatrix} = \begin{bmatrix} 1 & 1 & 0 & 1 & 0 \\ 1 & 1 & 0 & 1 & 0 \\ 1 & 1 & 0 & 1 & 0 \\ 1 & 1 & 0 & 0 & 1 \\ 1 & 1 & 0 & 0 & 1 \\ 1 & 1 & 0 & 0 & 1 \\ 1 & 0 & 1 & 0 & 1 \\ 1 & 0 & 1 & 0 & 1 \\ 1 & 0 & 1 & 0 & 1 \\ 1 & 0 & 1 & -1 & -1 \\ 1 & 0 & 1 & -1 & -1 \\ 1 & 0 & 1 & -1 & -1 \\ 1 & -1 & -1 & -1 & -1 \\ 1 & -1 & -1 & -1 & -1 \\ 1 & -1 & -1 & -1 & -1 \\ 1 & -1 & -1 & 1 & 0 \\ 1 & -1 & -1 & 1 & 0 \\ 1 & -1 & -1 & 1 & 0 \end{bmatrix} \begin{bmatrix} \mu \\ E_1 \\ E_2 \\ T_1 \\ T_2 \end{bmatrix}$$

The second and third columns of $\mathbf{X_z}$ pertain to the E_i effects, and the fourth and fifth columns pertain to the T_r (provenance) effects.

Results of Case 2

Recall from Eq. 11.4 that $\hat{\mathbf{g}} = \mathbf{C'V^{-1}(y-X\hat{\boldsymbol{\beta}})}$ given that all genotypes are from the same population, i.e. $E(\mathbf{g}) = \boldsymbol{\tau} = 0$. In the previous example in this chapter, we assumed $\boldsymbol{\tau}$ equal to **0**, but in this case we have assumed that the three different provenances have different fixed genetic effects. Offspring from tree 1 from provenance A in test 1 will have the same expected value as offspring from tree 2 from provenance A in test 1; however, they will have a different expected value than offspring from tree 1 from provenance B in test 1. Using Eqs. 11.4 and 11.5, and the matrices and vectors given above, we calculate $\hat{\boldsymbol{\beta}}^z$ as

$$\hat{\boldsymbol{\beta}}^z = \begin{bmatrix} 31.07 \\ 2.42 \\ -2.56 \\ 3.13 \\ -2.29 \end{bmatrix} = \begin{bmatrix} \mu^z \\ E_1^z \\ E_2^z \\ T_1^z \\ T_2^z \end{bmatrix}$$

Equation 11.4 uses $\mathbf{X}^z$ and $\hat{\boldsymbol{\beta}}^z$ to express observations as deviations from all fixed effects, both 'nuisance' and genetic, and thus, the resulting predicted breeding values are expressed as deviations from a particular fixed provenance effect. Thus, in $\hat{\mathbf{g}}$, the predicted breeding values for the three parents within a provenance sum to zero. Parental rankings can be compared within a provenance, but <u>not</u> across provenances.

Now consider that we wish to collect seed from certain of these mother trees to be planted in commercial plantations, and want to know from which trees to collect (i.e. which seed is expected to be the genetically the best). To compare trees from different provenances we need to add back the fixed genetic provenance effect. This is an application of Eq. 11.14,

$$\hat{\mathbf{w}} = \mathbf{P}_p\hat{\boldsymbol{\beta}}_p + \hat{\mathbf{g}}.$$

The elements of $\hat{\mathbf{w}}$ are used to compare trees among all the provenances. The vector $\hat{\boldsymbol{\beta}}_p$ is a vector consisting of only the fixed genetic effects (nuisance effects omitted), which for this case is a 2 x 1 vector containing the last two elements of $\hat{\boldsymbol{\beta}}^z$. The matrix $\mathbf{P}_p$ is an incidence matrix relating the fixed genetic effects to the trees for which genetic values are being predicted. For this case, $\mathbf{P}_p$ is 9 x 2 (specifically, the last two columns of $\mathbf{X}^z$), since we are predicting breeding values for 9 parents, and there are only two fixed genetic effects. There are three provenances, and therefore three T_r effects, but with the 'sum-to-zero' parameterization $T_3 = -T_1 - T_2$, and this is accounted for in the 2 columns of $\mathbf{P}_p$. The vector $\hat{\mathbf{w}}$ is a 9 x 1 vector, and the expected value of $\hat{\mathbf{w}}$ is

$$E(\hat{\mathbf{w}}) = E(\mathbf{P}_p\hat{\boldsymbol{\beta}}_p + \hat{\mathbf{g}}) = E(\mathbf{P}_p\hat{\boldsymbol{\beta}}_p) = \boldsymbol{\tau},$$

since $E(\hat{\mathbf{g}}) = \mathbf{0}$. The vector $\boldsymbol{\tau}$ contains the appropriate fixed provenance effects for each of the 9 parents. Using Eq. 11.14 to predict $\hat{\mathbf{w}}$ for this case results in

$$\hat{\mathbf{w}} = \mathbf{P}_p\hat{\boldsymbol{\beta}}_p + \mathbf{C'V}^{-1}(\mathbf{y}-\mathbf{X}\hat{\boldsymbol{\beta}})$$

$$\hat{\mathbf{w}} = \hat{\boldsymbol{\tau}} + \hat{\mathbf{g}}$$

$$\begin{bmatrix} 4.31 \\ 2.32 \\ 2.76 \\ -5.79 \\ 0.41 \\ -1.48 \\ -3.08 \\ -0.08 \\ 0.64 \end{bmatrix} = \begin{bmatrix} 3.13 \\ 3.13 \\ 3.13 \\ -2.29 \\ -2.29 \\ -2.29 \\ -0.84 \\ -0.84 \\ -0.84 \end{bmatrix} + \begin{bmatrix} 1.18 \\ -0.81 \\ -0.37 \\ -3.50 \\ 2.70 \\ 0.81 \\ -2.24 \\ 0.76 \\ 1.48 \end{bmatrix}$$

The first three elements of $\hat{\mathbf{g}}$, $\hat{\boldsymbol{\tau}}$, and $\hat{\mathbf{w}}$ pertain to the three parent trees from provenance A, the second three pertain to provenance B, and the last three to the trees from provenance C. Note that within each provenance, the elements of $\hat{\mathbf{g}}$ are centered on zero. The vector $\hat{\mathbf{w}}$ is calculated by adding the elements of $\hat{\mathbf{g}}$ to $\hat{\boldsymbol{\tau}}$, the fixed genetic values associated with each of the three provenances (as in Eq. 4.2). Thus, in $\hat{\mathbf{w}}$, genetic values within a provenance average to the fixed genetic effect for that provenance, and genetic values across all provenances average to zero.

SUMMARY AND CONCLUSIONS

The simple examples in this chapter illustrate how to apply BLUP, and how BLP and BLUP treat data differently. In general BLUP is the preferred method whenever fixed effect estimates obtained by simple averages or checklots are not both unbiased and precise. For the grand mean to be an unbiased precise estimate of the expected value of $\mathbf{y}$, 1) the parents in all tests should come from a single population (i.e. $E(\mathbf{g})=0$), and 2) the parents in any one test should be a large random sample of the parents in that single population. If the data on a test site do not come from a single population, then a simple average of all families cannot possibly be unbiased; the simple average cannot account for the fact that the genotypes come from different fixed groups (provenances, selection generations, etc.). If there are few families or checklots in the simple average, then the estimates of the fixed effects will be imprecise.

In either case, the GLS estimate of the fixed effects from BLUP will provide more precise estimates of the fixed effects and more precise prediction of the genetic values. How much more precise GLS estimates are (and therefore, the relative advantage of BLUP over BLP) depends upon the structure of the data. It is important to know what situations require BLUP *vs* BLP because BLP can often be simpler to compute (depending on the structure of **V**).

It is interesting that the increased precision of BLUP *vs* BLP is not reflected in the estimated error variances or correlations between the true and predicted breeding values. The formulae for estimates of precision of BLP do not 'recognize' that the fixed effects are estimated (i.e. all formulae assume that fixed effects are known). Because fixed effects do have to be estimated, the estimates of precision for BLP are too high.

While in this chapter we have focused on differences between BLP and BLUP, they share many similarities. For example, both treat genetic effects as random variables to be predicted, hence all of the examples in Chapters 5 - 8 about how to 1) specify linear models, 2) set up **V** and **C**, and 3) estimate second moments are identical for BLP and BLUP. Further, both combine genetic data of different precisions, different ages, from different mating designs, and from different traits in the same way. In fact, most of the concepts in Chapters 5 - 8 apply to both BLP and BLUP. There are many issues that need investigation (such as the effects of errors in estimating **C** and **V**); but compared to methods that treat genetic values as fixed, we believe that both of these methods provide a superior way of predicting genetic values when data are unbalanced, of different precisions and/or come from a variety of different sources. We believe these prediction methods will find more widespread use in plant and animal breeding programs.

PROBLEMS

Problem 12.1: For Case 1 of this chapter, set up the design matrix $\mathbf{X}^3$ with the constraint that $E_3 = 0$. Then,

a) calculate $\hat{\boldsymbol{\beta}}^3$,

b) set up $\mathbf{L}^3$ (i.e. find a matrix) such that

$$\mathbf{L^3X} = \begin{bmatrix} \mu^3 + E_1^3 \\ \mu^3 + E_2^3 \\ \mu^3 + E_3^3 \end{bmatrix}$$

c) confirm that $\mathbf{L}^3\hat{\boldsymbol{\beta}}^3$ is estimable by showing it equals $\hat{\boldsymbol{\beta}}^m$ presented in Figure 12.5.

Problem 12.2: Verify that $\mathrm{Var}(\mathbf{L}^*\hat{\boldsymbol{\beta}}^*) = \mathrm{Var}(\mathbf{L}^z\hat{\boldsymbol{\beta}}^z) = \mathrm{Var}(\hat{\boldsymbol{\beta}}^m) = \mathrm{Var}(\mathbf{L}^3\hat{\boldsymbol{\beta}}^3)$.

Problem 12.3: Consider the following situation: You have 100 open-pollinated families tested in a single progeny test (test 1). Based on the 5-year results from that test you selected 25 families to test in a second progeny test (test 2). You now have 15-year results from test 1 and 10-year results from test 2. You wish to use BLUP to predict breeding values for 15-year height for the 100 parents, using family means from a single site as the unit of observation.

a) For use in this instance, is BLP or BLUP more appropriate? Why?

b) What are the dimensions of **y**, **V**, **C**, **G**.

c) What effects should be in **X**? Show two rows of **X**: one row corresponding to an observation from test 1, and one from test 2.

d) What types of second moment estimates are needed?

Problem 12.4: Calculate $\mathbf{X}\hat{\boldsymbol{\beta}}$ for each of four different model parameterizations: $\hat{\boldsymbol{\beta}}^*$, $\hat{\boldsymbol{\beta}}^z$, $\hat{\boldsymbol{\beta}}^m$ in Case 1, and $\hat{\boldsymbol{\beta}}^3$ in Problem 12.1.

Problem 12.5: Verify that $\mathrm{Var}(\mathbf{X}^*\hat{\boldsymbol{\beta}}^*) = \mathrm{Var}(\mathbf{X}^z\hat{\boldsymbol{\beta}}^z) = \mathrm{Var}(\mathbf{X}^m\hat{\boldsymbol{\beta}}^m) = \mathrm{Var}(\mathbf{X}^3\hat{\boldsymbol{\beta}}^3)$.
Hint: $\mathrm{Var}(\mathbf{X}\hat{\boldsymbol{\beta}}) = \mathbf{X}\,\mathrm{Var}(\hat{\boldsymbol{\beta}})\,\mathbf{X}' = \mathbf{X}(\mathbf{X}'\mathbf{V}^{-1}\mathbf{X})^-\mathbf{X}'$.

Literature Cited

Allard, R.W. 1960. Principles of Plant Breeding. John Wiley and Sons, New York, NY. 485p.

Anderson, V.L. and R.A. McLean. 1974. Design of Experiments: A Realistic Approach. Marcel Dekker, Inc., New York, NY. 418p.

Baker, R.J. 1986. Selection Indices in Plant Breeding. CRC Press, Boca Raton, FL. 218p.

Baradat, P. 1976. Use of juvenile-mature relationships and information from relatives in combined multitrait selection. P. 121-138 in Proc. IUFRO Joint Meeting on Adv. Gen. Breeding, June 14-18, Bordeaux, France.

Becker, W.A. 1975. Manual of Quantitative Genetics. Washington State Univ. Press, Pullman, WA. 170p.

Binet, F.E. 1965. On the construction of an index for indirect selection. Biometrics 21:291-299.

Bridgwater, F.E. and A.E. Squillace. 1986. Selection indexes for forest trees. P. 17-20 in Advanced Generation Breeding of Forest Trees. Southern Coop. Series Bull. 309, Louisiana Ag. Exp. Stn., Baton Rouge, LA.

Bridgwater, F.E. and R.W. Stonecypher. 1979. Index selection for volume and straightness in a loblolly pine population. P. 132-139 in Proc. 15th. South. For. Tree Improve. Conf., June 19-21, Miss State, MS.

Bridgwater, F.E., J.T. Talbert and S. Jahromi. 1983. Index selection for increased dry weight production in a young loblolly pine population. Silvae Genet. 32:157-161.

Brunk, H.D. 1975. An Introduction to Mathematical Statistics. Xerox College Pub., Lexington, MA. 457p.

Bulmer, M.G. 1985. The Mathematical Theory of Quantitative Genetics. Clarendon Press, Oxford, GB. 254p.

Burdon, R.D. 1977. Genetic correlation as a concept for studying genotype-environment interaction in forest tree breeding. Silvae Genet. 26:168-175.

Burdon, R.D. 1979. Generalisation of multi-trait selection indices using information from several sites. New Zealand J. For. Sci. 9:145-152.

Burdon, R.D. 1982. Selection indices using information from multiple sources for the single-trait case. Silvae Genet. 31:81-85.

Busby, C.L. 1983. Crown-quality assessment and the relative economic importance of growth and crown characters in mature loblolly pine. P. 121-130 in Proc. 17th South. For. Tree Improve. Conf., June 7-9, Athens, GA.

Christophe, C. and Y. Birot. 1983. Genetic structures and expected genetic gains from multitrait selection in wild populations of Douglas-fir and Sitka spruce II. Practical application of index selection on several populations. Silvae Genet. 32:173-181.

Comstock, R.E and R.H Moll. 1963. Genotype-environment interactions. P. 164-194 in Statistical Genetics and Plant Breeding, R.E Hanson and H.F. Robinson (eds.). NAS-NRC Pub. 982, Washington, DC.

Cotterill, P.P. 1987. On estimating heritability according to practical applications. Silvae Genet. 36:46-48.

Cotterill, P.P. and C.A. Dean. 1989. Handbook of Applied Index Selection. CSIRO Div. Forestry, Canberra, Australia. (In Press).

Cotterill, P.P. and N. Jackson. 1981. Index selection with restrictions in tree breeding. Silvae Genet. 30:106-108.

Cotterill, P.P. and N. Jackson. 1985. On index selection: I. Methods of determining economic weights. Silvae Genet. 34:56-63.

Cotterill, P.P., R.L. Correll and R. Boardman. 1983. Methods of estimating the average performance of families across incomplete open-pollinated progeny tests. Silvae Genet. 32:28-32.

Dean, C.A., P.P. Cotterill and J.N. Cameron. 1983. Genetic parameters and gains expected from multiple trait selection of radiata pine in eastern Victoria. Aust. For. Res. 13:271-278.

Dean, C.A., P.P. Cotterill and R.L. Eisemann. 1986. Genetic parameters and gains expected from selection in *Pinus caribaea* var. *hondurensis* in northern Queensland, Australia. Silvae Genet. 35:229-236.

Eagles, H.A. and K.J. Frey. 1974. Expected and actual gains in economic value of oat lines from five selection methods. Crop Sci. 14:861-864.

Eisenhart, C. 1947. The assumptions underlying the analysis of variance. Biometrics 3:1-21.

Elgin, J.H., R.R. Hill and K.E. Zeiders. 1970. Comparison of four methods of multiple trait selection for five traits in alfalfa. Crop Sci. 10:190-193.

Falconer, D.S. 1981. Introduction to Quantitative Genetics. Longman & Co., New York, NY. 340p.

Freund, R.J. and R.C. Littell. 1981. SAS for Linear Models. SAS Institute, Inc., Cary, NC. 231p.

Gianola, D. 1986. On selection criteria and estimation of parameters when the variance is heterogeneous. Theor. Appl. Genet. 72:671-677.

Goddard, R.E. and R.K. Strickland. 1968. Volume and weight tables for five-year-old plantation grown slash pine. School of Forestry Res. Rep. No. 14. University of Florida, Gainesville, FL. 8p.

Graybill, F.A. 1976. Theory and Application of the Linear Model. Duxbury Press, North Scituate, MA. 704p.

Griffing, B. 1956. A generalised treatment of the use of diallel crosses in quantitative inheritance. Heredity 10:31-50.

Hallauer, A.R. and J.B. Miranda. 1981. Quantitative Genetics in Maize Breeding. Iowa State Univ. Press, Ames, IA. 468p.

Harris, D.L. 1963. The influence of errors of parameter estimation upon index selection. P. 491-500 in Statistical Genetics and Plant Breeding, R.E Hanson and H.F. Robinson (eds.). NAS-NRC Pub. 982, Washington, DC.

Harris, D.L. 1964. Expected and predicted progress from index selection involving estimates of the population parameters. Biometrics 20:46-72.

Harvey, W.R. and A.M. Townsend. 1985. Selection on a combination of individual, family, and stand merit in provenance tests. Forest Sci. 31:813-821.

Hatcher, A.V., F.E. Bridgwater and R.J. Weir. 1981. Performance level: Standardized score for progeny test performance. Silvae Genet. 30:184-187.

Hayes, J.F. and W.G. Hill. 1980. A reparameterization of a genetic selection index to locate its sampling properties. Biometrics 36:237-243.

Hazel, L.N. 1943. The genetic basis for constructing selection indexes. Genetics 28:476-490.

Henderson, C.R. 1963. Selection index and expected genetic advance. P. 141-163 in Statistical Genetics and Plant Breeding, W.D. Hanson and H.F. Robinson (eds.). NAS-NRC Pub. No. 982, Washington, DC.

Henderson, C.R. 1973. Sire evaluation and genetic trends. P. 10-41 in Animal Breeding and Genetics Symposium in Honor of J. Lush, Animal Sci. Assoc. Amer., Champaign, Ill.

Henderson, C.R. 1974. General flexibility of linear model techniques for sire evaluation. J. Dairy Sci. 57:963-972.

Henderson, C.R. 1975a. Best linear unbiased estimation and prediction under a selection model. Biometrics 31:423-447.

Henderson, C.R. 1975b. Comparison of alternative sire evaluation methods. J. Anim. Sci. 41:760-770.

Henderson, C.R. 1975c. Use of all relatives in intraherd prediction of breeding values and real producing abilities. J. Dairy Sci. 58:1910-1919.

Henderson, C.R. 1976. A simple method for computing the inverse of a numerator relationship matrix used in prediction of breeding values. Biometrics 32:69-83.

Henderson, C.R. 1977a. Best linear unbiased prediction of breeding values not in the model for records. J. Dairy Sci. 60:783-787.

Henderson, C.R. 1977b. Prediction of future records. P. 615-638 in Proc. Internat. Conf. Quan. Genet., E. Pollack, O. Kempthorne and T. Bailey (eds.). The Iowa State Univ. Press, Ames, IA.

Henderson, C.R. 1978. Undesirable properties of regressed least squares prediction of breeding values. J. Dairy Sci. 61:114-120.

Henderson, C.R. 1984. Applications of Linear Models in Animal Breeding. University of Guelph, Guelph, Ontario, CAN. 462p.

Henderson, C.R. 1988. Progress in statistical methods applied to quantitative genetics since 1976. P. 85-90 in Proc. Second Internat. Conf. on Quantitative Genet., B.S. Weir, E.J. Eisen, M.M. Goodman and G. Namkoong (eds.). Sinauer Assoc. Inc., Sunderland, Mass.

Henderson, C.R. and R.L. Quaas. 1976. Multiple trait evaluation using relatives' records. J. Anim. Sci. 43:1188-1197.

Hill, W.G. 1984. On selection among groups with heterogeneous variance. Anim. Prod. 39:473-477.

Iloeje, M.U. and C.J. Wilcox. 1981. Best linear unbiased prediction: A method of estimating breeding values. Dairy Sci. Infor. Sheet No. DY80-71. University of Florida, Gainesville, FL. 9p.

Kempthorne, O. and A.W. Nordskog. 1959. Restricted selection indices. Biometrics 15:10-19.

Kennedy, B.W. and D.A. Sorensen. 1988. Properties of mixed-model methods for prediction of genetic merit. P. 91-103 in Proc. Second Internat. Conf. Quantitative Genet., B.S. Weir, E.J. Eisen, M.M. Goodman and G. Namkoong (eds.). Sinauer Assoc. Inc., Sunderland, Mass.

Kmenta, J. 1971. Elements of Econometrics. Macmillan Co. Inc., New York, NY. 655p.

Lambeth, C.C. 1980. Juvenile-mature correlations in *Pinaceae* and implications for early selection. Forest Sci. 26:571-580.

Lambeth, C.C. 1983. Early testing - An overview with emphasis on loblolly pine. P. 297-311 in Proc. 17th South. For. Tree Improve. Conf., June 7-9, Athens, GA.

Lambeth, C.C., J.P. van Buijtenen, S.D. Duke and R.B. McCullough. 1983. Early selection is effective in 20-year-old genetic tests of loblolly pine. Silvae Genet. 32:210-215.

Land, S.B., B.C. Bongarten and J.R. Toliver. 1987. Genetic parameters and selection indices from provenance/progeny tests. P. 59-74 in Statistical Considerations in Genetic Testing of Forest Trees. South. Coop. Series Bull. 324, University of Florida, Gainesville, FL.

Lin, C.Y. 1978. Estimation of heritability of an index. Can. J. Genet. Cytol. 20:485-487.

Lindgren, D. 1986. How should breeders respond to breeding values? P. 361-372 in Proc. IUFRO Conf. Breeding Theory, Progeny Testing and Seed Orchards, October 13-17, Williamsburg, VA.

Lowe, W.J., R. Stonecypher and A.V. Hatcher. 1983. Progeny test data handling and analysis. P. 51-68 in Progeny Testing of Forest Trees. South. Coop. Series Bull. No. 275, Texas A&M Univ., College Station, TX.

McDaniel, B.T. 1974. Why new sire summaries are needed. J. Dairy Sci. 57:951-955.

Mendenhall, W. and R.L. Schaeffer. 1973. Mathematical Statistics with Applications. Duxbury Press, North Scituate, MA. 561p.

Milliken, G.A. and D.E. Johnson. 1984. Analysis of Messy Data I. Designed Experiments. Lifetime Learning Pub., Belmont, CA. 473p.

Mosteller, F. and J.W. Tukey. 1977. Data Analysis and Multiple Regression. Addison-Wesley Pub., London, GB. 588p.

Namkoong, G. 1976. A multiple-index selection strategy. Silvae Genet. 25:199-201.

Namkoong, G., E.B. Snyder and R.W. Stonecypher. 1966. Heritability and gain concepts for evaluating breeding systems such as seedling orchards. Silvae Genet. 15:76-84.

Neter, J. and W. Wasserman. 1974. Applied Linear Statistical Models. Richard D. Irwin & Co., Homewood, IL. 842p.

Noble, B. and J.W. Daniel. 1977. Applied Linear Algebra. Prentice-Hall, Inc., Englewood Cliffs, NJ. 477p.

Nordskog, A.W. 1978. Some statistical properties of an index of multiple traits. Theor. Appl. Genet. 52:91-94.

Openshaw, S.J. and H.H. Hadley. 1984. Selection indexes to modify protein concentration of soybean seeds. Crop Sci. 24:1-4.

Quaas, R.L. and E.J. Pollak. 1981. Modified equations for sire models with groups. J. Dairy Sci. 64:1868-1872.

Quaas, R.L., R.D. Anderson and A.R. Gilmour. 1984. BLUP School Handbook. Univ. New England, New South Wales, Auatralia. 158p.

Robinson, H.F., R.E. Comstock and P.H. Harvey. 1951. Genotypic and phenotypic correlations in corn and their implications in selection. Agron. J. 43:282-287.

Rockwood, D.L. and R.E. Goddard. 1973. Predicted gains for fusiform rust resistance in slash pine. P. 31-37 in Proc. Twelfth South. Forest Tree Improve. Conf., June 12-13, Baton Rouge, LA.

Sales, J. and W.G. Hill. 1976. Effect of sampling errors on efficiency of selection indices. Anim. Prod. 22:1-17.

SAS Institute Inc. 1985. SAS/IML User's Guide for Personal Computers, Version 6 Edition. SAS Insitute Inc., Cary, NC. 429p.

Searle, S.R. 1966. Matrix Algebra for the Biological Sciences. John Wiley and Sons, New York, NY. 296p.

Searle, S.R. 1971. Topics in variance component estimation. Biometrics 27:1-76.

Searle, S.R. 1974. Prediction, mixed models, and variance components. P. 229-266 in Reliability and Biometry, SIAM, Philadelphia, Penn.

Shelbourne, C.J.A. 1969. Tree breeding methods. For. Res. Inst. Tech. Pap. No. 55, New Zealand For. Serv., Wellington, New Zealand.

Shelbourne, C.J.A. and C.B. Low. 1980. Multi-trait index selection and associated genetic gains of *Pinus radiata* progenies at five sites. N. Z. J. For. Sci. 10:307-324.

Sluder, E.R. 1977. Fusiform rust in loblolly and slash pine plantations on high-hazard sites in Georgia. U.S. Dep. Agric. For. Serv. Res. Pap. SE-160, Southeast. For. Exp. Stn., Asheville, NC.

Smith, H. 1936. A discriminant function for plant selection. Ann. Eug. 7:240-250.

Snedecor, G.W. and W.G Cochran. 1967. Statistical Methods. Iowa State Univ. Press, Ames, IA. 593p.

Sohn, S. and R.E. Goddard. 1979. Influence of infection percent on improvement of fusiform rust resistance in slash pine. Silvae Genet. 28:173-180.

Stonecypher, R. and M. Arbez. 1976. Methods of selection. P. 31-46 in Proc. IUFRO Conf. Advanced Generation Breeding, June 14-18, Bordeaux, France.

Talbert, C.B. 1984. An analysis of several approaches to multiple-trait index selection in loblolly pine (*Pinus taeda* L.). Ph.D. Dissertation, North Carolina St. Univ., Raleigh, NC. 105p.

Talbert, C.B. 1986. Multi-criterion index selection as a tool for operational tree improvement. P. 228-238 in Proc. IUFRO Conf. on Breeding Theory, Progeny Testing and Seed Orchards, October 13-17, Williamsburg, VA.

Thompson, R. 1979. Sire evaluation. Biometrics 35:339-353.

van Vleck, L.D. 1970. Index selection for direct and maternal genetic components of economic traits. Biometrics 26:477-483.

van Vleck, L.D. 1976. Selection for direct, maternal and grandmaternal genetic components of economic traits. Biometrics 32:173-181.

White, T.L. and G.R. Hodge. 1988. Best linear prediction of breeding values in a forest tree improvement program. Theor. Appl. Genet. 76:719-727.

White, T.L., G.R. Hodge and M.A. Delorenzo. 1986. Best linear prediction of breeding values in forest tree improvement. P. 99-122 in Proc. Statistical Considerations in Genetic Testing of Forest Trees. South. Coop. Series Bull. 324, Univ. of Florida, Gainesville, FL.

Wilcox, C.J. and M.A. Delorenzo. 1983. Progeny testing in dairy cattle. P. 101-113 in Proc. First Internat. Symp. on Animal Production, Nov. 15-19, Riberao Preto, S.P., Brazil.

Williams, J.S. 1962a. The evaluation of a selection index. Biometrics 18:375-393.

Williams, J.S. 1962b. Some statistical properties of a genetic selection index. Biometrika 49:325-337.

Appendices

Appendix 1: One minus the cumulative standard normal distribution. Each entry is α, the area from z to ∞ under the standard normal distribution.

z	.00	.01	.02	.03	.04	.05	.06	.07	.08	.09
.0	.5000	.4960	.4920	.4880	.4840	.4801	.4761	.4721	.4681	.4641
.1	.4602	.4562	.4522	.4483	.4443	.4404	.4364	.4325	.4286	.4247
.2	.4207	.4168	.4129	.4090	.4052	.4013	.3974	.3936	.3897	.3859
.3	.3821	.3783	.3745	.3707	.3669	.3632	.3594	.3557	.3520	.3483
.4	.3446	.3409	.3372	.3336	.3300	.3264	.3228	.3192	.3156	.3121
.5	.3085	.3050	.3015	.2981	.2946	.2912	.2877	.2843	.2810	.2776
.6	.2743	.2709	.2676	.2643	.2611	.2578	.2546	.2514	.2483	.2451
.7	.2420	.2389	.2358	.2327	.2296	.2266	.2236	.2206	.2177	.2148
.8	.2119	.2090	.2061	.2033	.2005	.1977	.1949	.1922	.1894	.1867
.9	.1841	.1814	.1788	.1762	.1736	.1711	.1685	.1660	.1635	.1611
1.0	.1587	.1562	.1539	.1515	.1492	.1469	.1446	.1423	.1401	.1379
1.1	.1357	.1335	.1314	.1292	.1271	.1251	.1230	.1210	.1190	.1170
1.2	.1151	.1131	.1112	.1093	.1075	.1056	.1038	.1020	.1003	.0985
1.3	.0968	.0951	.0934	.0918	.0901	.0885	.0869	.0853	.0838	.0823
1.4	.0808	.0793	.0778	.0764	.0749	.0735	.0721	.0708	.0694	.0681
1.5	.0668	.0655	.0643	.0630	.0618	.0606	.0594	.0582	.0571	.0559
1.6	.0548	.0537	.0526	.0516	.0505	.0495	.0485	.0475	.0465	.0455
1.7	.0446	.0436	.0427	.0418	.0409	.0401	.0392	.0384	.0375	.0367
1.8	.0359	.0351	.0344	.0336	.0329	.0322	.0314	.0307	.0301	.0294
1.9	.0287	.0281	.0274	.0268	.0262	.0256	.0250	.0244	.0239	.0233
2.0	.0228	.0222	.0217	.0212	.0207	.0202	.0197	.0192	.0188	.0183
2.1	.0179	.0174	.0170	.0166	.0162	.0158	.0154	.0150	.0146	.0143
2.2	.0139	.0136	.0132	.0129	.0125	.0122	.0119	.0116	.0113	.0110
2.3	.0107	.0104	.0102	.0099	.0096	.0094	.0091	.0089	.0087	.0084
2.4	.0082	.0080	.0078	.0075	.0073	.0071	.0069	.0068	.0066	.0064
2.5	.0062	.0060	.0059	.0057	.0055	.0054	.0052	.0051	.0049	.0048
2.6	.0047	.0045	.0044	.0043	.0041	.0040	.0039	.0038	.0037	.0036
2.7	.0035	.0034	.0033	.0032	.0031	.0030	.0029	.0028	.0027	.0026
2.8	.0026	.0025	.0024	.0023	.0023	.0022	.0021	.0021	.0020	.0019
2.9	.0019	.0018	.0018	.0017	.0016	.0016	.0015	.0015	.0014	.0014
3.0	.0013	.0013	.0013	.0012	.0012	.0011	.0011	.0011	.0010	.0010
3.1	.0010	.0009	.0009	.0009	.0008	.0008	.0008	.0008	.0007	.0007
3.2	.0007	.0007	.0006	.0006	.0006	.0006	.0006	.0005	.0005	.0005
3.3	.0005	.0005	.0005	.0004	.0004	.0004	.0004	.0004	.0004	.0003
3.4	.0003	.0003	.0003	.0003	.0003	.0003	.0003	.0003	.0003	.0002
3.6	.0002	.0002	.0001	.0001	.0001	.0001	.0001	.0001	.0001	.0001
3.9	.0000									

Appendix 2: Selection intensity (i) for proportion selected (p) when the population size is infinite. Abridged from Becker (1975).

p	i	p	i	p	i	p	i
.001	3.367	.20	1.400	.48	0.830	.76	0.409
.002	3.170	.21	1.372	.49	0.814	.77	0.394
.003	3.050	.22	1.346	.50	0.798	.78	0.380
.004	2.962	.23	1.320	.51	0.782	.79	0.365
.005	2.892	.24	1.295	.52	0.766	.80	0.350
.006	2.834	.25	1.271	.53	0.751	.81	0.335
.007	2.784	.26	1.248	.54	0.735	.82	0.320
.008	2.740	.27	1.225	.55	0.720	.83	0.305
.009	2.701	.28	1.202	.56	0.704	.84	0.290
.01	2.665	.29	1.180	.57	0.689	.85	0.274
.02	2.421	.30	1.159	.58	0.674	.86	0.259
.03	2.268	.31	1.138	.59	0.659	.87	0.243
.04	2.154	.32	1.118	.60	0.644	.88	0.227
.05	2.063	.33	1.097	.61	0.629	.89	0.211
.06	1.985	.34	1.078	.62	0.614	.90	0.195
.07	1.918	.35	1.058	.63	0.599	.91	0.178
.08	1.858	.36	1.039	.64	0.585	.92	0.162
.09	1.804	.37	1.020	.65	0.570	.93	0.144
.10	1.755	.38	1.002	.66	0.555	.94	0.127
.11	1.709	.39	0.984	.67	0.541	.95	0.109
.12	1.667	.40	0.966	.68	0.526	.96	0.090
.13	1.627	.41	0.948	.69	0.511	.97	0.070
.14	1.590	.42	0.931	.70	0.497	.98	0.049
.15	1.554	.43	0.913	.71	0.482	.99	0.027
.16	1.521	.44	0.896	.72	0.468		
.17	1.489	.45	0.880	.73	0.453		
.18	1.458	.46	0.863	.74	0.438		
.19	1.428	.47	0.846	.75	0.424		

Appendix 3: Program for matrix construction and BLP breeding value calculation written using SAS Interactive Matrix Language (SAS/IML).

/* MATRIX.PGM - Written by Gary R. Hodge, March 27, 1987.
MATRIX.PGM is written for SAS/IML on a PC. This program automates the process of Best Linear Prediction of breeding values for slash pine using open-pollinated progeny test data. The program uses a SAS data set with ten variables:

SAS VAR. NAME	MATRIX & COLUMN	DESCRIPTION
UFNUM	FC[I,1]	Parent tree identification code.
COMPID	FC[I,2]	Company identification code.
TESTNUM	FC[I,3]	Test identification number.
TRAITID	FN[I,1]	Identifies the age of the test, and the measurement as either rust or volume (15 = 5-year volume, 110 = 10-year volume, 115 = 15-year volume, 25 = 5-year rust, 210 = 10-year rust, 215 = 15-year rust).
DEV	FN[I,2]	Family mean expressed as a deviation from the test mean.
RUSTMN	FN[I,3]	The test mean rust infection, $0 \leq \bar{y} \leq 1$.
VOLMN	FN[I,4]	The test mean volume in ft^3/tree.
CVVOL	FN[I,5]	The coefficient of variation for volume.
VFBARR	FN[I,6]	Variance of family means for rust.
VFBARV	FN[I,7]	Variance of family means for volume.

Note that VFBARR and VFBARV are the final estimates of the variances of family means, and will go directly into the diagonals of the $\mathbf{V}_k$ matrices. Thus, only the covariances in the $\mathbf{V}_k$ and $\mathbf{C}_k$ matrices will be calculated in MATRIX.PGM. The SAS data set includes one line for each observation from a progeny test. For example, the data set will include two lines for each parent in a test where both rust and volume are measured. The data set should be sorted by family (UFNUM) before use of MATRIX.PGM.

Appendix 3 cont.

The program has two main sections. Section 1, which is quite short, does two things. First it reads the data from the SAS data set for one family at a time into two matrices, FC (character variables) and FN (numeric variables). Then it creates the y vector and the correct size V and C matrices which are filled with zeros. Section 2, which makes up the bulk of the program is divided into three subsections. Section 2.1 replaces the zeros in the the V matrix with the appropriate second moments, Section 2.2 does the same thing with the C matrix, and Section 2.3 calculates the breeding values. */

```
/*--Section 1 --*/
/*--Initialization--*/
PROC IML WORKSIZE=100;
LIBNAME BLP 'C:\BLP';
USE BLP.MASTER;
BVHATS={NAME BV1 BV2};
EVPHATS={NAME EVP1 EVP2};
K=1;

/*--Read the first observation into FC and FN--*/
READ POINT 1 VAR{A,B,C} INTO FC; READ POINT 1 INTO FN;

DO WHILE (K<8000);
  K=K+1;
  LIST POINT K;
/*--Read the next observation into CHECKC and CHECKN--*/
  READ POINT K VAR{A,B,C} INTO CHECKC;
  READ POINT K INTO CHECKN;

/*--If observation is for current family, add it to CHECKC and FC--*/
  IF FC[1,1]=CHECKC[1,1] THEN DO;
    FC=FC//CHECKC; FN=FN//CHECKN;
    FREE CHECKC; FREE CHECKN;
  END;

/*If last observation for current family has been read, construct matrices and vectors,
  and calculate breeding values*/
  ELSE DO;
    N=NROW(FN);
    Y=FN[,2];                                  /*--Create y vector--*/
    V=SHAPE(0,N,N);          /*--Create nxn matrix of zeros for V matrix.--*/
    C=SHAPE(0,N,2);          /*--Create nx2 matrix of zeros for C matrix.--*/
```

Appendix 3 cont.

```
/*--SECTION 2.1 - Construction of the V matrix--*/
  DO I=1 TO N;
    DO J=I TO N;

/*--Measurements from the same test--*/
      IF FC[I,{2,3}]=FC[J,{2,3}] THEN DO;

/*--Fill in the diagonals with variances of family means--*/
        IF I=J THEN DO;
          IF FN[I,1]=25 | FN[I,1]=210 | FN[I,1]=215 THEN V[I,J]=FN[I,6];
          ELSE V[I,J]=FN[I,7];
        END;

/*-Fill in off-diagonals with Type A rust-volume covariances-*/
        ELSE DO;
          IF FN[I,1]=115 | FN[I,1]=215 THEN                  /*-A 15-year test-*/
              V[I,J]=0.137*SQRT(FN[I,6]*FN[I,7]);
          IF FN[I,1]=110 | FN[I,1]=210 THEN                  /*-A 10-year test-*/
              V[I,J]=0.134*SQRT(FN[I,6]*FN[I,7]);
          IF FN[I,1]=15 | FN[I,1]=25 THEN                     /*-A 5-year test-*/
              V[I,J]=0.057*SQRT(FN[I,6]*FN[I,7]);
          V[J,I]=V[I,J];       /*-Set corresponding element equal for symmetry-*/
        END;

      END;

/*--Measurements from different tests--*/
      ELSE DO;

/*--Fill in off-diagonals with Type B rust-rust covariances--*/
        IF (FN[I,1]=25 | FN[I,1]=210 | FN[I,1]=215) &
           (FN[J,1]=25 | FN[J,1]=210 | FN[J,1]=215) THEN DO;
          V[I,J]=(.2741+.6414*(FN[I,3]+FN[J,3])
               -.2680*((FN[I,3]+FN[J,3])##2)
               -.5006*ABS(FN[I,3]-FN[J,3]))*SQRT(FN[I,6]*FN[J,6]);
          V[J,I]=V[I,J];          /*Set corresponding element equal for symmetry*/
        END;
```

Appendix 3 cont.

```
/*--Fill in off-diagonals with Type B vol-vol covariances--*/
        ELSE DO;

            CVS=FN[I,5]+FN[J,5]; IF CVS>.65 THEN CVS=.65;

          IF FN[I,1]=15 & FN[J,1]=15 THEN DO;              /*-Two 5-year tests-*/
            V[I,J]=(.7289-(1.0291*CVS)-(.0931*ABS(FN[I,4]-FN[J,4])))
                  *SQRT(FN[I,7]*FN[J,7]);
            IF V[I,J] < 0 THEN V[I,J] = 0;
            V[J,I]=V[I,J];         /*--Set corresponding element equal for symmetry--*/
          END;

          IF FN[I,1]=110 & FN[J,1]=110 THEN DO;           /*-Two 10-year tests-*/
            V[I,J]=(.8213-(0.6924*CVS)-(.2110*ABS(FN[I,4]-FN[J,4])))
                  *SQRT(FN[I,7]*FN[J,7]);
            IF V[I,J] < 0 THEN V[I,J] = 0;
            V[J,I]=V[I,J];          /*-Set corresponding element equal for symmetry-*/
          END;

          IF FN[I,1]=115 & FN[J,1]=115 THEN DO;           /*-Two 15-year tests-*/
            V[I,J]=(.8700-(0.6924*CVS)-(.2110*ABS(FN[I,4]-FN[J,4])))
                  *SQRT(FN[I,7]*FN[J,7]);
            IF V[I,J] < 0 THEN V[I,J] = 0;
            V[J,I]=V[I,J];          /*-Set corresponding element equal for symmetry-*/
          END;

          IF FN[I,1]=15 & FN[J,1]=110 THEN DO;
            V[I,J]=(.7289-(1.0291*CVS)-(.0931*ABS(FN[I,4]-FN[J,4])))
                  *SQRT(FN[I,7]*FN[J,7]);                /*-A 5 and a 10-year test-*/
            IF V[I,J] < 0 THEN V[I,J] = 0;
            V[J,I]=V[I,J];          /*-Set corresponding element equal for symmetry-*/
          END;

          IF FN[I,1]=110 & FN[J,1]=15 THEN DO;
            V[I,J]=(.7289-(1.0291*CVS)-(.0931*ABS(FN[I,4]-FN[J,4])))
                  *SQRT(FN[I,7]*FN[J,7]);                /*-A 10 and a 5-year test-*/
            IF V[I,J] < 0 THEN V[I,J] = 0;
            V[J,I]=V[I,J];          /*-Set corresponding element equal for symmetry-*/
          END;
```

Appendix 3 cont.

```
          IF FN[I,1]=15 & FN[J,1]=115 THEN DO;
            V[I,J]=(.7289-(1.0291*CVS)-(.0931*ABS(FN[I,4]-FN[J,4])))
                   *SQRT(FN[I,7]*FN[J,7]);              /*-A 5 and a 15-year test-*/
            IF V[I,J] < 0 THEN V[I,J] = 0;
            V[J,I]=V[I,J];          /*-Set corresponding element equal for symmetry-*/
          END;

          IF FN[I,1]=115 & FN[J,1]=15 THEN DO;
            V[I,J]=(.7289-(1.0291*CVS)-(.0931*ABS(FN[I,4]-FN[J,4])))
                   *SQRT(FN[I,7]*FN[J,7]);              /*-A 15 and a 5-year test-*/
            IF V[I,J] < 0 THEN V[I,J] = 0;
            V[J,I]=V[I,J];          /*-Set corresponding element equal for symmetry-*/
          END;

          IF FN[I,1]=110 & FN[J,1]=115 THEN DO;
            V[I,J]=(.8547-(0.6924*CVS)-(.2110*ABS(FN[I,4]-FN[J,4])))
                   *SQRT(FN[I,7]*FN[J,7]);              /*-A 10 and a 15-year test-*/
            IF V[I,J] < 0 THEN V[I,J] = 0;
            V[J,I]=V[I,J];          /*-Set corresponding element equal for symmetry-*/
          END;

          IF FN[I,1]=115 & FN[J,1]=110 THEN DO;
            V[I,J]=(.8547-(0.6924*CVS)-(.2110*ABS(FN[I,4]-FN[J,4])))
                   *SQRT(FN[I,7]*FN[J,7]);              /*-A 15 and a 10-year test-*/
            IF V[I,J] < 0 THEN V[I,J] = 0;
            V[J,I]=V[I,J];          /*-Set corresponding element equal for symmetry-*/
          END;

        END;

      END;

    END;
  END;
```

Appendix 3 cont.

```
/*--SECTION 2.2 - Construction of the C matrix--*/
   DO I=1 TO N;

/*--Measurement on rust--*/
     IF FN[I,1]=25 | FN[I,1]=210 | FN[I,1]=215 THEN DO;
       C[I,2]=(.2741+.6414*(FN[I,3]+.50)-.2680*((FN[I,3]+.50)##2)
            -.5006*ABS(FN[I,3]-.50))*SQRT(FN[I,6]*.02184)*2;
     END;

/*--Measurement on volume--*/
     ELSE DO;
       CVS=FN[I,5]+.1924; IF CVS>.65 THEN CVS=.65;

       IF FN[I,1]=15 THEN DO;                          /*--A 5-year test--*/
         C[I,1]=(.7289-(1.0291*CVS)-(.0931*ABS(FN[I,4]+.3921)))
                *SQRT(FN[I,7]*.32221)*2;
         IF C[I,1] < 0 THEN C[I,1] = 0;
       END;

       IF FN[I,1]=110 THEN DO;                         /*-A 10-year test-*/
         C[I,1]=(.8547-(0.6924*CVS)-(.2110*ABS(FN[I,4]+.3921)))
                *SQRT(FN[I,7]*.32221)*2;
         IF C[I,1] < 0 THEN C[I,1] = 0;
       END;

       IF FN[I,1]=15 THEN DO;                          /*-A 15-year test-*/
         C[I,1]=(.8700-(0.6924*CVS)-(.2110*ABS(FN[I,4]+.3921)))
                *SQRT(FN[I,7]*.32221)*2;
         IF C[I,1] < 0 THEN C[I,1] = 0;
       END;
     END;

   END;
```

Appendix 3 cont.

```
/*--SECTION 2.3 - Calculation of breeding values and EVP--*/
   G={0.904   0,
      0   .06223};                                   /*--Genetic variances--*/

   CPVINV=C'*INV(V); FREE V;
   BV=CPVINV*Y;                                        /*--Calculate BV's--*/
   EVP=G-CPVINV*C;                                     /*--Calculate EVP--*/
   EVPHAT=SHAPE(0,1,2)
   DO I=1 TO 2;
     EVPHAT[1,I]=EVP[I,I]                /*--Get diagonals of EVP matrix--*/
   END;

   BVHAT=BV'||EVPHAT;                    /*--Put BV's and EVP's together--*/

   BVHAT=FC[1,1]||BVHAT;                    /*--Add family identification--*/

   BVHATS=BVHATS//BVHAT;                           /*Add current BV to
                                            previously calculated BV's.-*/

   IF CHECKC[1,1]=99 THEN DO;  /* Have we reached dummy variable at the */
     K=8000;                   /* end of the data set?                  */
   END;                        /*  If yes, K=8000 so program will quit. */

/*If we have more families, clear intermediate matrices and continue*/
   ELSE DO;
     K=K+1;
     FC=CHECKC; FN=CHECKN;        /*--Set CHECKC and CHECKN (first
                          observation of new family) as new FC and FN--*/

     FREE CPVINV C EVP EVPHAT BVHAT CHECKC CHECKN;
                                                /*-Clear other matrices-*/
   END;

  END;
END;

/*All families finished, print breeding values and error variances*/

  PRINT BVHATS[FORMAT=5.4]; PRINT EVPHATS[FORMAT=5.4];
```

Answers to Problems

Problem 1.1

Recall that a row vector has dimension 1 x c, and a column vector has dimension r x 1.

a) (1 x n) • (n x 1) = (1 x 1)
b) (r x 1) • (1 x c) = (r x c)
c) (r x n) • (n x 1) = (r x 1)
d) (1 x n) • (n x c) = (1 x c)

Problem 1.2

$z_{11} = 15 \cdot 25 + 10 \cdot 22 + 30 \cdot 33 + 8 \cdot 15 = 1705$

$z_{12} = 15 \cdot 36 + 10 \cdot 33 + 30 \cdot 44 + 8 \cdot 22 = 2366$

$z_{13} = 15 \cdot 22 + 10 \cdot 18 + 30 \cdot 20 + 8 \cdot 18 = 1254$

$z_{21} = 25 \cdot 25 + 15 \cdot 22 + 29 \cdot 33 + 12 \cdot 15 = 2092$

$z_{22} = 25 \cdot 36 + 15 \cdot 33 + 29 \cdot 44 + 12 \cdot 22 = 2935$

$z_{23} = 25 \cdot 22 + 15 \cdot 18 + 29 \cdot 20 + 12 \cdot 18 = 1616$

$z_{31} = 10 \cdot 25 + 5 \cdot 22 + 14 \cdot 33 + 9 \cdot 15 = 957$

$z_{32} = 10 \cdot 36 + 5 \cdot 33 + 14 \cdot 44 + 9 \cdot 22 = 1339$

$z_{33} = 10 \cdot 22 + 5 \cdot 18 + 14 \cdot 20 + 9 \cdot 18 = 752$

Problem 1.4

a) dim(**Z**) = dim(**X'X**) = (3 x 3). dim(**W**) = dim(**XX'**) = (4 x 4)

b)

$$\mathbf{Z} = \begin{bmatrix} 1289 & 1491 & 692 \\ 1491 & 1835 & 839 \\ 692 & 839 & 402 \end{bmatrix} \qquad \mathbf{W} = \begin{bmatrix} 950 & 575 & 1315 & 510 \\ 575 & 350 & 805 & 305 \\ 1315 & 805 & 1937 & 714 \\ 510 & 305 & 714 & 289 \end{bmatrix}$$

Problem 1.5

All answers are equal to **D** below:

$$\mathbf{D} = \begin{bmatrix} 2631618 & 201042 & 1823307 \\ 2019452 & 132534 & 1328454 \\ 1522646 & 68310 & 900381 \\ 4573017.6 & 215519.4 & 2735139.6 \end{bmatrix}$$

Problem 1.6

a)

$$\mathbf{A}^{-1} = \begin{bmatrix} -0.006509 & -0.007264 & 0.013807 \\ 0.003217 & 0.020880 & -0.006896 \\ 0.000549 & -0.000101 & -0.000104 \end{bmatrix}$$

b) $\mathbf{I}_3$

c) $\mathbf{A}$

d) $\mathbf{A}$

e) $\mathbf{I}_3$

f) $(\mathbf{A}^{-1})' = \mathbf{A}'^{-1}$

Problem 1.7

a)

$$(\mathbf{AB})^{-1} = \begin{bmatrix} 0.000031 & -0.000279 & -0.000006 \\ -0.001125 & -0.007325 & 0.002459 \\ 0.003061 & 0.022277 & -0.006867 \end{bmatrix}$$

b) $\mathbf{I}_3$

c) $(\mathbf{AB})^{-1}$

d) $\mathbf{I}_3$

Problem 1.8

a) $\mathbf{L}^{-1}$ does not exist.

$$\mathbf{M}^{-1} = \begin{bmatrix} 0.006349 & 0.090675 & -0.040675 \\ -0.017027 & -0.086923 & 0.055105 \\ 0.015873 & 0.007937 & -0.007937 \end{bmatrix}$$

c) Using a Moore-Penrose generalized inverse:

$$\mathbf{L}^{-} = \begin{bmatrix} -0.031044 & 0.051359 & -0.010730 \\ 0.013180 & -0.016524 & 0.009836 \\ 0.026556 & -0.038269 & 0.014843 \end{bmatrix}$$

$$\mathbf{M}^{-} = \mathbf{M}^{-1}$$

d) $\mathbf{LL}^{-} \neq \mathbf{I}_3$, $\mathbf{MM}^{-} = \mathbf{I}_3$

e) $\mathbf{LL}^{-}\mathbf{L} = \mathbf{L}$, $\mathbf{MM}^{-}\mathbf{M} = \mathbf{M}$

Problem 2.1

$E(z) = 1(p) + 0(q) = p.$

$E(z^2) = 1^2(p) + 0^2(q) = p.$

$Var(z) = E(z^2) - [E(z)]^2 = p - p^2 = p(1 - p) = pq.$

Problem 2.2

$$\begin{aligned} E(by + c) &= \Sigma\,(by + c)/N \\ &= \Sigma\, by/N + \Sigma\, c/N \\ &= b\,\Sigma\, y/N + Nc/N \\ &= bE(y) + c, \end{aligned}$$

where all summation (Σ) is from $i=1$ to N.

Problem 2.3

$$\begin{aligned} E[y - E(y)]^2 &= E\{y^2 - 2y\,E(y) + [E(y)]^2\} \\ &= E(y^2) - 2E(y)\,E(y) + [E(y)]^2 \\ &= E(y^2) - [E(y)]^2. \end{aligned}$$

Problem 2.4

a) $4\,Var(x) = 1600$ ft^2.

b) $6\,Cov(x,y) = 960$ ft-in.

c) $Corr(x,y) = 0.8$.

d) $4\,Var(x) + 9\,Var(y) + 12\,Cov(x,y) = 4420$ ft-in.

Problem 2.5

$$
\begin{aligned}
\mathrm{Cov}(bx,bx) &= E\{[bx - E(bx)][bx - E(bx)]\} \\
&= E\{b^2x^2 - 2bx\,E(bx) + [E(bx)]^2\} \\
&= E(b^2x^2) - [E(bx)]^2 \\
&= b^2\,E(x^2) - [b\,E(x)]^2 \\
&= b^2\,\{E(x^2) - [E(x)]^2\} \\
&= b^2\,\mathrm{Var}(x).
\end{aligned}
$$

Problem 2.6

$$
\begin{aligned}
\mathrm{Corr}(x,x) &= \mathrm{Cov}(x,x)/[\mathrm{Var}(x)\ \mathrm{Var}(x)]^{½} \\
&= \mathrm{Var}(x)/\mathrm{Var}(x) = 1.
\end{aligned}
$$

Problem 2.7

$$
\begin{aligned}
E(z) &= E[(y - \mu)/\sigma] \\
&= [E(y) - \mu]/\sigma \\
&= (\mu - \mu)/\sigma = 0.
\end{aligned}
$$

$$
\begin{aligned}
E(z^2) &= E[(y - \mu)/\sigma]^2 \\
&= [E(y^2 - 2y\,E(y) + \mu^2]/\sigma^2 \\
&= \{E(y^2) - [E(y)]^2\}/\sigma^2 \\
&= \mathrm{Var}(y)/\sigma^2 = \sigma^2/\sigma^2 = 1.
\end{aligned}
$$

$$
\begin{aligned}
\mathrm{Var}(z) &= E(z^2) - [E(z)]^2 \\
&= 1 - 0 \\
&= 1.
\end{aligned}
$$

Problem 2.8

For the effects μ, E_i, B_{ij}, f_k, fe_{ik}, p_{ijk}, and w_{ijkl}, respectively, the number of levels = 1, 1, 4, 30, 30, 120, and 1200.

Problem 2.9

a) $\bar{y}_{ijk\cdot} = \mu + E_i + B_{ij} + f_k + fe_{ik} + p_{ijk} + w_{ijk\cdot}/10$

b) $\bar{y}_{\cdot\cdot k\cdot} = \mu + f_k + fe_{\cdot k}/3 + p_{\cdot\cdot k}/12 + w_{\cdot\cdot k\cdot}/120$

c) $\mathrm{Var}(\bar{y}_{\cdot\cdot k\cdot}) = \sigma^2_f + \sigma^2_{fe}/3 + \sigma^2_p/12 + \sigma^2_w/120$

Problem 2.11

a)

$$\mathbf{A} = \begin{bmatrix} 1 & 0 & 2 & 0 \\ 0 & 1 & 0 & 2 \end{bmatrix}$$

b)

$$\mathbf{S} = \mathrm{Var}(\mathbf{Ay}) = \begin{bmatrix} \mathrm{Var}(y_1+2y_3) & \mathrm{Cov}(y_1+2y_3,\ y_2+2y_4) \\ \mathrm{Cov}(y_1+2y_3,\ y_2+2y_4) & \mathrm{Var}(y_1+2y_4) \end{bmatrix}$$

Problem 2.12

$$\begin{aligned}
\mathrm{Cov}(p_{i\cdot k}/b,\ p_{i\cdot\cdot}/sb) &= (1/sb^2)\ \mathrm{Cov}(p_{i\cdot k},\ p_{i\cdot\cdot}) \\
&= (1/sb^2)\ \mathrm{Cov}(p_{i\cdot k},\ \sum_k^s p_{i\cdot k}) \\
&= (1/sb^2)\ \mathrm{Cov}(p_{i\cdot k},\ p_{i\cdot 1} + p_{i\cdot 2} + \ldots + p_{i\cdot k} + \ldots + p_{i\cdot s}) \\
&= (1/sb^2)\ \mathrm{Cov}(p_{i\cdot k},\ p_{i\cdot k}) \\
&= (1/sb^2)\ \mathrm{Cov}(\sum_j^b p_{ijk},\ \sum_j^b p_{ijk}) \\
&= (1/sb^2)\ \mathrm{Cov}(p_{i1k} + p_{i2k} + \ldots + p_{ibk},\ p_{i1k} + p_{i2k} + \ldots + p_{ibk}) \\
&= (1/sb^2)\ b\ \mathrm{Cov}(p_{ijk},\ p_{ijk}) \\
&= (1/sb)\ \mathrm{Var}(p_{ijk}) \\
&= \sigma^2_p/sb
\end{aligned}$$

Problem 3.1

$\mathrm{Var}(\bar{y}_{A\cdot k\cdot}) = 13.78$, $z_A = 1.35$, will select approximately 9 families.

$\mathrm{Var}(\bar{y}_{B\cdot k\cdot}) = 20.53$, $z_B = 1.10$, will select approximately 14 families.

Problem 3.2

$\mathrm{Var}(\bar{y}_{C\cdot k\cdot}) = 25.00$, $z_C = 1.20$, will select approximately 12 families.

$\mathrm{Var}(\bar{y}_{D\cdot k\cdot}) = 156.25$, $z_D = 0.48$, will select approximately 32 families.

Problem 3.3

$z_C = 1.20$, will select approximately 12 families.

$z_E = 0.81$, will select approximately 21 families.

Problem 3.4

a,b) With t=1, $\mathrm{Var}(\bar{y}_{..k.}) = 16.53$, $z_1 = 1.23$, will select approximately 22 families.

With t=4, $\mathrm{Var}(\bar{y}_{..k.}) = 8.63$, $z_4 = 1.70$, will select approximately 9 families.

With t=8, $\mathrm{Var}(\bar{y}_{..k.}) = 7.32$, $z_8 = 1.85$, will select approximately 6 families.

c) Given the level of σ^2_{fe}, four tests appear to be adequate to sample the range of possible sites. In other words, since $\bar{y}_{..k.}$ has approximately the same variance across 4 tests as across 8 tests (and the same mean of zero as a result of standardization) approximately the same number of families would be expected above any given truncation point in the 4 test group as in the 8 test group.

Problem 3.5

a,b) With t=1, $\mathrm{Var}(\bar{y}_{..k.}) = 16.53$, $z_1 = 1.23$, will select approximately 22 families.

With t=4, $\mathrm{Var}(\bar{y}_{..k.}) = 10.13$, $z_4 = 1.57$, will select approximately 12 families.

With t=8, $\mathrm{Var}(\bar{y}_{..k.}) = 9.07$, $z_8 = 1.66$, will select approximately 10 families.

c) The lower the value of σ^2_{fe}, the less $\bar{y}_{..k.}$ across all tests will be affected by family x environment effects, or the lower the marginal advantage of additional tests.

Problem 4.1

Assume we wish to express heights in units of feet, and volumes in units of ft^3. Then elements of **y** will be in feet, elements of $\boldsymbol{\alpha}$ will be in feet, elements of **g** in ft^3, elements of **V** in units of ft^2, and some elements in **C** will be in $ft(ft^3) = ft^4$, and some in units of $ft(ft) = ft^2$.

Problem 4.2

a) $\mathrm{Var}(\hat{\mathbf{g}})$ has dimension 4 x 4.

$$\mathrm{Var}(\hat{\mathbf{g}}) = \begin{bmatrix} \hat{g}_{11} & \hat{g}_{12} & \hat{g}_{13} & \hat{g}_{14} \\ g_{21} & g_{22} & g_{23} & g_{24} \\ g_{31} & g_{32} & g_{33} & g_{34} \\ g_{41} & g_{42} & g_{43} & g_{44} \end{bmatrix}$$

where

$\hat{g}_{11} = \hat{g}_{33} =$ variance of the predicted breeding value for 25-year volume for parents 1 and 2, respectively: units of $(ft^3)^2 = ft^6$.

$\hat{g}_{22} = \hat{g}_{44} =$ variance of the predicted breeding value for 5-year height for parents 1 and 2, respectively: units of ft^2.

$\hat{g}_{12} = \hat{g}_{21} =$ covariance of the predicted breeding values for 25-year volume and 5-year height for parent 1: units of $(ft^3)(ft) = ft^4$.

$\hat{g}_{34} = \hat{g}_{43} =$ covariance of the predicted breeding values for 25-year volume and 5-year height for parent 2: units of $(ft^3)(ft) = ft^4$.

$\hat{g}_{13} = \hat{g}_{31} =$ covariance of the predicted breeding values for 25-year volume for parents 1 and 2: units of $(ft^3)(ft^3) = ft^6$.

$\hat{g}_{24} = \hat{g}_{42} =$ covariance of the predicted breeding values for 5-year height for parents 1 and 2: units of $(ft)(ft) = ft^2$.

$\hat{g}_{14} = \hat{g}_{41} =$ covariance of the predicted breeding values for 25-year volume for parent 1 and 5-year height for parent 2: units of $(ft^3)(ft) = ft^4$.

$\hat{g}_{23} = \hat{g}_{32} =$ covariance of the predicted breeding values for 25-year volume for parent 2 and 5-year height for parent 1: units of $(ft^3)(ft) = ft^4$.

b) $Var(\hat{\mathbf{g}} - \mathbf{g})$ has dimension 4 x 4.

$$Var(\hat{\mathbf{g}} - \mathbf{g}) = \begin{bmatrix} e_{11} & e_{12} & e_{13} & e_{14} \\ e_{21} & e_{22} & e_{23} & e_{24} \\ e_{31} & e_{32} & e_{33} & e_{34} \\ e_{41} & e_{42} & e_{43} & e_{44} \end{bmatrix}$$

where

$e_{11} = e_{33} =$ error variance of prediction for 25-year volume for parents 1 and 2, respectively: units of $(ft^3)^2 = ft^6$.

$e_{22} = e_{44} =$ error variance of prediction for 5-year height for parents 1 and 2, respectively: units of ft^2.

$e_{12} = e_{21} =$ covariance of the errors of prediction for 25-year volume and 5-year height for parent 1: units of $(ft^3)(ft) = ft^4$.

$e_{34} = e_{43} =$ covariance of the errors of prediction for 25-year volume and 5-year height for parent 2: units of $(ft^3)(ft) = ft^4$.

$e_{13} = e_{31} =$ covariance of the errors of prediction for 25-year volume for parents 1 and 2: units of $(ft^3)(ft^3) = ft^6$.

$e_{24} = e_{42} =$ covariance of the errors of prediction for 5-year height for parents 1 and 2: units of (ft)(ft) = ft^2.

$e_{14} = e_{41} =$ covariance of the errors of prediction for 25-year volume for parent 1 and for 5-year height for parent 2: units of $(ft^3)(ft) = ft^4$.

$e_{23} = e_{32} =$ covariance of the errors of prediction for 25-year volume for parent 2 and for 5-year height for parent 1: units of $(ft^3)(ft) = ft^4$.

Problem 5.1

$$\mathbf{V} = \begin{bmatrix} .0220 & .0142 & .0205 & 0 & 0 & 0 \\ .0142 & .0771 & .0635 & 0 & 0 & 0 \\ .0205 & .0635 & .2390 & 0 & 0 & 0 \\ 0 & 0 & 0 & .0220 & .0142 & .0205 \\ 0 & 0 & 0 & .0142 & .0771 & .0635 \\ 0 & 0 & 0 & .0205 & .0635 & .2390 \end{bmatrix}$$

$$\mathbf{C} = \begin{bmatrix} .0409 & 0 \\ .1271 & 0 \\ .3000 & 0 \\ 0 & .0409 \\ 0 & .1271 \\ 0 & .3000 \end{bmatrix} \qquad \mathbf{G} = [.6000]$$

$$\begin{aligned} Var(\hat{g}_i) &= 0.417 \text{ m}^2 \\ Var(g_i-\hat{g}_i) &= 0.183 \text{ m}^2 \\ Corr(\hat{g}_i,g_i) &= 0.83 \\ \hat{g}_A &= 1.43 \text{ m} \\ \hat{g}_B &= 1.16 \text{ m} \end{aligned}$$

Problem 5.2

$$\mathbf{C} = \begin{bmatrix} .0284 & 0 \\ .0800 & 0 \\ .1271 & 0 \\ 0 & .0284 \\ 0 & .0800 \\ 0 & .1271 \end{bmatrix} \qquad \mathbf{G} = [.1600]$$

$$\begin{aligned} Var(\hat{g}_i) &= 0.109 \text{ m}^2 \\ Var(g_i-\hat{g}_i) &= 0.051 \text{ m}^2 \\ Corr(\hat{g}_i,g_i) &= 0.83 \\ \hat{g}_A &= 0.78 \text{ m} \\ \hat{g}_B &= 1.25 \text{ m} \end{aligned}$$

Problem 5.3

Three 5-year-old tests: $Var(\hat{g}) = 0.136$, $Var(g-\hat{g}) = 0.464$, $Corr(\hat{g},g) = 0.48$.

One 10-year-old test: $Var(\hat{g}) = 0.209$, $Var(g-\hat{g}) = 0.391$, $Corr(\hat{g},g) = 0.81$.

Problem 5.4

Three 5-year-old tests: $Var(\hat{g}) = 0.147$, $Var(g-\hat{g}) = 0.453$, $Corr(\hat{g},g) = 0.50$.

One 10-year-old test: $Var(\hat{g}) = 0.166$, $Var(g-\hat{g}) = 0.434$, $Corr(\hat{g},g) = 0.53$.

Problem 5.5

With t=3, z_3 = 1.36, will select approximately 43 families from 500 (8.7%).
With t=2, z_2 = 1.49, will select approximately 34 families from 500 (6.8%).
With t=1, z_1 = 1.81, will select approximately 18 families from 500 (3.5%).

As the truncation point increases, relatively more selections will come from the three-test group. As the truncation point decreases toward 50% selected, approximately equal numbers of selections will be made from all three groups.

Problem 5.7

All trees in the orchard are both male and female. The average expected breeding value of offspring from the orchard is

$$BV_{offspring} = (1/2)\ (BV_{♂} + BV_{♀}).$$

All the genes in orchard seed which are contributed by females come from orchard trees, but 30% of the genes contributed by males are from unimproved trees. If these are assumed to have an average BV = 0 m³, then

$$BV_{offspring} = (1/2)\ [(0.7 \cdot 1.0\ m^3) + (0.3 \cdot 0\ m^3) + (1.0 \cdot 1.0\ m^3)]$$
$$= 0.85\ m^3.$$

Problem 6.1

a) If $\mathbf{y}' = [\bar{y}_{1\cdot AB\cdot}\ \ \bar{y}_{2\cdot AB\cdot}\ \ \bar{y}_{3\cdot AB\cdot}\ \ \bar{y}_{4\cdot AB\cdot}\ \ \bar{y}_{1\cdot AC\cdot}\ \ \cdots\ \ \bar{y}_{3\cdot CD\cdot}\ \ \bar{y}_{4\cdot CD\cdot}]$, then

$$\mathbf{V} = \begin{bmatrix} \mathbf{V1} & \mathbf{V2} & \mathbf{V2} & \mathbf{V2} & \mathbf{V2} & \mathbf{Z_v} \\ \mathbf{V2} & \mathbf{V1} & \mathbf{V2} & \mathbf{V2} & \mathbf{Z_v} & \mathbf{V2} \\ \mathbf{V2} & \mathbf{V2} & \mathbf{V1} & \mathbf{Z_v} & \mathbf{V2} & \mathbf{V2} \\ \mathbf{V2} & \mathbf{V2} & \mathbf{Z_v} & \mathbf{V1} & \mathbf{V2} & \mathbf{V2} \\ \mathbf{V2} & \mathbf{Z_v} & \mathbf{V2} & \mathbf{V2} & \mathbf{V1} & \mathbf{V2} \\ \mathbf{Z_v} & \mathbf{V2} & \mathbf{V2} & \mathbf{V2} & \mathbf{V2} & \mathbf{V1} \end{bmatrix} \qquad \mathbf{C} = \begin{bmatrix} \mathbf{C1} & \mathbf{C1} & \mathbf{Z_c} & \mathbf{Z_c} \\ \mathbf{C1} & \mathbf{Z_c} & \mathbf{C1} & \mathbf{Z_c} \\ \mathbf{C1} & \mathbf{Z_c} & \mathbf{Z_c} & \mathbf{C1} \\ \mathbf{Z_c} & \mathbf{C1} & \mathbf{C1} & \mathbf{Z_c} \\ \mathbf{Z_c} & \mathbf{C1} & \mathbf{Z_c} & \mathbf{C1} \\ \mathbf{Z_c} & \mathbf{Z_c} & \mathbf{C1} & \mathbf{C1} \end{bmatrix}$$

where the sub-matrices **V1**, **V2**, $\mathbf{Z_v}$, **C1**, and $\mathbf{Z_c}$ are as follow:

$$\mathbf{V1} = \begin{bmatrix} 8.51 & 4.50 & 4.50 & 4.50 \\ 4.50 & 8.51 & 4.50 & 4.50 \\ 4.50 & 4.50 & 8.51 & 4.50 \\ 4.50 & 4.50 & 4.50 & 8.51 \end{bmatrix} \qquad \mathbf{V2} = \begin{bmatrix} 2.70 & 1.80 & 1.80 & 1.80 \\ 1.80 & 2.70 & 1.80 & 1.80 \\ 1.80 & 1.80 & 2.70 & 1.80 \\ 1.80 & 1.80 & 1.80 & 2.70 \end{bmatrix}$$

$$\mathbf{Z_v} = \begin{bmatrix} 0 & 0 & 0 & 0 \\ 0 & 0 & 0 & 0 \\ 0 & 0 & 0 & 0 \\ 0 & 0 & 0 & 0 \end{bmatrix} \qquad \mathbf{C1} = \begin{bmatrix} 3.60 \\ 3.60 \\ 3.60 \\ 3.60 \end{bmatrix} \qquad \mathbf{Z_c} = \begin{bmatrix} 0 \\ 0 \\ 0 \\ 0 \end{bmatrix}$$

The matrix **G** is a 4 x 4 diagonal matrix with 7.20 on the diagoanls, and 0's on the off-diagonals.

b) $\mathrm{Var}(\hat{\mathbf{g}})$ is a 4 x 4 matrix with 4.96 on the diagonals, and 0.25 on all off-diagonals. $\mathrm{Corr}(\hat{g},g) = 0.83$ for all parents.

Problem 6.2

a) $\mathrm{Var}(\bar{y}_{i\cdot kl\cdot}) = 10.76$.

b)

$$\mathbf{V} = \begin{bmatrix}
\mathbf{D} & \mathbf{F} & \mathbf{F} & \mathbf{F} & \mathbf{F} & \mathbf{F} & \mathbf{F} & \mathbf{F} & \mathbf{F} & \mathbf{Z} & \mathbf{Z} & \mathbf{Z} & \mathbf{Z} & \mathbf{Z} & \mathbf{Z} \\
\mathbf{F} & \mathbf{D} & \mathbf{F} & \mathbf{F} & \mathbf{F} & \mathbf{F} & \mathbf{Z} & \mathbf{Z} & \mathbf{Z} & \mathbf{F} & \mathbf{F} & \mathbf{F} & \mathbf{Z} & \mathbf{Z} & \mathbf{Z} \\
\mathbf{F} & \mathbf{F} & \mathbf{D} & \mathbf{F} & \mathbf{F} & \mathbf{Z} & \mathbf{F} & \mathbf{Z} & \mathbf{Z} & \mathbf{F} & \mathbf{Z} & \mathbf{Z} & \mathbf{F} & \mathbf{F} & \mathbf{Z} \\
\mathbf{F} & \mathbf{F} & \mathbf{F} & \mathbf{D} & \mathbf{F} & \mathbf{Z} & \mathbf{Z} & \mathbf{F} & \mathbf{Z} & \mathbf{Z} & \mathbf{F} & \mathbf{Z} & \mathbf{F} & \mathbf{Z} & \mathbf{F} \\
\mathbf{F} & \mathbf{F} & \mathbf{F} & \mathbf{F} & \mathbf{D} & \mathbf{Z} & \mathbf{Z} & \mathbf{Z} & \mathbf{F} & \mathbf{Z} & \mathbf{Z} & \mathbf{F} & \mathbf{Z} & \mathbf{F} & \mathbf{F} \\
\mathbf{F} & \mathbf{F} & \mathbf{Z} & \mathbf{Z} & \mathbf{Z} & \mathbf{D} & \mathbf{F} & \mathbf{F} & \mathbf{F} & \mathbf{F} & \mathbf{F} & \mathbf{F} & \mathbf{Z} & \mathbf{Z} & \mathbf{Z} \\
\mathbf{F} & \mathbf{Z} & \mathbf{F} & \mathbf{Z} & \mathbf{Z} & \mathbf{F} & \mathbf{D} & \mathbf{F} & \mathbf{F} & \mathbf{F} & \mathbf{Z} & \mathbf{Z} & \mathbf{F} & \mathbf{F} & \mathbf{Z} \\
\mathbf{F} & \mathbf{Z} & \mathbf{Z} & \mathbf{F} & \mathbf{Z} & \mathbf{F} & \mathbf{F} & \mathbf{D} & \mathbf{F} & \mathbf{Z} & \mathbf{F} & \mathbf{Z} & \mathbf{F} & \mathbf{Z} & \mathbf{F} \\
\mathbf{F} & \mathbf{Z} & \mathbf{Z} & \mathbf{Z} & \mathbf{F} & \mathbf{F} & \mathbf{F} & \mathbf{F} & \mathbf{D} & \mathbf{Z} & \mathbf{Z} & \mathbf{F} & \mathbf{Z} & \mathbf{F} & \mathbf{F} \\
\mathbf{Z} & \mathbf{F} & \mathbf{F} & \mathbf{Z} & \mathbf{Z} & \mathbf{F} & \mathbf{F} & \mathbf{Z} & \mathbf{Z} & \mathbf{D} & \mathbf{F} & \mathbf{F} & \mathbf{F} & \mathbf{F} & \mathbf{Z} \\
\mathbf{Z} & \mathbf{F} & \mathbf{Z} & \mathbf{F} & \mathbf{Z} & \mathbf{F} & \mathbf{Z} & \mathbf{F} & \mathbf{Z} & \mathbf{F} & \mathbf{D} & \mathbf{F} & \mathbf{F} & \mathbf{Z} & \mathbf{F} \\
\mathbf{Z} & \mathbf{F} & \mathbf{Z} & \mathbf{Z} & \mathbf{F} & \mathbf{F} & \mathbf{Z} & \mathbf{Z} & \mathbf{F} & \mathbf{F} & \mathbf{F} & \mathbf{D} & \mathbf{Z} & \mathbf{F} & \mathbf{F} \\
\mathbf{Z} & \mathbf{Z} & \mathbf{F} & \mathbf{F} & \mathbf{Z} & \mathbf{Z} & \mathbf{F} & \mathbf{F} & \mathbf{Z} & \mathbf{F} & \mathbf{F} & \mathbf{Z} & \mathbf{D} & \mathbf{F} & \mathbf{F} \\
\mathbf{Z} & \mathbf{Z} & \mathbf{F} & \mathbf{Z} & \mathbf{F} & \mathbf{Z} & \mathbf{F} & \mathbf{Z} & \mathbf{F} & \mathbf{F} & \mathbf{Z} & \mathbf{F} & \mathbf{F} & \mathbf{D} & \mathbf{F} \\
\mathbf{Z} & \mathbf{Z} & \mathbf{Z} & \mathbf{F} & \mathbf{F} & \mathbf{Z} & \mathbf{Z} & \mathbf{F} & \mathbf{F} & \mathbf{Z} & \mathbf{F} & \mathbf{F} & \mathbf{F} & \mathbf{F} & \mathbf{D}
\end{bmatrix}$$

where the sub-matrices **D**, **F**, and **Z** are as follow:

$$\mathbf{D} = \begin{bmatrix} 10.76 & 4.50 \\ 4.50 & 10.76 \end{bmatrix} \qquad \mathbf{F} = \begin{bmatrix} 2.70 & 1.80 \\ 1.80 & 2.70 \end{bmatrix} \qquad \mathbf{Z} = \begin{bmatrix} 0 & 0 \\ 0 & 0 \end{bmatrix}$$

The matrix **C** is constructed of sub-matrices **X** and **Z** as follow:

$$\mathbf{C} = \begin{bmatrix} \mathbf{X} & \mathbf{X} & \mathbf{Z} & \mathbf{Z} & \mathbf{Z} & \mathbf{Z} \\ \mathbf{X} & \mathbf{Z} & \mathbf{X} & \mathbf{Z} & \mathbf{Z} & \mathbf{Z} \\ \mathbf{X} & \mathbf{Z} & \mathbf{Z} & \mathbf{X} & \mathbf{Z} & \mathbf{Z} \\ \mathbf{X} & \mathbf{Z} & \mathbf{Z} & \mathbf{Z} & \mathbf{X} & \mathbf{Z} \\ \mathbf{X} & \mathbf{Z} & \mathbf{Z} & \mathbf{Z} & \mathbf{Z} & \mathbf{X} \\ \mathbf{Z} & \mathbf{X} & \mathbf{X} & \mathbf{Z} & \mathbf{Z} & \mathbf{Z} \\ \mathbf{Z} & \mathbf{X} & \mathbf{Z} & \mathbf{X} & \mathbf{Z} & \mathbf{Z} \\ \mathbf{Z} & \mathbf{X} & \mathbf{Z} & \mathbf{Z} & \mathbf{X} & \mathbf{Z} \\ \mathbf{Z} & \mathbf{X} & \mathbf{Z} & \mathbf{Z} & \mathbf{Z} & \mathbf{X} \\ \mathbf{Z} & \mathbf{Z} & \mathbf{X} & \mathbf{X} & \mathbf{Z} & \mathbf{Z} \\ \mathbf{Z} & \mathbf{Z} & \mathbf{X} & \mathbf{Z} & \mathbf{X} & \mathbf{Z} \\ \mathbf{Z} & \mathbf{Z} & \mathbf{X} & \mathbf{Z} & \mathbf{Z} & \mathbf{X} \\ \mathbf{Z} & \mathbf{Z} & \mathbf{Z} & \mathbf{X} & \mathbf{X} & \mathbf{Z} \\ \mathbf{Z} & \mathbf{Z} & \mathbf{Z} & \mathbf{X} & \mathbf{Z} & \mathbf{X} \\ \mathbf{Z} & \mathbf{Z} & \mathbf{Z} & \mathbf{Z} & \mathbf{X} & \mathbf{X} \end{bmatrix} \qquad \mathbf{X} = \begin{bmatrix} 3.6 \\ 3.6 \end{bmatrix} \qquad \mathbf{Z} = \begin{bmatrix} 0 \\ 0 \end{bmatrix}$$

c) Var($\hat{\mathbf{g}}$) is a 6 x 6 matrix with 4.40 on the diagonals, and 0.13 on all off-diagonals. Corr($\hat{g}$,g) = 0.78 for all parents.

Problem 6.3

a)

$$\mathbf{V} = \begin{bmatrix} 8.51 & 4.50 & 2.70 & 1.80 & 2.70 & 1.80 & 0.00 & 0.00 & 2.70 & 1.80 & 0.00 & 0.00 \\ 4.50 & 8.51 & 1.80 & 2.70 & 1.80 & 2.70 & 0.00 & 0.00 & 1.80 & 2.70 & 0.00 & 0.00 \\ 2.70 & 1.80 & 8.51 & 4.50 & 0.00 & 0.00 & 2.70 & 1.80 & 0.00 & 0.00 & 2.70 & 1.80 \\ 1.80 & 2.70 & 4.50 & 8.51 & 0.00 & 0.00 & 1.80 & 2.70 & 0.00 & 0.00 & 1.80 & 2.70 \\ 2.70 & 1.80 & 0.00 & 0.00 & 8.51 & 4.50 & 2.70 & 1.80 & 2.70 & 1.80 & 0.00 & 0.00 \\ 1.80 & 2.70 & 0.00 & 0.00 & 4.50 & 8.51 & 1.80 & 2.70 & 1.80 & 2.70 & 0.00 & 0.00 \\ 0.00 & 0.00 & 2.70 & 1.80 & 2.70 & 1.80 & 8.51 & 4.50 & 0.00 & 0.00 & 2.70 & 1.80 \\ 0.00 & 0.00 & 1.80 & 2.70 & 1.80 & 2.70 & 4.50 & 8.51 & 0.00 & 0.00 & 1.80 & 2.70 \\ 2.70 & 1.80 & 0.00 & 0.00 & 2.70 & 1.80 & 0.00 & 0.00 & 8.51 & 4.50 & 2.70 & 1.80 \\ 1.80 & 2.70 & 0.00 & 0.00 & 1.80 & 2.70 & 0.00 & 0.00 & 4.50 & 8.51 & 1.80 & 2.70 \\ 0.00 & 0.00 & 2.70 & 1.80 & 0.00 & 0.00 & 2.70 & 1.80 & 2.70 & 1.80 & 8.51 & 4.50 \\ 0.00 & 0.00 & 1.80 & 2.70 & 0.00 & 0.00 & 1.80 & 2.70 & 1.80 & 2.70 & 4.50 & 8.51 \end{bmatrix}$$

$$\mathbf{C} = \begin{bmatrix} 3.6 & 0 & 0 & 3.6 & 0 \\ 3.6 & 0 & 0 & 3.6 & 0 \\ 3.6 & 0 & 0 & 0 & 3.6 \\ 3.6 & 0 & 0 & 0 & 3.6 \\ 0 & 3.6 & 0 & 3.6 & 0 \\ 0 & 3.6 & 0 & 3.6 & 0 \\ 0 & 3.6 & 0 & 0 & 3.6 \\ 0 & 3.6 & 0 & 0 & 3.6 \\ 0 & 0 & 3.6 & 3.6 & 0 \\ 0 & 0 & 3.6 & 3.6 & 0 \\ 0 & 0 & 3.6 & 0 & 3.6 \\ 0 & 0 & 3.6 & 0 & 3.6 \end{bmatrix}$$

G is a 5 x 5 diagonal matrix with 7.20 on the diagonals, and 0's on the off-diagonals.

b)

$$\mathrm{Var}(\hat{\mathbf{g}}) = \begin{bmatrix} 3.31 & -.68 & -.68 & 0.98 & 0.98 \\ -.68 & 3.31 & -.68 & 0.98 & 0.98 \\ -.68 & -.68 & 3.31 & 0.98 & 0.98 \\ 0.98 & 0.98 & 0.98 & 3.69 & -.75 \\ 0.98 & 0.98 & 0.98 & -.75 & 3.69 \end{bmatrix}$$

$\mathrm{Corr}(\hat{g},g)$ = 0.678, 0.678, 0.678, 0.716 and 0.716 for parents A, B, C, D, and E, respectively.

Problem 6.4

a) Delete the 1st and 2nd rows and columns of **V**, and the 1st and 2nd rows of **C** in the answer to Problem 6.3.

b)

$$\mathrm{Var}(\hat{\mathbf{g}}) = \begin{bmatrix} 2.46 & -.50 & -.50 & 0.35 & 1.11 \\ -.50 & 3.27 & -.71 & 1.11 & 0.95 \\ -.50 & -.71 & 3.27 & 1.11 & 0.95 \\ 0.35 & 1.11 & 1.11 & 3.22 & -.66 \\ 1.11 & 0.95 & 0.95 & -.66 & 3.67 \end{bmatrix}$$

$\mathrm{Corr}(\hat{g},g)$ = 0.584, 0.674, 0.674, 0.669 and 0.713 for parents A, B, C, D, and E, respectively.

Problem 6.5

a) dim(**V**) = 8 x 8, dim(**C**) = 8 x 2, dim (**G**) = 2 x 2.

b) seven types of elements in **V**.

c) five types of elements in **C**.

d) two types of elements in **G**.

e) $\frac{1}{2} : \frac{1}{4} : 1$

Problem 7.1

a)

$$\mathbf{V}_k = \begin{bmatrix} 1.91 & 0.56 & 0.56 & 0.56 \\ 0.56 & 1.91 & 0.56 & 0.56 \\ 0.56 & 0.56 & 1.91 & 0.56 \\ 0.56 & 0.56 & 0.56 & 1.91 \end{bmatrix} \quad \mathbf{C}_k = \begin{bmatrix} 1.88 \\ 1.88 \\ 1.88 \\ 1.88 \end{bmatrix} \quad \mathbf{G}_k = [7.20]$$

b) $\text{Corr}(\hat{g},g) = 0.74$

c) $\mathbf{G}_k = [7.20]$

$$\mathbf{V}_k = \begin{bmatrix} 5.63 & 1.50 & 1.80 & 0.94 & 1.80 & 0.94 & 1.80 & 0.94 \\ 1.50 & 1.91 & 0.94 & 0.56 & 0.94 & 0.56 & 0.94 & 0.56 \\ 1.80 & 0.94 & 5.63 & 1.50 & 1.80 & 0.94 & 1.80 & 0.94 \\ 0.94 & 0.56 & 1.50 & 1.91 & 0.94 & 0.56 & 0.94 & 0.56 \\ 1.80 & 0.94 & 1.80 & 0.94 & 5.63 & 1.50 & 1.80 & 0.94 \\ 0.94 & 0.56 & 0.94 & 0.56 & 1.50 & 1.91 & 0.94 & 0.56 \\ 1.80 & 0.94 & 1.80 & 0.94 & 1.80 & 0.94 & 5.63 & 1.50 \\ 0.94 & 0.56 & 0.94 & 0.56 & 0.94 & 0.56 & 1.50 & 1.91 \end{bmatrix} \quad \mathbf{C}_k = \begin{bmatrix} 3.60 \\ 1.88 \\ 3.60 \\ 1.88 \\ 3.60 \\ 1.88 \\ 3.60 \\ 1.88 \end{bmatrix}$$

d) $\text{Corr}(\hat{g},g) = 0.85$

Problem 7.2

a)

$$\mathbf{C}_k = \begin{bmatrix} 15.12 & 4.32 & 9.00 \\ 15.12 & 4.32 & 9.00 \\ 4.32 & 6.72 & 4.80 \\ 4.32 & 6.72 & 4.80 \end{bmatrix}$$

b) $\text{Corr}(\hat{g}_A,g_A) = 0.78$, $\text{Corr}(\hat{g}_B,g_B) = 0.63$, $\text{Corr}(\hat{g}_C,g_C) = 0.78$

c) $(\mathbf{y}_D - \boldsymbol{\alpha}_D)' = [12\ 12\ 0\ 0]$; $(\mathbf{y}_E - \boldsymbol{\alpha}_E)' = [0\ 0\ 8\ 8]$; $(\mathbf{y}_F - \boldsymbol{\alpha}_F)' = [6\ 6\ 4\ 4]$

d) Parent D: $\hat{g}_A = 13.52$, $\hat{g}_B = 7.20$, $\hat{g}_C = 1.77$.
Parent E: $\hat{g}_A = 2.66$, $\hat{g}_B = 4.94$, $\hat{g}_C = 9.02$.
Parent F: $\hat{g}_A = 8.09$, $\hat{g}_B = 6.07$, $\hat{g}_C = 5.39$.

e) Parent D: Δ_A = 22.5%, Δ_B = 14.4%, Δ_C = 4.4%.
Parent E: Δ_A = 4.4%, Δ_B = 9.9%, Δ_C = 22.6%.
Parent F: Δ_A = 13.5%, Δ_B = 12.1%, Δ_C = 13.5%.

f) Parent D: $\hat{h}$ = 7.49.
Parent E: $\hat{h}$ = 5.53.
Parent F: $\hat{h}$ = 6.51.

g) Parent D: Δ_A = 15.0% = (7.49/50)
Parent E: Δ_A = 11.1% = (5.53/50)
Parent F: Δ_A = 13.0% = (6.51/50)

Problem 9.1

The dimension of all results of Eqs. 9.1 - 9.11 are 1 x 1, except for Eq. 9.6 where dim(**b**) = m x 1, and m = number of observations pertaining to a single candidate.

Problem 9.2

a) units of elements in **V** and **C** will all be either ft^2, (degrees)^2, or (ft-degrees).
b) magnitude of variances of heights and covariances of heights and straightness will decrease, variances of straightness will remain constant.

Problem 10.1

a, b) For tree i1AB2, $(\mathbf{y} - \boldsymbol{\alpha})$ = [2.0 1.2 0.4 0.1 0.1 -0.4 -1.0], and $\hat{g}$=0.68.
For tree i2AB4, $(\mathbf{y} - \boldsymbol{\alpha})$ = [-2.0 1.2 0.4 0.1 0.1 -0.4 -1.0], and $\hat{g}$=0.19.

c) For tree i1AB2, $\hat{g}$ = .69. For tree i2AB4, $\hat{g}$ = 0.20.

d) For case 3a, Var($\hat{g}$) = 0.184. For case 3b, Var($\hat{g}$) = 0.166.

Problem 10.2

a)

$$C = \begin{bmatrix} 0.5320 & 0.2660 & 0.2660 & 0.2660 \\ 0.2734 & 0.2660 & 0.2660 & 0.2660 \\ 0.1330 & 0.2660 & 0 & 0.1330 \\ 0.1330 & 0.2660 & 0 & 0.1330 \\ 0.1330 & 0 & 0.2660 & 0.1330 \\ 0.1330 & 0 & 0.2660 & 0.1330 \\ 0 & 0 & 0 & 0 \end{bmatrix}$$

b)

$$G = \begin{bmatrix} 0.5320 & 0.2660 & 0.2660 & 0.2660 \\ 0.2660 & 0.5320 & 0 & 0.1330 \\ 0.2660 & 0 & 0.5320 & 0.1330 \\ 0.2660 & 0.1330 & 0.1330 & 0.5320 \end{bmatrix}$$

c) The predicted breeding values for the individuals are different, but $\hat{g}_A$, $\hat{g}_B$, and $\hat{g}_{AxB}$ are identical regardless of which individual observation is used.

d) $\hat{g}_A = 0\ (y_{i1Ab2}) + 0.344\ (\bar{y}_{i\cdot AB\cdot}) + 0.344\ (\bar{y}_{i\cdot AC\cdot}) + 0.344\ (\bar{y}_{i\cdot AD\cdot})$
$- 0.134\ (\bar{y}_{i\cdot BC\cdot}) - 0.134\ (\bar{y}_{i\cdot BD\cdot}) - 0.134\ (\bar{y}_{i\cdot CD\cdot})$

There is a zero coefficient on the individual tree observation from the AxB family because, heuristically, it provides no additional information about parent A other than what is already contained in the AxB family mean.

e) $\hat{g}_{AxB} = 0.567 = ½\ (\hat{g}_A + \hat{g}_B) = ½\ (0.758 + 0.376)$

Problem 10.3

a) $\mathrm{Var}(\bar{y}_{ijkl\cdot}) = 2\sigma^2_f + 2\sigma^2_{fe} + \sigma^2_d + \sigma^2_{de} + \sigma^2_p + \sigma^2_w/n = 1.1281.$

b) $\mathrm{Cov}(y_{ijklm}, \bar{y}_{ijkl\cdot}) = \mathrm{Var}(\bar{y}_{ijkl\cdot}) = 1.1281.$

c) $\mathrm{Cov}(\bar{y}_{ijkl\cdot}, \bar{y}_{i\cdot kl\cdot}) = \mathrm{Var}(\bar{y}_{i\cdot kl\cdot}) = 0.5574.$

d) $\mathrm{Cov}(\bar{y}_{ijkl\cdot}, \bar{y}_{i\cdot k'l\cdot}) = \sigma^2_f + \sigma^2_{fe} = 0.1773.$

e) $\mathrm{Cov}(\bar{y}_{ijkl\cdot}, \bar{y}_{i\cdot k'l'\cdot}) = 0.$

f) $\mathrm{Cov}(\bar{y}_{ijkl\cdot}, g_{ijklm}) = ½\ \sigma^2_A\ (1 + 1/n) = 0.3103.$

g) **V** is now 8 x 8, insert **p** and **p'** as the second row and second column in the **V** described for Case 3a, where

$$\mathbf{p} = [1.1281 \quad 1.1281 \quad 0.5574 \quad 0.1773 \quad 0.1773 \quad 0.1773 \quad 0.1773 \quad 0].$$

C is now 8 x 1, insert 0.3103 as the second element in **C** described for Case 3a.

h) The coefficient on plot is negative (-0.078); it adjusts for the micro-site environmental effects associated with the individual tree.

i) $\mathrm{Corr}(\hat{g},g) = 0.588$ and $\Delta I = 0.885$ (9.5%) if the plot mean is not used; $\mathrm{Corr}(\hat{g},g) = 0.592$ and $\Delta I = 0.891$ (9.6%) if the plot mean is used.

Problem 11.1

$$\begin{aligned}
\mathbf{QV} &= (\mathbf{V}^{-1} - \mathbf{V}^{-1}\mathbf{X}(\mathbf{X}'\mathbf{V}^{-1}\mathbf{X})^{-}\mathbf{X}'\mathbf{V}^{-1})\mathbf{V} \\
&= (\mathbf{V}^{-1}\mathbf{V} - \mathbf{V}^{-1}\mathbf{X}(\mathbf{X}'\mathbf{V}^{-1}\mathbf{X})^{-}\mathbf{X}'\mathbf{V}^{-1}\mathbf{V}) \\
&= (\mathbf{I} - \mathbf{V}^{-1}\mathbf{X}(\mathbf{X}'\mathbf{V}^{-1}\mathbf{X})^{-}\mathbf{X}')
\end{aligned}$$

$$\begin{aligned}
\mathbf{QVQ} &= (\mathbf{I} - \mathbf{V}^{-1}\mathbf{X}(\mathbf{X}'\mathbf{V}^{-1}\mathbf{X})^{-}\mathbf{X}')\ \mathbf{Q} \\
&= (\mathbf{I} - \mathbf{V}^{-1}\mathbf{X}(\mathbf{X}'\mathbf{V}^{-1}\mathbf{X})^{-}\mathbf{X}')\ (\mathbf{V}^{-1} - \mathbf{V}^{-1}\mathbf{X}(\mathbf{X}'\mathbf{V}^{-1}\mathbf{X})^{-}\mathbf{X}'\mathbf{V}^{-1}) \\
&= \mathbf{V}^{-1} - \mathbf{V}^{-1}\mathbf{X}(\mathbf{X}'\mathbf{V}^{-1}\mathbf{X})^{-}\mathbf{X}'\mathbf{V}^{-1} - \mathbf{V}^{-1}\mathbf{X}(\mathbf{X}'\mathbf{V}^{-1}\mathbf{X})^{-}\mathbf{X}'\mathbf{V}^{-1} \\
&\qquad + \mathbf{V}^{-1}\mathbf{X}(\mathbf{X}'\mathbf{V}^{-1}\mathbf{X})^{-}\mathbf{X}'\mathbf{V}^{-1}\mathbf{X}(\mathbf{X}'\mathbf{V}^{-1}\mathbf{X})^{-}\mathbf{X}'\mathbf{V}^{-1} \\
&= \mathbf{V}^{-1} - \mathbf{V}^{-1}\mathbf{X}(\mathbf{X}'\mathbf{V}^{-1}\mathbf{X})^{-}\mathbf{X}'\mathbf{V}^{-1} - \mathbf{V}^{-1}\mathbf{X}(\mathbf{X}'\mathbf{V}^{-1}\mathbf{X})^{-}\mathbf{X}'\mathbf{V}^{-1} \\
&\qquad + \mathbf{V}^{-1}\mathbf{X}(\mathbf{X}'\mathbf{V}^{-1}\mathbf{X})^{-}\mathbf{X}'\mathbf{V}^{-1}. \\
&= \mathbf{V}^{-1} - \mathbf{V}^{-1}\mathbf{X}(\mathbf{X}'\mathbf{V}^{-1}\mathbf{X})^{-}\mathbf{X}'\mathbf{V}^{-1} \\
&= \mathbf{Q}
\end{aligned}$$

Problem 11.2

$$\begin{aligned}
\mathrm{Var}(\mathbf{V}^{-1}\mathbf{X}\hat{\boldsymbol{\beta}}) &= \mathbf{V}^{-1}\mathbf{X}\ \mathrm{Var}(\hat{\boldsymbol{\beta}})\ \mathbf{X}'\mathbf{V}^{-1} \\
&= \mathbf{V}^{-1}\mathbf{X}\ \{\mathrm{Var}[(\mathbf{X}'\mathbf{V}^{-1}\mathbf{X})^{-}\mathbf{X}'\mathbf{V}^{-1}\mathbf{y}]\}\ \mathbf{X}'\mathbf{V}^{-1} \\
&= \mathbf{V}^{-1}\mathbf{X}\ [(\mathbf{X}'\mathbf{V}^{-1}\mathbf{X})^{-}\mathbf{X}'\mathbf{V}^{-1}\ \mathrm{Var}(\mathbf{y})\ \mathbf{V}^{-1}\mathbf{X}(\mathbf{X}'\mathbf{V}^{-1}\mathbf{X})^{-}]\ \mathbf{X}'\mathbf{V}^{-1} \\
&= \mathbf{V}^{-1}\mathbf{X}\ [(\mathbf{X}'\mathbf{V}^{-1}\mathbf{X})^{-}\mathbf{X}'\mathbf{V}^{-1}\mathbf{V}\mathbf{V}^{-1}\mathbf{X}(\mathbf{X}'\mathbf{V}^{-1}\mathbf{X})^{-}]\ \mathbf{X}'\mathbf{V}^{-1} \\
&= \mathbf{V}^{-1}\mathbf{X}\ [(\mathbf{X}'\mathbf{V}^{-1}\mathbf{X})^{-}\mathbf{X}'\mathbf{V}^{-1}\mathbf{X}(\mathbf{X}'\mathbf{V}^{-1}\mathbf{X})^{-}]\ \mathbf{X}'\mathbf{V}^{-1} \\
&= \mathbf{V}^{-1}\mathbf{X}\ (\mathbf{X}'\mathbf{V}^{-1}\mathbf{X})^{-}\ \mathbf{X}'\mathbf{V}^{-1}
\end{aligned}$$

Problem 12.1

a)

$$\mathbf{X} = \begin{bmatrix} 1 & 1 & 0 \\ 1 & 1 & 0 \\ 1 & 1 & 0 \\ 1 & 1 & 0 \\ 1 & 1 & 0 \\ 1 & 1 & 0 \\ 1 & 0 & 1 \\ 1 & 0 & 1 \\ 1 & 0 & 1 \\ 1 & 0 & 1 \\ 1 & 0 & 1 \\ 1 & 0 & 1 \\ 1 & 0 & 0 \\ 1 & 0 & 0 \\ 1 & 0 & 0 \\ 1 & 0 & 0 \\ 1 & 0 & 0 \\ 1 & 0 & 0 \end{bmatrix} \qquad \hat{\boldsymbol{\beta}}^3 = \begin{bmatrix} 31.58 \\ 2.06 \\ -3.57 \end{bmatrix}$$

b)

$$\mathbf{L}^3 = \begin{bmatrix} 1 & 1 & 0 \\ 1 & 0 & 1 \\ 1 & 0 & 0 \end{bmatrix}$$

Problem 12.3

a) BLUP would be more appropriate since the population in test 2 is not a random group, but is selected.

b) If we use the 10-year measurements from test 2, and the 15-year measurements from test 1, then dim(**y**) = 125 x 1, dim(**V**) = 125 x 125, dim(**C**) = 125 x 100, and dim(**G**) = 100 x 100.

c) Using the 'means' approach, **X** will contain 2 effects which can be designated μ_1 for the mean from test 1, and μ_2 for the mean from test 2. Then two rows of **X** for observations from different tests would be

$$\begin{bmatrix} 1 & 0 \\ 0 & 1 \end{bmatrix}$$

d) Second moments needed are: $\mathrm{Var}(\bar{y}_{1\cdot k\cdot})$, $\mathrm{Var}(\bar{y}_{2\cdot k\cdot})$, $\mathrm{Cov}(\bar{y}_{1\cdot k\cdot},\bar{y}_{2\cdot k\cdot})$, $\mathrm{Cov}(\bar{y}_{1\cdot k\cdot},g_k)$, and $\mathrm{Cov}(\bar{y}_{2\cdot k\cdot},g_k)$.

Problem 12.4

For all different approaches,

$$\mathbf{X}\hat{\boldsymbol{\beta}} = \begin{bmatrix} 33.63 \\ 33.63 \\ 33.63 \\ 33.63 \\ 33.63 \\ 33.63 \\ 28.01 \\ 28.01 \\ 28.01 \\ 28.01 \\ 28.01 \\ 28.01 \\ 31.58 \\ 31.58 \\ 31.58 \\ 31.58 \\ 31.58 \\ 31.58 \end{bmatrix}$$

Index

D

E

F

G

H

I

L

M

O

P

R

S

T

V

Y

FORESTRY SCIENCES

Baas P, ed: New Perspectives in Wood Anatomy. 1982. ISBN 90–247–2526–7
Prins CFL, ed: Production, Marketing and Use of Finger-Jointed Sawnwood. 1982. ISBN 90–247–2569–0
Oldeman RAA, et al., eds: Tropical Hardwood Utilization: Practice and Prospects. 1982. ISBN 90–247–2581–X
Den Ouden P and Bloom BK: Manual of Cultivated Conifers: Hardy in Cold and Warm-Temperate Zone. 1982. ISBN 90–247–2148–2
Bonga JM and Durzan DJ, eds: Tissue Culture in Forestry. 1982. ISBN 90–247–2660–3
Satoo T and Magwick HAI: Forest Biomass. 1982. ISBN 90–247–2710–3
Van Nao T, ed: Forest Fire Prevention and Control. 1982. ISBN 90–247–3050–3
Douglas J: A Re-appraisal of Forestry Development in Developing Countries. 1983. ISBN 90–247–2830–4
Gordon JC and Wheeler CT, eds: Biological Nitrogen Fixation in Forest Ecosystems: Foundations and Applications. 1983. ISBN 90-247–2849–5
Németh MV: The Virus-Mycoplasma and Rikettsia Disease of Fruit Trees. 1984. ISBN 90–247–2868–1
Duryea ML and Landis TD, eds: Forest Nursery Manual: Production of Bareroot Seedlings. 1984. ISBN 90–247–2913–0
Hummel FC, ed: Forest Policy: A Contribution to Resource Development. 1984. ISBN 90–247–2883–5
Manion PD, ed: Scleroderris Canker of Conifers. 1984. ISBN 90–247–2912–2
Duryea ML and Brown GN, eds: Seedling Physiology and Reforestation Success. 1984. ISBN 90–247–2949–1
Staaf KAG and Wiksten NA: Tree Harvesting Techniques. 1984. ISBN 90–247–2994–7
Boyd JD: Biophysical Control of Microfibril Orientation in Plant Cell Walls. 1985. ISBN 90–247–3101–1
Findlay WPK, ed: Preservation of Timber in the Tropics. 1985. ISBN 90–247–3112–7
Samset I: Winch and Cable Systems. 1985. ISBN 90–247–3205–0
Leary RA: Interaction Theory in Forest Ecology and Management. 1985. ISBN 90–247–3220–4
Gessel SP: Forest Site and Productivity. 1986. ISBN 90–247–3284–0
Hennessey TC, Dougherty PM, Kossuth SV and Johnson JD, eds: Stress Physiology and Forest Productivity. 1986. ISBN 90–247–3359–6
Shepherd KR: Plantation Silviculture. 1986. ISBN 90–247–3379–0
Sohlberg S and Sokolov VE, eds: Practical Application of Remote Sensing in Forestry. 1986. ISBN 90–247–3392–8
Bonga JM and Durzan DJ, eds: Cell and Tissue Culture in Forestry. Volume 1: General Principles and Biotechnology. 1987. ISBN 90–247–3430–4
Bonga JM and Durzan DJ, eds: Cell and Tissue Culture in Forestry. Volume 2: Specific Principles and Methods: Growth and Developments. 1987. ISBN 90–247–3431–2
Bonga JM and Durzan DJ, eds: Cell and Tissue Culture in Forestry. Volume 3: Case Histories: Gymnosperms, Angiosperms and Palms. 1987. ISBN 90–247–3432–0
Richards EG, ed: Forestry and the Forest Industries: Past and Future. 1987. ISBN 90–247–3592–0
Kossuth SV and Ross SD, eds: Hormonal Control of Tree Growth. 1987. ISBN 90–247–3621–8
Sundberg U and Silversides CR: Operational Efficiency in Forestry. Volume 1: Analysis. 1988. ISBN 90–247–3683–8
Silversides CR and Sundberg U: Operational Efficiency in Forestry. Volume 2: Practice. 1989. ISBN 0–7923–0063–7
White, TL and Hodge GR: Predicting Breeding Values with Applications in Forest Tree Improvement. 1989. ISBN 0–7923–0460–8